Collins and Lyne's
Microbiological Methods

Sixth edition

C. H. Collins MBE, DSc, FRCPath, FIBiol, FIMLS
Senior Visiting Research Fellow, Department of Microbiology, National Heart and Lung Institute, University of London

Patricia M. Lyne CBiol, MIBiol
The Ashes, Hadlow, Kent

J. M. Grange MD, MSc
Director, Department of Microbiology, National Heart and Lung Institute, University of London and Honorary Consultant Microbiologist, National Heart and Lung Hospital, London

BUTTERWORTH
HEINEMANN

Butterworth–Heinemann
Halley Court, Jordan Hill,
Oxford OX2 8EJ

 PART OF REED INTERNATIONAL P.L.C.

Oxford London Guildford Boston Munich New Delhi Singapore Sydney Tokyo Toronto Wellington

First published 1964
Reprinted 1966
Second Edition 1967
Reprinted 1970

Third Edition 1970
Fourth Edition 1976
Reprinted with additions 1979
Fifth Edition 1984

Revised reprint 1985
Reprinted 1987
Sixth Edition 1989
Revised reprint 1991

British Library Cataloguing in Publication Data

Collins, C. H. (Christopher Herbert)
 Collins and Lyne's microbiological methods.
 – 6th. ed.
 1. Microbiology. Laboratory techniques
 I. Title   II. Lyne, Patricia M. (Patricia
 Mary), *1934–*   III. Grange, John M.   IV.
 Collins, C.H. (Christopher Herbert).
 Microbiological methods
 576'.028

 ISBN 0–7506 142 85

Library of Congress Cataloging-in-Publication Data

Collins and Lyne's microbiological methods.

 Rev. ed. of: Microbiological methods/C.H. Collins,
Patricia M. Lyne.   5th ed. 1984.
 Includes index.
 1. Microbiology—Technique.   I. Collins, C.H.
(Christopher Herbert)   II. Lyne, Patricia M.
III. Grange, John M.   IV. Collins, C.H. (Christopher
Herbert) Microbiological methods.   V. Title. [DNLM:
1. Bacteriological Technics.   2. Microbiology—methods.
QW 25 M6265]
QR65.C64   1989   576'.028   88–8674
ISBN 0–7506 142 85

Printed and bound by Hartnolls Ltd, Bodmin, Cornwall

Preface to the sixth edition

This book first appeared in 1964 and has passed through five editions and several reprints. It has therefore outgrown its original authors who needed more and more advice and assistance from younger and more active microbiologists.

As we wished to retain the original, and apparently successful, style we resisted the temptation to hand over separate chapters to individual authors. Instead we invited a number of our colleagues to look at those parts of the fifth edition which interested them particularly and to make suggestions, corrections, improvements and amendments. These we have incorporated on a 'peer review' principle so that in the end no chapter is the sole product of any one individual.

As in earlier editions, the names of some organisms have changed, and so has their position in the book. Although we now accept that 'kit' and 'automated' methods have come to stay we have retained many of the manual techniques for the use of students and those who work in less well-developed areas.

We wish to thank the microbiologists who have provided so much useful information for this edition and for allowing us to mould their material into the original, and admittedly somewhat telegraphic, style.

C. H. Collins   J. M. Grange
Patricia M. Lyne   London
Hadlow

Contributors

J. S. Brazier, MSc, PhD, FIMLS, Public Health Laboratory, Luton, Beds

E. Y. Bridson, CBiol, FIBiol, FIMLS, 3 Bellever Hill, Camberley, Surrey

C. K. Campbell, MSc, PhD, Mycology Reference Laboratory, Central Public Health Laboratory, Colindale, London

C. H. Collins, MBE, DSc, FRCPath, Microbiology Department, National Heart and Lung Institute, University of London

A. W. Cremer, CBiol, MIBiol, FIMLS, Department of Clinical Microbiology, University College Hospital, London

T. J. Donovan, PhD, CBiol, FIMLS, Public Health Laboratory, Ashford, Kent

G. F. Down, FIMLS, Public Health Laboratory, Dulwich, London

J. M. Grange, MD, MSc, Microbiology Department, National Heart and Lung Institute, University of London

J. D. Jarvis, FIMLS, Department of Medical Microbiology, The London Hospital, London

Patricia M. Lyne, CBiol, MIBiol, The Ashes, Hadlow, Kent

D. N. Petts, MSc, CBiol, FIMLS, Microbiology Department, Basildon Hospital, Basildon, Essex

Contents

Chapter 1

Safety in microbiology

In spite of the attention given to laboratory-acquired infections in recent years and the publication of codes of practice, guidelines and safety manuals (see below), these infections still occur.

More than 4000 laboratory-associated infections have been reported so far this century (Pike, 1979; Collins, 1988a). Many more have probably gone unreported. Most of the victims worked in research laboratories with organisms whose potential pathogenicity was unknown at the time, and/or used techniques now known to be hazardous. As a result of investigations over the last 30 years the causes of many of these infections have been discovered and methods for preventing them have been developed. There is now no need for microbiological laboratory workers (with certain notable exceptions which are outside the scope of this book) to feel at risk from the organisms they handle providing that they are aware of:

(1) the potential hazards of the materials and organisms they handle,
(2) the routes by which these organisms may enter the body and cause infections,
(3) the correct methods of 'containing' these organisms so that they do not have access to those routes.

Classification of microorganisms on the basis of risk

Microorganisms vary in their ability to infect individuals and cause disease. Some are harmless; some may be responsible for diseases with mild symptoms; others can cause serious illnesses; a few have the potential for spreading in the community and causing serious epidemic disease. Experience and research into laboratory-acquired infections have enabled investigators to classify organisms and viruses into four Classes, Risk or Hazard Groups (1–4 or I–IV) in increasing order of hazard to laboratory workers and to the community. The three systems in current use are shown in Table 1.1. The most recent lists of microorganisms within each group have been published by the US Public Health Service (USPHS, 1981) and the UK Advisory Committee on Dangerous Pathogens (ACDP, 1990).

Most of the organisms discussed in this book are in Risk/Hazard Groups 1 and 2. Warnings (in italic type) are given in the sections on those in Risk/Hazard Group 3 which require Containment Level 3 facilities. No Risk/Hazard Group 4 agents are mentioned in the text.

Table 1.1 Summary of systems for classifying microorganisms on the basis of hazards to laboratory workers and the community

	Hazard			
	Low			High
USPHS (1974)	Class 1 none or minimal	Class 2 ordinary potential	Class 3 special, to individual	Class 4 high, to individual
WHO (1979)	Risk Group I low individual low community	Risk Group II moderate individual low community	Risk Group III high individual low community	Risk Group IV high, to individual and community
ACDP (1990)	Hazard Group 1 unlikely to cause human disease	Hazard Group 2 possibly to laboratory workers, unlikely to community	Hazard Group 3 some hazard to laboratory workers. May spread to community	Hazard Group 4 Serious hazard to laboratory workers. High risk to community

Routes of infection

Microorganisms can enter the body through the mouth (ingestion), through the lungs (inhalation), through the skin (injection) and through the eyes.

They may be ingested during mouth pipetting and may also enter the mouth from contaminated fingers and articles which have been contaminated from the laboratory benches, e.g. cigarettes, food, pencils. Such environmental contamination may be the result of spills and splashes, either unrecognized or inadequately disinfected.

Inhalation of infected airborne particles (aerosols) released during many common laboratory manipulations have probably caused the largest number of laboratory-associated diseases (Pike, 1979; Collins, 1988a).

Injection may result from accidental stabbing with hypodermic needles, pasteur pipettes or broken and infected glassware. Organisms may also enter the body through cuts and abrasions in the skin, some of which may be too small to be obvious to the worker himself.

Splashing of infected liquids into the eyes has been the cause of a number of infections, some fatal.

The routes of infection in laboratory-acquired infections are not necessarily the same as those of 'naturally'-acquired infections. In addition, the infecting dose may be much greater and the symptoms may therefore be different.

Aerosols: infected airborne particles

These are likely to be released and may be inhaled or may contaminate the hands, bench, etc. during the following manipulations: work with loops and syringes; pipetting, centrifuging, blending and homogenizing; pouring; opening culture tubes and dishes. (For details see Collins, 1988a.)

Levels of containment

The Risk/Hazard Group number allocated to an organism indicates the appropriate Containment (UK) or Biosafety (US) Level, i.e. the necessary accommodation, equipment, techniques and precautions. Thus, there are four of these levels (see Table 1.2). In the WHO classification the Basic Laboratory equates with Levels 1 and 2, the Containment Laboratory with Level 3 and the Maximum Containment Laboratory with Level 4.

Table 1.2 Summary of Biosafety/Containment level requirements

Level	Facilities	Laboratory practice	Safety equipment
1	Basic	GMT[a]	None. Work on open bench
2	Basic	GMT plus protective clothing, biohazard signs	Open bench plus safety cabinet for aerosol potential
3	Containment	Level 2 plus special clothing, controlled access	Safety cabinet for all activities
4	Maximum Containment	Level 3 plus air lock entry, shower exit, special waste disposal	Class III safety cabinet, pressure gradient, double-ended autoclave

[a] GMT, Good Microbiological Technique

Levels 1 and 2: the Basic laboratory

This is intended for work with organisms in Risk/Hazard Groups 1 and 2. Ample space should be provided. Walls, ceilings and floors should be smooth, non-absorbent, easy to clean and disinfect and resistant to the chemicals that are likely to be used. Floors should also be slip resistant. Lighting and heating should be adequate. Hand basins, other than laboratory sinks, are essential.

Bench tops should be wide, the correct height for work at a comfortable sitting position, smooth, easy to clean and disinfect and resistant to chemicals likely to be used. Adequate storage facilities should be provided.

Access should be restricted to authorized persons.

Level 3: the Containment laboratory

This is intended for work with organisms in Risk/Hazard Group 3. All the features of the Basic laboratory should be incorporated, plus the following.

The room should be physically separated from other rooms, with no communication (e.g. by pipe ducts, false ceilings) with other areas, apart from the door, which should be lockable, and transfer grilles for ventilation (see below).

Ventilation should be one way, achieved by having a lower pressure in the Containment laboratory than in other, adjacent, rooms and areas. Air should be removed to atmosphere (total dump, not recirculated to other parts of the building) by an exhaust system coupled to a microbiological safety cabinet (see p. 13) so that during working hours air is continually extracted by the cabinet or directly from the room and airborne particles cannot be moved around the building. Replacement air enters through transfer grilles.

Access should be strictly regulated. The international biohazard warning sign, with appropriate wording, should be fixed to the door (Figure 1.1).

Level 4: the Maximum Containment laboratory

This is required for work with Risk/Hazard Group 4 materials and is outside the

BIOHAZARD   Figure 1.1 International biohazard sign

scope of this book. Construction and use generally require government licence or supervision.

More information about the design of laboratories at the various levels is given by WHO (1983), USPHS (1984), Collins (1988a), ACDP (1990a).

Prevention of laboratory-acquired infections: a code of practice

The principles of containment involve Good Microbiological Technique (GMT), with which this book is particularly concerned, and the provision of:

(1) primary barriers around the organisms to prevent their dispersal into the laboratory,
(2) secondary barriers around the worker to act as a safety net should the primary barriers be breached,
(3) tertiary barriers to prevent the escape into the community of any organisms which are not contained by the primary and secondary barriers.

This Code of Practice draws very largely on the requirements and recommendations published in other works (WHO, 1983; Collins, 1988a,b; ACDP, 1990a).

Primary barriers

These are techniques and equipment designed to contain microorganisms and prevent their direct access to the worker and their dispersal as aerosols.

(1) All body fluids and pathological material should be regarded as potentially infected. 'High Risk' specimens, i.e. those that may contain agents in Risk/Hazard Group 3 or originate from patients in the AIDS and hepatitis risk categories should be so labelled (see below).
(2) Mouth pipetting should be *banned in all circumstances*. Pipetting devices (p.26) should be provided.
(3) No article should be placed in or at the mouth. This includes mouthpieces of rubber tubing attached to pipettes, pens, pencils, labels, smoking materials, fingers, food and drink.

(4) The use of hypodermic needles should be restricted. Cannulas are safer.

(5) Sharp glass pasteur pipettes should be replaced by soft plastic varieties (p.25).

(6) Cracked and chipped glassware should be replaced.

(7) Centrifuge tubes should be filled only to within 2 cm of the lip. All Risk/Hazard Group 3 materials should be centrifuged in sealed centrifuge buckets (p.11).

(8) Bacteriological loops should be completely closed, with a diameter not more than 3 mm and a shank not more than 5 cm and be held in metal, not glass, handles (see Collins, 1988).

(9) Homogenizers should be inspected regularly for defects which might disperse aerosols (p.13). Only the safest models should be used. Glass Griffith's tubes and tissue homogenizers should be held in a pad of wadding in a gloved hand when operated.

(10) All Risk/Hazard Group 3 materials should be processed in a Containment Laboratory (see Tertiary barriers below) and in a Microbiological Safety Cabinet (p.13) unless exempted by agreed national regulations.

(11) A supply of suitable disinfectant (p.36) at use dilution should be available at every work station.

(12) Benches and work surfaces should be disinfected regularly and after spillage of potentially infected material.

(13) Discard jars for small objects and for re-usable pipettes should be provided at each work station. They should be emptied, decontaminated and refilled with freshly prepared disinfectant daily (p.36).

(14) Discard bins or bags supported in bins should be placed near to each work station. They should be removed and autoclaved daily (p.30).

(15) Broken culture vessels, e.g. after accidents, should be covered with a cloth. Suitable disinfectant should be poured over the cloth and the whole left for 30 min. The debris should then be cleared up into a suitable container (tray or dustpan) and autoclaved. The hands should be gloved and stiff cardboard used to clear up the debris.

(16) Infectious and pathological material to be sent through the post or by airmail should be packed in accordance with government, post office and airline regulations, obtainable from those authorities. Full instructions are also given elsewhere (WHO, 1983; USPHS, 1984; Collins, 1988a; HSAC, 1986).

(17) Specimen containers should be robust and leak proof.

(18) Discarded infectious material should not leave the laboratory until it has been sterilized or otherwise made safe.

Secondary barriers

These are intended to protect the worker should the primary barriers fail. They should, however, be observed as strictly as the latter.

(1) Proper overalls should be worn at all times and fastened. They should be kept apart from outdoor and other clothing. When agents in Risk/Hazard Group 3 and material containing hepatitis B or the AIDS viruses are handled plastic aprons should be worn over the normal protective clothing. Surgeons' gloves should also be worn.

(2) Laboratory protective clothing should be removed when the worker leaves the laboratory and not worn in any other area such as canteens and rest rooms.
(3) Hands should be washed after handling infectious material and always before leaving the laboratory.
(4) Any obvious cuts, scratches and abrasions on exposed parts of the worker's body should be covered with waterproof dressings.
(5) In laboratories where pathogens are handled there should be medical supervision.
(6) All illnesses should be reported to the medical or occupational health supervisors. They may be laboratory associated. Pregnancy should also be reported. It is inadvisable to work with certain organisms during pregnancy.
(7) Any member of the staff who is receiving steroids or immunosuppressive drugs should report this to the medical supervisor.
(8) Workers should be immunized where possible and practicable against likely infections (Collins, 1988; Wright, 1988).
(9) In laboratories where tuberculous materials are handled the staff should have received BCG or have evidence of a positive skin reaction before starting work. They should have annual chest X-rays thereafter (Wright, 1988).

Tertiary barriers

These are intended to offer additional protection to the worker and to prevent the escape into the community of microorganisms that are under investigation in the laboratory.

They concern the provision of accommodation and facilities at the Biosafety/ Containment levels appropriate for the organisms handled, and of microbiological safety cabinets.

Precautions against infection with hepatitis B (HBV) and the human immunodeficiency virus (HIV)

Laboratory workers who may be exposed to these viruses in clinical material are advised to follow the special precautions described by one or more of the following: HSAC, 1984; USPHS, 1984; Collins, 1988a,b; ACDP, 1990a,b.

Control of substances hazardous to health regulations (COSHH) 1988

These apply to microorganisms as well as to chemicals. Assessments in relation to microorganisms should be made according to the Hazard Groups in which they are placed (ACDP, 1990), the level of containment applied, and the techniques used. Hazard Group 3 organisms are identified as such in this book.

Some of the chemicals used in microbiology are known or thought to be hazardous to health. They are identified in the text by the word 'caution'.

Instruction in safety

This instruction should be part of the general training given to all microbiology laboratory workers and should be included in what is known as Good Microbiological Technique. In general, methods that protect cultures from contamination also protect workers from infection but this should not be taken for granted. Personal protection can be achieved only by good training and careful work. Programmes for training in safety in microbiology have been published (WHO, 1983; Collins, 1988a).

For a review of the history, incidence, causes and prevention of laboratory-acquired infection see Collins (1988a).

References

ACDP (1990a) *Categorization of Pathogens According to Hazard and Categories of Containment*, 2nd edn, Advisory Committee on Dangerous Pathogens, HMSO, London

ACDP (1990b) *HIV – the Causative Agent of AIDS and Related Conditions*, Advisory Committee on Dangerous Pathogens, HMSO, London

Collins, C. H. (1988a) *Laboratory-acquired Infections*, 2nd edn, Butterworths, London

Collins, C. H. (188b) Microbiological hazards. In *Safety in Clinical and Biomedical Laboratories* (ed C. H. Collins) Chapman and Hall, London

DHSS (1978) *Code of Practice for the Prevention of Infection in Clinical Laboratories and Post-mortem Rooms*, Department of Health and Social Security, HMSO, London

HSAC (1986) *Safety in Health Services Laboratories: the Labelling, Transport and Reception of Specimens*, Health Services Advisory Committee, Health and Safety Commission, HMSO, London

Pike, R. M. (1979) Laboratory-associated infections; incidence, fatalities, causes and prevention. *Annual Review of Microbiology*, **33**, 41–66

USPHS (1974) *Classification of Etiologic Agents on the Basis of Hazard*, Centers for Disease Control, Atlanta, Government Printing Office, Washington DC

USPHS (1981) Classification of etiologic agents on the basis of hazard. *Federal Register*, **46**, 59379–59380

USPHS (1984) *Biosafety in Microbiological and Biomedical Laboratories.* Centers for Disease Control and National Institutes of Health, Government Printing Office, Washington DC

WHO (1983) *Laboratory Biosafety Manuel*, World Health Organization, Geneva

Wright, A. E. (1988) Health care in the laboratory. In *Safety in Clinical and Biomedical Laboratories* (ed. C. H. Collins), Chapman and Hall, London

Chapter 2

Laboratory equipment

It is important to choose the correct equipment for microbiological work. A wide range of laboratory apparatus, glassware and plastic goods is available from scientific equipment suppliers, but unfortunately not all of it is suitable for microbiological use. Before purchasing any new piece of equipment, it is best to obtain the personal advice of microbiologists who have had experience in its use. It is bad policy to rely entirely on advertisements, catalogues, extravagant claims of representatives and the opinions of purchasing officers who are mainly concerned with balancing budgets. The best is not always the most expensive, but it is rarely the cheapest. Few microbiologists require very expensive research-type micro-scopes, but centrifuges should be the very best available. Laboratory autoclaves should be designed for laboratory, not pharmacy or hospital, use. Thermostatic equipment designed for chemical work is usually suboptimal for microbiological use.

In this chapter, we have indicated the types of equipment that we and our associates and friends have found adequate for use in medical, veterinary and food microbiology (except virology) work.

The operation of some microbiological equipment may result in the release of infectious aerosols (see Chapter 1 and Collins, 1988a).

Microscopes

Although the microscope is an essential piece of microbiological apparatus, it is not used as often as might be expected and the purchase of elaborate and expensive (prestige) instruments is not justified. The medium priced instruments are ade-quate for most basic laboratory work.

Laboratory workers should be familiar with the general mechanical and optical principles, but a detailed knowledge of either is unnecessary and, apart from superficial cleaning, maintenance should be left to the manufacturers, who will arrange for periodic visits by their technicians. Most manufacturers publish handbooks containing useful explanations and information.

Wide-field compensating eyepieces should be fitted; they are less tiring than other eyepieces and are more convenient for spectacle wearers. For low power work, × 5 and × 10 objectives are most useful and for high power (oil immersion) microscopy, × 50 and × 100 fluorite objectives are desirable. These fluorite lenses

8

are much better than achromatics and much less expensive than the hardly justifiable apochromats.

Careful attention to critical illumination, centring, and the position of diaphragms is essential for adequate microscopy.

Microscopes for fluorescence microscopy

Suitable equipment is described in Chapter 7.

Low power microscopes

Although many observations on colony forms can be made with a hand-lens, a low power binocular 'dissecting' microscope is very useful for examining very small colonies and doing subcultures. An instrument that can be illuminated from above and below is desirable. Paired $\times$ 5 eyepieces and 59-mm objectives give a satisfactory image.

Incubators and water-baths

Incubators

Incubators are available in various sizes. In general, it is best to obtain the largest possible model that can be accommodated, although this may create space problems in laboratories where several different incubation temperatures are employed. The small incubators suffer wider fluctuations in temperature when their doors are opened than do the larger models and most laboratory workers find that incubator space, like refrigerator space, is subject to 'Parkinson's law'.

Although incubators rarely develop faults, it is advisable, before choosing one, to ascertain that service facilities are available. The circuits are not complex but require expert technical knowledge to repair. Transporting incubators back to the manufacturers is most inconvenient.

In medical and veterinary laboratories, incubators are usually operated only at 35–37°C. The food or industrial laboratory usually requires machines operated at 15–20, 28–32 and 55°C. For temperatures above ambient there are no problems, but lower temperatures may need a cooling as well as a heating device.

Incubation in an atmosphere of 5–8% carbon dioxide is preferable for the cultivation of almost all bacteria of medical importance. Automatic control of carbon dioxide and humidity is now possible but as a back-up 'candle jars' should be available.

Water used to maintain humidity may become contaminated with fungi, especially aspergilli, causing problems with cultures that require prolonged incubation.

Cooled incubators
For incubation at temperatures below the ambient, incubators must be fitted with modified refrigeration systems with heating and cooling controls. These need to be correctly balanced.

Automatic temperature changes

These are necessary when cultures are to be incubated at different temperatures for varying lengths of time, as in the examination of water by membrane filtration, and when it is not convenient to move them from one incubator to another. Two thermostats are required, wired in parallel, and a time switch wired to control the thermostat set at the higher temperature. These are now built in to some models.

Incubator rooms

A 'hot room' or ' walk-in incubator' is not difficult to adapt from a small room or a corner of a large room. Windows must, of course, be blocked up.

The walls need two layers of building paper on which is glued 48-mm thick slabs of expanded polystyrene (cork is much more expensive) between battens at 600-mm centres. The inner lining can be ordinary plaster or insulation board. The ceiling must be lagged in the same way and in high rooms lower false ceilings are preferable. The floor can be covered with insulation board and hardboard and the doors lagged on their inner surfaces in the same way, and fitted in their jambs on sponge rubber draught-prevention strips.

Two methods of heating are possible. Tubular heaters around the walls are satisfactory and a power of 3 kW is more than adequate for a room of 5–7 m^3 that is to be maintained at 37°C. A large circulating fan, to avoid hot and cold spots, must be fitted on one wall and should operate constantly. An alternative arrangement is the greenhouse- or space-heater in which 2.5–3.0 kW heater and a fan are built into a steel casing. The wiring must be altered so that the fan is always on and the heater connected to a sensitive thermostat (the thermostats built into these heaters are not sensitive enough for this purpose). A second, independent thermostat with a manual re-set should be fitted to switch off the electricity supply if the main thermostat fails and the temperature rises above the required level.

These two methods of heating may be combined. Smaller tubular heaters, permanently switched on, will supply background heat and the space heater will maintain the required temperature.

Ideally, there should be a recording thermometer, with one or two sensing devices, fitted at approximately 0.5 and 2 m from floor level.

Wooden shelving and racking is undesirable. If a high humidity is maintained, fungi may grow on the wood. Steel or aluminium racks are preferable and can be tailored by a racking specialist. We have used steel racking to accommodate the usual aluminium culture bottle racks, thus avoiding solid shelving. All shelves should be free, i.e. easily removed for cleaning, and there should be air spaces between the shelves and the walls so as to allow for circulation of air.

Complete incubator rooms can be purchased from several companies.

Rooms in which a constant temperature at or below ambient has to be maintained need a refrigerator coil through which a fan blows cold air into the room. Such a device can be thermostatically operated.

Plastic food containers with snap-on lids, containing wet filter papers, are useful for maintaining humidity in cultures in hot rooms and incubators.

Water-baths

The contents of a test-tube placed in a water-bath are raised to the required temperature much more rapidly than in an incubator. These instruments are therefore used for short-term incubation. If the level of water in the bath comes one-half to two-thirds of the way up that of the column of liquid in the tube, convection currents are caused which keep the contents of the tube well mixed, and hasten reactions such as agglutination.

All modern water-baths are equipped with electrical stirrers and in some the heater, thermostat and stirrer are in one unit, easily detached from the bath for use elsewhere or for servicing. Water-baths must also be lagged to prevent heat loss through the walls. A bath that has not been lagged by the manufacturers can be insulated with slabs of expanded polystyrene.

Water-baths should be fitted with lids in order to prevent heat loss and evaporation. These lids must slope so that condensation water does not drip on the contents. To avoid chalky deposits on tubes and internal surfaces, only distilled water should be used, except, of course, in those baths which operate at or around ambient temperature, which are connected to the cold water supply and fitted with a constant level and overflow device. This kind of bath offers the same problems as the low-temperature incubators unless the water arrives at a much lower temperature than that required in the water-bath. In premises that have an indirect cold water supply from a tank in the roof, the water temperature in summer may be near or even above the required water-bath temperature, giving the thermostat no leeway to operate. In this case, unless it is possible to make direct connection with the rising main, the water supply to the bath must be passed through a refrigerating coil. Alternatively, the water-bath can be placed in a refrigerator and its electricity supply adapted from the refrigerator light or brought in through a purpose-made hole in the cabinet. The problem is best overcome by fitting laboratory refrigeration units, designed for this purpose.

Metal block heaters
These give better temperature control than water-baths and do not dry out, but they can be used only for tubes and thin glass bottles which must fit neatly into the holes in the steel or aluminium blocks.

Centrifuges

The ordinary laboratory centrifuge is capable of exerting a force of up to 3000 g and this is the force necessary to deposit bacteria within a reasonable time.

For general microbiology, a centrifuge capable of holding 15- and 50-ml buckets and working at a maximum speed of 4000 rev/min is adequate. The swing-out head is safer than the angle head as it is less likely to distribute aerosols (Collins, 1988).

For maximum microbiological safety, sealed buckets should be fitted. These confine aerosols if breakage occurs and are safer than sealed rotors. Sealed buckets ('safety cups') are supplied by several companies. Windshields offer no protection. There should be an electrically operated safety catch so that the lid cannot be opened when the rotors are spinning.

Centrifuge buckets are usually made of stainless steel and are fitted with rubber buffers. They are paired and their weight is engraved on them. They are always used in pairs, opposite to one another, and it is convenient to paint each pair with different coloured patches so as to facilitate recognition. If the buckets fit in the centrifuge head on trunnions, these are also paired.

Centrifuge tubes to fit the buckets are made of glass, plastic, nylon or spun aluminium. Conical and round-bottomed tubes are made. Although the conical tubes concentrate the deposit into a small button, they break much more readily than do the round-bottomed tubes. When the supernatant fluid, not the deposit, is required, round-bottomed tubes should be used.

Instructions for using the centrifuge

(1) Select two centrifuge tubes of identical lengths and thicknesses. Place liquid to be centrifuged in one tube and water in the other to within about 2 cm of the top.
(2) Place the tubes in paired centrifuge buckets and place the buckets on the pans of the centrifuge balance. This can be made by boring holes large enough to take the buckets in small blocks of hard wood and fitting them on the pans of a simple balance.
(3) Balance the tubes and buckets by adding 70% alcohol to the lightest bucket. Use a pasteur pipette and allow the alcohol to run down between the tube and the bucket.
(4) Place the paired buckets and tubes in diametrically opposite positions in the centrifuge head.
(5) Close the centrifuge lid and make sure that the speed control is at zero before switching on the current. (Many machines are fitted with a 'no-volt' release to prevent the machine starting unless this is done.)
(6) Move the speed control slowly until the speed indicator shows the required number of revolutions per minute.

Precautions

(1) Make sure that the rubber buffers are in the buckets, otherwise the tubes will break.
(2) Check the balancing carefully. Improperly balanced tubes will cause 'head wobble', spin-off accidents and wear out the bearings.
(3) Check that the balanced tubes are really opposite one another in multi-bucket machines.
(4) Never start or stop the machine with a jerk.
(5) Observe the manufacturer's instructions about the speed limits for the various loads.
(6) Open sealed centrifuge buckets in a microbiological safety cabinet.

Maintenance

The manufacturers will, by contract, arrange periodic visits for inspection and maintenance in the interests of safety and efficiency.

Blenders, tissue grinders and shakers

These range from instruments suitable for grinding and emulsifying large samples, e.g. of food, to glass devices for homogenizing small pieces of tissue.

Most electrically driven machines are efficient but suffer from the disadvantage of requiring a fresh, sterile cup for each sample. These cups are expensive and it takes time to clean and re-sterilize them.

Some of them may release aerosols during operation and when they are opened. A heavy Perspex or metal cover should be placed over them during use. This should be decontaminated afterwards. All blenders, etc. should be opened in a microbiological safety cabinet because the contents will be warm and under pressure. Aerosols will be released.

The Stomacher Lab-Blender overcomes these problems. Samples are emulsified in heavy-duty sterile plastic bags by the action of paddles. This is an efficient machine, capable of processing large numbers of samples in a short time without pauses for washing and re-sterilizing. There is little risk of aerosol dispersal. The bags are automatically sealed while the machine is working.

Tissue grinders (Griffith's tubes), essential for emulsifying small pieces of tissue for microbiological examination, are available in several sizes. A heavy glass tube is constricted near to its closed end and a pestle, usually made of glass covered with PTFE or of stainless steel, is ground into the constriction. With the pestle in place, the tissue and some fluid are put into the tube. The pestle is rotated by hand. The tissue is ground through the constriction and the emulsion collects in the bottom of the tube.

These grinders present some hazards. The tubes, even though made of borosilicate glass, may break and disperse infected material. They should be used inside a microbiological safety cabinet and be held in the gloved hand in a wad of absorbent material.

Small pieces of soft tissue, e.g. curettings, may be emulsified by shaking in a screw-capped bottle (Bijou) with a few glass beads and 1 or 2 ml of broth on a vortex mixer.

Shaking machines

Conventional shaking machines are sold by almost all laboratory suppliers. They are useful for mixing and shaking cultures and for serological tests, but should be fitted with racks, preferably of polypropylene, which hold the bottles or tubes firmly, and covered with a stout Perspex box when in use to prevent the dispersal of aerosols.

Vortex mixers are useful for mixing the contents of single bottles, e.g. emulsifying sputum, but should always be operated in a microbiological safety cabinet.

We believe that all bottles that contain infected material and which are shaken in any machine should be placed inside individual self-sealing or heat-sealed plastic bags in order to minimize the dispersal of infected airborne particles.

Microbiological safety cabinets

Microbiological safety cabinets are intended to capture and retain infected airborne particles released in the course of certain manipulations and to protect the laboratory worker from infection which may arise from inhaling them.

There are three kinds: Classes I, II and III. Class I and Class II cabinets are used in diagnostic and Containment laboratories for work with Risk/Hazard Group 3 organisms. Class III cabinets are used for Risk/Hazard Group 4 viruses.

A Class I cabinet is shown in Figure 2.1. The operator sits at the cabinet, works with his hands inside and sees what he is doing through the glass screen. Any aerosols released from his cultures are retained because a current of air passes in at

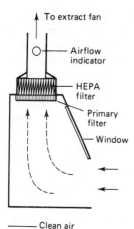

Clean air
Contaminated air   **Figure 2.1** Class I microbiological safety cabinet

the front of the cabinet. This sweeps the aerosols up through the filters which remove all or most of the organisms. The clean air then passes through the fan, which maintains the airflow, and is exhausted to atmosphere, where any particles or organisms that have not been retained on the filter are so diluted that they are no longer likely to cause infection if inhaled. A minimum airflow of 0.75 m/s must be maintained through the front of the cabinet and modern cabinets have airflow indicators and warning devices. The filters must be changed when the airflow falls below this level.

The Class II cabinet (Figure 2.2) is more complicated and is sometimes called a laminar flow cabinet but as this term is also used for clean air cabinets which do not protect the worker it should be avoided. In the Class II cabinet about 70% of the air

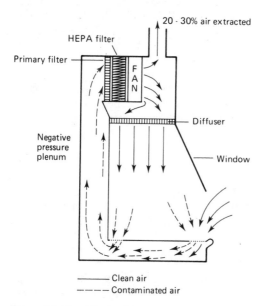

Clean air
Contaminated air

Figure 2.2 Class II microbiological safety cabinet

is recirculated through filters so that the working area is bathed in clean (almost sterile) air. This entrains any aerosols produced in the course of the work and these are removed by the filters. Some of the air (about 30%) is exhausted to atmosphere and is replaced by a 'curtain' of room air which enters at the working face. This prevents the escape of any particles or aerosols released in the cabinet.

There are other types of Class II cabinets, with different air flows and exhaust systems, which may also be used for toxic chemicals and volatile chemicals. Manufacturers should be consulted.

Class III cabinets are totally enclosed and are tested under pressure to ensure that no particles can leak from them into the room. The operator works with gloves which are an integral part of the cabinet. Air enters through a filter and is exhausted to atmosphere through one or two more filters (Figure 2.3).

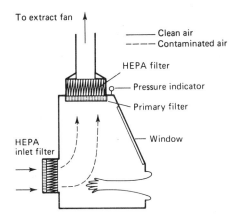

Figure 2.3 Class III microbiological safety cabinet

Purchasing

Safety cabinets should comply with national standards, e.g. British Standards Institution (BSI, 1979), Standards Association of Australia (SAA, 1980), National Sanitation Foundation (USA) (NSF, 1984).

Use of safety cabinets

These cabinets are intended to protect the worker from airborne infection. They will not protect him from spillage and the consequences of poor techniques. The cabinet should not be loaded with unnecessary equipment or it will not do its job properly. Work should be done in the middle to the rear of the cabinet, not near the front and the worker should avoid bringing his hands and arms out of the cabinet while he is working. After each set of manipulations and before withdrawing his hands he should wait for 2–3 min to allow any aerosols to be swept into the filters. The hands and arms may be contaminated and should be washed immediately after ceasing work in a safety cabinet. Do not use bunsen burners, even micro-incinerators. They disturb the airflow. Use disposable plastic loops (p.27).

Siting

The efficiency of a safety cabinet depends also on correct siting and proper maintenance. Possible sites for cabinets in a room are shown in Figure 2.4. A is a

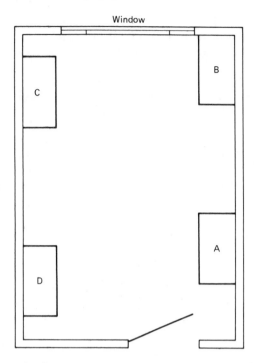

Figure 2.4 Possible sites for safety cabinets in relation to cross draughts from door and window and movement of staff. A is bad; B is poor; C is better; D is best

poor site, as it is near to the door and airflow into the cabinet will be disturbed every time the door is opened and someone walks past the cabinet into the room. Site B is not much better, as it is in almost a direct line between door and window, although no-one is likely to walk past it. Its left side is also close to the wall and airflow into it on that side may be affected by the 'skin effect', i.e. the slowing down of air when it passes parallel and close to a surface. Air passing across the window may be cooled, and will meet warm air from the rest of the room at the cabinet face, when turbulence may result.

Site C is better and site D is best of all. If two cabinets are required in the same room, sites C and D would be satisfactory, but they should not be too close together, or one may disturb the airflow of the other. We remember a laboratory where a Class I cabinet was placed next to a Class II cabinet. When the former was in use it extracted air from the latter and rendered that cabinet quite ineffective.

Care must also be taken in siting any other equipment that might generate air currents, e.g. fans and heaters. Mechanical room ventilation may be a problem if (rarely) it is efficient, but this can be overcome by linking it to the electric circuits of the cabinets so that either, but not both, are extracting air from the room at any one time. Alternatively baffles may be fitted to air inlets and outlets to avoid conflicting air currents near to the cabinet. Tests with smoke generators (see below) will establish the directions of air currents in rooms.

None of these considerations need apply to Class III cabinets which are in any case usually operated in more controlled environments.

Testing airflows

The presence and direction of air currents and draughts are determined with 'smoke'. This may indeed be from burning material in a device like that used by bee keepers, but is usually a chemical which produces a dense, visible vapour. Titanium tetrachloride is commonly used. If a cotton wool tipped stick, e.g. a throat swab, is dipped into this liquid and then waved in the air a white cloud is formed which responds to quite small air movements.

Commercial airflow testers are more convenient. They are small glass tubes, sealed at each end. Both ends are broken off with the gadget provided and a rubber bulb fitted to one end. Pressing the bulb to pass air through the tube causes it to emit white smoke. These methods are suitable for ascertaining air movements indoors.

To measure airflow an anemometer is required. Small, vane anemometers, timed by a stopwatch, are useful for occasional work but for serious activities electronic models, with a direct reading scale, are essential. The electronic vane type has a diameter of about 10 cm. It has a satisfactory time constant and responds rapidly enough to show the changes in velocity that are constantly occurring when air is passing into a Class I safety cabinet. Hot wire or thermistor anemometers may be used but they may show very rapid fluctuations and need damping. Both can be connected to recorders.

It is necessary to measure the airflow into a Class I cabinet in at least five places in the plane of the working face (Figure 2.5). An average is then calculated, but at no place should there be a reading that is 20 lfm (0.1 m/s) more or less than any of the others. If there is such a difference there will be turbulence within the cabinet. Usually some piece of equipment, inside or outside the cabinet, or the operator himself, is influencing the airflow.

Although airflow into Class I cabinets are usually much the same at all points on the working face this is not true for Class II cabinets, where the flow is greater at the bottom than at the top. The average inward flow can be calculated by measuring the velocity of air leaving the exhaust and the area of the exhaust vent. From this the volume per minute is found and this is also the amount entering the cabinet. Divided by the area of the working face it gives the average velocity. A rough and ready way, suitable for day-to-day use requires a sheet of plywood or

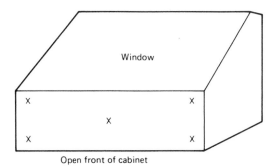

Figure 2.5 Testing airflow into a Class I safety cabinet. Anemometer readings should be taken at five places, marked with an X, in the plane of the working face with no-one working at the cabinet

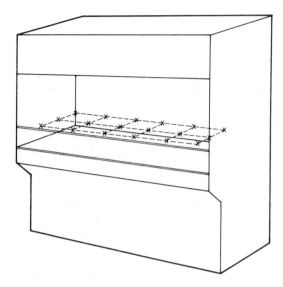

Figure 2.6 Testing the vertical airflow in a Class II safety cabinet. Anemometer readings should be taken at points marked X on an imaginary grid 6 inches within the cabinet walls and just above the level of the bottom of the glass window

metal which can be fitted over the working face. In the centre of this an aperture is made which is 2 × 2.5 cm. The inward velocity of air through this is measured and the average velocity over the whole face calculated from this figure and the area of the working face. Whichever way is used a check should also be made with smoke to ensure that air is in fact entering the cabinet all the way round its perimeter and not just at the lower edge.

The downward velocity of air in a Class II cabinet should be measured with an anemometer at 18 points in the horizontal plane 10 cm above the top edge of the working face (Figure 2.6). No reading should differ from the mean by more than 20%.

It is usual, when testing airflows with an anemometer, to observe or record the readings at each position for several minutes, because of possible fluctuations.

To measure the airflow through a Class III cabinet the gloves should be removed and readings taken at each glove port. Measurements should also be taken at the inlet filter face when the gloves are attached.

Recommended airflows for each class of cabinet are shown in Table 2.1.

Table 2.1 Mean airflows (m/s) for microbiological safety cabinets according to British Standard 5726

	Class I	Class II	Class III
Inflow at working face	0.7–1.0[a]	0.4	—
Downflow		0.25–0.5	
Inflow at glove ports			0.75
Inflow at filter			3

[a] With unused filters
Data from British Standard 5726 (1979) by permission of the British Standards Institution.

Decontamination

As safety cabinets are used to contain aerosols which may be released during work with Risk/Hazard Group 3 microorganisms the inside surfaces and the filters will become contaminated. The working surface and the walls may be decontaminated on a day-to-day basis by swabbing them with disinfectant. Glutaraldehyde is probably the best disinfectant for this purpose as phenolics may leave sticky residues and hypochlorites may, in time, corrode the metal. For thorough decontamination, however, and before filters are changed formaldehyde should be used and precautions should be taken before use. The installation should be checked to ensure that none of the gas can escape to the room or to other rooms. The front closure (night door) of the cabinet should seal properly onto the carcase, or masking tape should be available to seal it. Any service holes in the carcase should be sealed and the HEPA filter seating examined to ensure that there are no leaks. If the filters are to be changed and the primary or roughing filter is accessible from inside the cabinet it should be removed and left in the working area. The supply filter on a Class III cabinet should be sealed with plastic film.

Formaldehyde is generated by:
(1) boiling formalin (*caution*), which is a 40% solution of the gas in water;
(2) by heating paraformaldehyde, which is its solid polymer; or
(3) by mixing formalin or paraformaldehyde (*caution*) and water with potassium permanganate.

Boiling formalin
The volume used is important. Too little will be ineffective; too much leads to deposits of the polymer, which is persistent and which may contribute to the natural blocking of the filters. The British Standard 5726 specifies a concentration of 0.05 g/m^3 (0.0014 g/ft^3) which is approximately 25 ml of neat formalin for a Class I cabinet with a capacity of 0.34 m^3.

The formalin may be boiled in several ways. Some cabinets have built-in devices in which the liquid is dripped on to a hot-plate. Other manufacturers supply small boilers. A laboratory hot-plate, connected to a timing clock is adequate. The correct amount of formalin is placed in a flask. The formalin need not be diluted. There is enough water to achieve the humidity required already in the solution. The addition of water may diminish the efficiency of the gas and will result in a very wet cabinet.

Heating paraformaldehyde
Tablets (1 g) are available from chemical manufacturers. Three or four tablets are adequate for an average size cabinet. They may be heated on an electric frying pan connected to a time clock or placed in a dish and heated by an electric hair drier.

Oxidation with permanganate
Formaldehyde is released when potassium permanganate is mixed with formalin solution or with paraformaldehyde in the presence of water. The reaction may be violent but is not particularly dangerous unless too much permanganate is used. For most cabinets a mixture of 35 ml of formalin and 7.5 g of permanganate will suffice. The formalin is placed in a large beaker or bowl, in the cabinet, the permanganate, previously weighed out, is added and the front closure put in place immediately. The mixture boils almost at once. If paraformaldehyde is preferred 4 g of this and

8 g of permanganate are mixed immediately before they are required and placed in a large beaker or bowl in the cabinet and the 10 ml of water added. The front closure must be put on at once.

Exposure time
It is convenient to start decontamination in the late afternoon and let the gas act overnight. In the morning the fan should be switched on and the front closure 'cracked' open very slightly to allow air to enter and purge the cabinet. Some new cabinets have a hole in the front closure, fitted with a stopper which can be removed for this purpose. With Class III cabinets the plastic film is taken off the supply filter to allow air to enter the cabinet and purge it of formaldehyde. After several minutes the gloves may be removed. The fan is then allowed to run for about 30 min which should remove all formaldehyde.

Summary of decontamination procedure

The prefilter should be removed from inside the cabinet (where possible) and placed on the working suface. It is usually pushed or clipped into place and is easily removed, but as it is likely to be contaminated gloves should be worn. Neat formalin, 2 ml/ft^3 (70 ml/m^3), which is about 25 ml for a normal size Class I cabinet, should be placed in its container on the heater in the cabinet or in the reservoir. The front closure is then put in place and sealed if necessary. Then the heater is switched on and the formalin boiled away. After switching off the heater the cabinet is left closed overnight. The next morning the cabinet fan is switched on and then the front closure is opened very slightly to allow air to pass in and purge the cabinet of formaldehyde. After several minutes the front closure is removed and the cabinet fan allowed to run for about 30 min. Any obvious moisture remaining on the cabinet walls and floor may then be wiped away and the filters changed.

Changing the filters and maintenance

The cabinet must be decontaminated before filters are changed and any work is done on the motor and fans. If these are to be done by an outside contractor, e.g. by the manufacturer's service engineer the front closure should be sealed on again after the initial purging of the gas and a notice: *Cabinet decontaminated but not to be used* placed on it awaiting the engineer's arrival. He will also require a certificate stating that the cabinet has been decontaminated.

The primary, or prefilters, should be changed when the cabinet airflow approaches its agreed local minimum. Used filters should be placed in plastic or tough paper bags, which are then sealed and burned. If the airflow is not restored to at least the middle of the range then arrangements should be made to replace the HEPA filter. This is usually done by the service engineer, but may be done by laboratory staff if they have received instruction. Unskilled operators often place the new filter in upside down or fail to set it securely and evenly in its place. Used filters should be placed in plastic bags, which are then sealed for disposal. They are not combustible. Some manufacturers accept used filters and recover the cases, but this is not usually a commercial practice – no refund is given. When manufacturers replace HEPA filters they may offer a testing service.

Testing; further information

This is beyond the scope of this book and the reader is referred to national standards and other publications (Clark, 1983; Collins, 1988).

Preparation room equipment

Sterilizers are discussed in Chapter 3.

Inspissators

In the preparation of slopes of medium consisting of egg or serum, the amount of heating required to coagulate the protein must be carefully controlled. A steamer heats the medium too rapidly and raises the temperature too high. The modern inspissator is thermostatically controlled at 75–85°C and fitted with a large internal circulating fan. The shelves on which the tubes are sloped are made of wire mesh so that circulation is not impeded. It is convenient to have wire mesh or aluminium racks made which hold tubes or bottles at the correct angle (5–10°) and which slide on the shelf brackets. These facilitate rapid loading and unloading while the instrument is hot.

An egg or serum medium is usually coagulated in 45–60 min at 80°C. Whether the inspissator is loaded hot or cold is a matter of personal choice, but better control, required for media that contain drugs, e.g. mycobacterial sensitivity tests, is obtained by raising the temperature first and then putting in a standard load for a constant time.

Laminar flow clean air work stations

These cabinets are designed to protect the work from the environment and are most useful for aseptic distribution of certain media and plate pouring. A stream of sterile (filtered) air is directed over the working area, either horizontally into the room or vertically downwards when it is usually re-circulated. They are particularly useful for preventing contamination when distributing sterile fluids. They are not 'safety cabinets' and should not be used for handling bacteria or tissue cultures, but a Class II microbiology safety cabinet (p.14) will also protect against contamination.

Glassware-washing machines

These are useful in large laboratories which use enough of any one size of tube or bottle to give an economic load.

Glassware drying cabinets

A busy wash-up room requires a drying cabinet with wire-rack shelves and a 3-kW electric heater in the base, operated through a three-heat control. An extractor fan on the top is a refinement, otherwise the sides near the top should be louvred.

Media distributors

Although time-honoured methods such as the funnel, rubber tube and clip are still useful in laboratories where only a small amount of culture media is distributed or tubed, automatic machines are now available and are indispensable in the larger media room.

Two types of machines are available; one is based on a peristaltic pump and the other on a syringe pump.

Peristaltic pump models deliver a pre-determined volume from a reservoir (large media bottle) and can be operated manually, by a pedal switch or by a timing device that permits slow or rapid delivery to suit the operator. The fluid is transferred entirely through autoclavable plastic tubing by a peristaltic pump and the cycle ensures that some fluid is returned to the reservoir to ensure mixing. As some media stain the tubing, separate 'manifolds' can be kept for each medium.

Pump models deliver accurate, pre-determined volumes over a wide range automatically or by foot or hand control. Measurement is in glass or stainless steel syringes with stainless steel or nylon valve assemblies and plastic tubing. For sterile distribution, the syringe assemblies and the tubing may be autoclaved.

Fully automatic machines for pouring petri dishes will sterilize reconstituted media, pour and then stack the poured petri dishes.

Glassware, plastics and small equipment

For ordinary bacteriological work, soda-glass tubes and bottles are satisfactory. In assay work, the more expensive resistance glass might be justified. An important consideration is whether glassware should be washed or discarded. Purchasing cheaper glassware in bulk and using plastic disposable petri dishes and culture tubes may, in some circumstances, be more economical than employing labour to clean them.

Plastics fall into two categories:

(1) disposable items, such as petri dishes, specimen containers and plastic loops, which are destroyed when autoclaved and cannot be sterilized by ordinary laboratory methods (but see Chapter 3), and
(2) recoverable material, which must be sterilized in the autoclave.

Non-autoclavable plastics
These include polyethylene, styrene, acrylonitrile, polystyrene and rigid polyvinyl chloride.

Autoclavable plastics
These withstand a temperature of 121°C and include polypropylene, polycarbonate, nylon, PTFE (Teflon), polyallomer, TPX (methylpentene polymer), Viken and vinyl tubing.

It is best to purchase plastic apparatus from specialist firms who will give advice on the suitability of their products for specific purposes. Most recoverable plastics used in microbiology can be washed in the same way as glassware. Some plastics soften during autoclaving and may become distorted unless packed carefully.

Petri dishes

Disposable plastic petri dishes are used in most laboratories in developed countries. They are supplied already sterilized and packed in batches in polythene bags. They can be stored indefinitely, are cheap when purchased in bulk and are well made and easy to handle. Two kinds are available, vented and unvented. Vented dishes have one or two nibs that raise the top slightly from the bottom and are to be preferred for anaerobic and carbon dioxide cultures.

Glass petri dishes are, however, still popular in some areas. In general, two qualities are obtainable. The thin, blown dishes, usually made of borosilicate glass, are pleasant to handle, their tops and bottoms are flat and they stack safely. They do not become etched through continued use but are fragile, must be washed with care and are expensive. The thick pressed glass dishes are often convex, cannot always be stacked safely and are easily etched and scratched with use and washing. On the other hand, they are cheap and not fragile.

Glass petri dishes are sterilized by hot air in aluminium boxes.

Test-tubes and bottles, plugs, caps and stoppers

Culture media can be tubed in either test-tubes or small bottles. Apart from personal choice, test-tubes are easier to handle in busy laboratories and take up less space in storage receptacles and incubators, but bottles are more convenient in the smaller workroom where media is kept for longer periods before use. Media in test-tubes may dry up during storage. Screw-capped test-tubes are available.

The most convenient sizes of test-tube are: 127×12.5 mm, holding 4 ml of medium: 152×16 mm, holding 5–10 ml; 152×19 mm, holding 10–15 ml; and 178×25 mm, holding 20 ml. Rimless test-tubes of heavy quality are made for bacteriological work. The lipped, thin glass chemical test-tubes are useless.

Cotton wool plugs have been used for many years to stopper test-tube cultures, but have largely been replaced by metal or plastic caps, or in some laboratories by soft, synthetic sponge bungs.

Aluminium test-tube caps were introduced some years ago but they have a limited tolerance and, in spite of alleged standard specifications of test-tubes, a laboratory very soon accumulates many tubes that will not fit the caps. Caps that are too loose are useless. These caps are cheap, last a long time, are available in many colours and save a great deal of time and labour. Those held in place by a small spring have a wider tolerance and fit most tubes. Polypropylene caps in several sizes and colours and which stand up to repeated autoclaving are obtainable.

Temporary closures can be made from kitchen aluminium foil.

Aluminium capped test-tubes are sterilized in the hot air oven in baskets. Polypropylene capped tubes must be autoclaved.

Rubber stoppers of the orthodox shape are useless as they are blown out of the tubes in the autoclave. The Astell rubber seals are designed to avoid this and to allow steam to penetrate into the tubes as readily as with cotton wool plugs or aluminium or polypropylene closures. These stoppers fit 16-mm tubes and also the Astell culture bottles used for roll-tube work, etc. They have a very long laboratory life.

Several sizes of small culture bottles are made for microbiological work. Some of these bottles are of strong construction and are intended to be reused. Others, although tough enough for safe handling, are disposable. The most useful sizes are: the 'bijou', holding 7 ml; the 'small McCartney' 14 ml; the 'McCartney', 28 ml; and the 'Universal Container', 28 ml, which has a larger neck than the others and is also used as a specimen container. These bottles usually have aluminium screw-caps with rubber liners. The liners should be made of black rubber; some red rubbers are thought to give off bactericidal substances. Polypropylene caps are also used; they need no liners but in our experience they may loosen spontaneously during long incubation or storage. The medium dries up. They are satisfactory in the short term. Astell culture bottles, holding about 20 ml and closed by Astell seals, are popular for roll-tube counting.

All of these bottles can be sterilized by autoclaving.

Media storage bottles

'Medical flats' or 'rounds' are made in sizes from 60 ml upwards. The most convenient sizes are 110 ml, holding 50–100 ml of medium, and 560 ml, for storing 250–500 ml. The flat bottles are easiest to handle and store but the round ones are more robust.

Specimen containers

We favour the screw-capped glass or plastic 'Universal Containers' which hold about 28 ml. There are many other containers, mostly made of plastic. There are too many different containers, many of plastic; some are satisfactory, others are not; some leak easily and others do not stand up to handling by patients. Only screw-capped containers with more than one and a half threads on their necks should be used. Those with 'push-in' or 'pop-up' stoppers are dangerous. They generate aerosols when opened. Waxed paper pots should not be used as they invariably leak.

For larger specimens, there is a variety of strong screw-capped jars.

Before any containers are purchased in bulk, it is advisable to test samples by filling them with coloured water and standing them upside down on blotting paper for several days after screwing the caps on moderately well. Patients and nurses may not screw caps on as tightly as possible. Similar bottles should be sent through the post, wrapped in absorbent material in accordance with postal regulations. Leakage will be evident by the staining of the blotting paper or wrapping. More severe tests include filling the containers with coloured water and centrifuging them upside down on a wad of blotting paper. Specimen container problems and tests are reviewed by Collins (1988a).

There is a British Standard (BSI, 1975) for medical specimen containers.

Sample jars and containers

Containers for food samples, water, milk, etc., should conform to local or national requirements. In general, large screw-capped jars are suitable but are rather heavy if many samples are taken, and should not be used in food factories. Plastic containers may be used as leakage and spillage is not such a problem as with pathological material. Strong plastic bags are useful, particularly if they are of the self-sealing type. Otherwise 'quick-ties' may be used.

Pasteur pipettes

These pipettes are probably the most dangerous pieces of laboratory equipment in unskilled hands. They are used, with rubber teats, to transfer liquid cultures, serum dilutions, etc.

Very few laboratories make their own pasteur pipettes nowadays; they are obtainable, plugged and unplugged, from most laboratory suppliers and are best purchased in bulk. Long and short forms are available and most are made of 6–7 mm diameter glass tubing.

After plugging, they can be sterilized in aluminium boxes made for this purpose.

New and safer pasteur pipettes with integral teats and made of low-density polypropylene are now available and are supplied ready sterilized.

Pasteur pipettes are used once only.

Graduated pipettes

Straight-sided blow-out pipettes, 1–10-ml capacity are used. They must be plugged with cotton wool at the suction end to prevent bacteria entering from the pipettor or teat and contaminating the material in the pipette. These plugs must be tight enough to stay where they are during pipetting but not so tight that they cannot be removed during cleaning. About 25 mm of non-absorbent cotton wool is pushed into the end with a piece of wire. The ends are then passed through a bunsen flame to tidy them. Wisps of cotton wool which get between the glass and the pipettor or teat may permit air to enter and the contents to leak.

Pipettes are sterilized in the hot air oven in square section aluminium containers similar to those used for pasteur pipettes. A wad of glass wool at the bottom of the container prevents damage to the tips.

Disposable 1 ml and 10 ml pipettes are available. Some firms supply them already plugged, wrapped and sterilized.

Pipetting aids

Rubber teats and pipetting devices provide an alternative to the highly dangerous practice of mouth pipetting.

Rubber teats
Choose teats with a capacity greater than that of the pipettes for which they are intended, i.e. a 1-ml teat for pasteur pipettes, a 2-ml teat for 1-ml pipette, otherwise the teat must be used fully compressed, which is tiring. Most beginners compress the teat completely, then suck up the liquid and try to hold it at the mark while transferring it. This is unsatisfactory and leads to spilling and inaccuracy. Compress the teat just enough to suck the liquid a little way past the mark on the pipette. Withdraw the pipette from the liquid, press the teat slightly to bring the fluid to the mark and then release it. The correct volume is now held in the pipette without tiring the thumb and without risking loss. To discharge the pipette, press the teat slowly and gently and then release it in the same way. Violent operation usually fails to eject all the liquid; bubbles are sucked back and aerosols are formed.

'Pipettors'

A large number of devices that are more sophisticated than simple rubber teats are now available. Broadly speaking there are four kinds of these:

(1) rubber bulbs with valves that control suction and dispensing;
(2) syringe-like machines that hold pipettes more rigidly than rubber bulbs and have a plunger operated by a rack and pinion or a lever;
(3) electrically operated pumps fitted with flexible tubes in which pipettes can be inserted;
(4) mechanical plunger devices which take small plastic pipette tips and are capable of repeatedly delivering very small volumes with great accuracy.

It is extremely difficult to give advice on the relative merits of the various devices. That which suits one operator, or is best for one purpose may not be suitable for others. Choice should therefore be made by the operators, not by managers or administrators who will not use them. None of those in categories 1, 2 and 4 above are expensive, and it should be possible for several different models to be available. What is necessary is some system of instruction in their use and in their maintenance. None should be expected to last forever.

Micro-slides and cover-glasses

Unless permanent preparations are required, the cheaper microscope slides are satisfactory. Slides with ground and polished edges are much more expensive. Most micro-slides are sold in boxes of 100 slides. They should not be washed and re-used but discarded.

Cover-glasses are sold in several thicknesses and sizes. Those of thickness grade No. 1, 16-mm square, are the most convenient. They are sold in boxes containing about 100 glasses. Plastic cover-glasses are available.

Durham's (fermentation) tubes

These are small glass tubes, usually 25–30 × 5–6 mm and closed at one end, which are placed inverted in tubes of culture media to detect gas formation. They are very cheap and are not worth washing.

Inoculating loops and wires

These are usually made of 25 SWG Nicrome wire, although this is more springy than platinum iridium. They should be short (not more than 5 cm long) in order to minimize vibration and therefore involuntary discharge of contents. Loops should be small (not more than 3 mm in diameter). Large loops are also inclined to empty spontaneously and scatter infected airborne particles. They should be completely closed, otherwise they will not hold fluid cultures. This can be carried out by twisting the end of the wire round the shank, or by taking a piece of wire 12 cm long, bending the centre round a nail or rod of appropriate diameter and twisting the ends together in a drill chuck. Ready-made loops of this kind are available.

Loops and wires should not be fused into glass rods. Aluminium holders are sold by most laboratory suppliers.

Plastic disposable loops are excellent. There are two sizes, 1 and 10 µl, and both are useful. They are sold sealed in packs of about 25, sterilized ready for use.

Spreaders

Cut a glass rod of 3–4 mm diameter into 180 mm lengths and round off the cut ends in a flame. Hold each length horizontally across a bat's wing flame so that it is heated and bends under its own weight approximately 36 mm from one end, and an L shape is obtained. Plastic spreaders are on the market, sterilized ready for use.

Racks and baskets

Test-tube and culture bottle racks should be made of polypropylene or of metal covered with polypropylene or nylon so that they can be autoclaved. These racks also minimize breakage, which is not uncommon when metal racks are used. Wooden racks are unhygienic.

Aluminium trays for holding from 10–100 bottles, according to size are widely employed in the UK. They can be autoclaved, and are easily taken apart for cleaning.

The traditional wire baskets are unsafe for holding test-tubes. They contribute to breakage hazards and do not retain spilled fluids. For non-infective work, these baskets, covered with polypropylene or nylon, are satisfactory, but autoclavable plastic boxes of various sizes are safer for use with cultures.

Bunsen burners

The usual bunsen, with a bypass, is satisfactory for most work, but for material that may spatter or that is highly infectious a hooded bunsen should be used. There are several versions of these, but we recommend the Kampff and Bactiburner types which enclose the flame in a borosilicate tube.

Other bench equipment

A hand magnifier, forceps and a knife or scalpel should be provided, and also a supply of tissues for mopping up spilled material and general cleaning. Swab sticks, usually made of wood and about 6 in long, are useful for handling some specimens, and wooden throat spatulas are useful for food samples. These can be sterilized in large test-tubes or in the aluminium boxes used for pasteur pipettes.

Some form of bench 'tidy' or rack is desirable to keep loops and other small articles together.

Discard jars and disinfectant pots are considered in Chapter 3.

References

BSI (1975) BS 5213 *Medical Specimen Containers for Microbiology*, British Standards Institution, London

BSI (1979) *BS 5726: Specification for Microbiological Safety Cabinets*, British Standards Institution, London

Clark, R. P. (1983) *The Performance, Testing and Limitations of Microbiological Safety Cabinets*. Science Reviews, Norwood

Collins, C. H. (1988) *Laboratory-acquired Infections,* 2nd edn, Butterworths, London
NSF (1984) *Standard No 49. Class II (Laminar Flow) Biohazard Cabinetry.* National Sanitation
    Foundation, Ann Arbor, MI, USA
SAA (1980) *Biological Safety Cabinets*, AS 2552, Standards Association of Australia, Sydney

Chapter 3

Sterilization, disinfection and the treatment of infected materials

'Sterilization' implies the complete destruction of all microorganisms, including spores.

'Disinfection' implies the destruction of vegetative organisms which might cause disease, or, in the context of the food industries, which might cause spoilage. Disinfection does not necessarily kill spores.

The two terms are not synonymous.

Sterilization

The methods commonly used in microbiological laboratories are:

(1) red heat (flaming);
(2) dry heat (hot air);
(3) steam under pressure (autoclaving);
(4) steam not under pressure (Tyndallization); and
(5) filtration.

Incineration is also a method of sterilization but as it is applied outside the laboratory for the ultimate disposal of laboratory waste it is considered separately (p.49).

Red heat

Instruments such as inoculating wires and loops are sterilized by holding them in a bunsen flame until they are red hot. Hooded bunsens (p.27) are recommended for sterilizing inoculating wires contaminated with highly infectious material (e.g. tuberculous sputum) to avoid the risk of spluttering contaminated particles over the surrounding areas.

Dry heat

This is applied in an electrically heated oven which is thermostatically controlled and is fitted with a large circulating fan to ensure even temperatures in all parts of the load. Modern equipment has electronic controls which can be set to raise the temperature to the required level, hold it there for a pre-arranged time and then switch off the current. A solenoid lock is incorporated in some models to prevent the oven being opened before the cycle is complete. This safeguards sterility and protects the staff from accidental burns.

Materials which can be sterilized by this method include glass petri dishes, flasks, pipettes and metal objects. Various metal canisters and cylinders which conveniently hold glassware during sterilization and keep it sterile during storage are available from laboratory suppliers.

Loading
Air is not a good conductor of heat so oven loads must be loosely arranged, with plenty of spaces to allow the hot air to circulate.

Holding times and temperatures
When calculating processing times for hot air sterilizing equipment, there are three time periods which must be considered.

(1) *the heating-up period*, which is the time taken for the entire load to reach the sterilization temperature; this may take about 1 h;
(2) *the holding periods* at different sterilization temperatures recommended by the UK Department of Health (DHSS, 1980) which are 160°C for 45 min, 170°C for 18 min, 180°C for $7\frac{1}{2}$ min and 190°C for $1\frac{1}{2}$ min;
(3) *the cooling-down period*, which is carried out gradually to prevent glassware from cracking as a result of a too rapid fall in temperature; this period may take up to 2 h.

Control of hot air sterilizers

Tests for electrically operated fan ovens are described in British Standard 3421 (BSI, 1961).

Hot air sterilization equipment should be calibrated with thermocouples when the apparatus is first installed and checked with thermocouples afterwards when necessary.

Ordinary routine control can be effected simply with the aid of Browne's tubes. The Browne's Type 3 tubes (Green Spot) are used for fan-operated ovens.

Steam under pressure

This is done by autoclaving. Bacteria are more readily killed by moist heat (saturated steam) than by dry heat. Steam kills bacteria by denaturing their protein. An agreed safe condition for sterilization is to use steam at 121°C for 15 min (DHSS, 1980). This is suitable for culture media, aqueous solutions, treatment of discarded cultures and specimens, etc.

Air has an important influence on the efficiency of steam sterilization because its presence changes the pressure–temperature relationship. For example the temperature of saturated steam at 15 lb/in^2 is 121°C, provided that all of the air is first removed from the vessel. With only half of the air removed, the temperature of the resulting air–steam mixture at the same pressure is only 112°C. In addition, the presence of air in mixed loads will prevent penetration by steam.

All the air that surrounds and permeates the load must first be removed before steam sterilization can commence.

Loads in autoclaves

As successful autoclaving depends on the removal of all the air from the chamber and the load, the materials to be sterilized should be packed loosely. 'Clean' articles may be placed in wire baskets, but contaminated material (e.g. discarded cultures) should be in solid bottomed containers not more than 8 in deep (see Disposal of Infected Waste, below). Large air spaces should be left around each container and none should be covered.

Types of autoclave

Only autoclaves designed for laboratory work and capable of dealing with a 'mixed load' should be used. 'Porous load' and 'bottled fluid sterilizers' are rarely satisfactory for laboratory work. There are two varieties of laboratory autoclave:

(1) pressure cooker types; and
(2) gravity displacement models with automatic air and condensate discharge.

'Pressure cooker' laboratory autoclaves

These are still in use in many parts of the world. The most common type is a device for boiling water under pressure. It has a vertical metal chamber with a strong metal lid which can be fastened down and sealed with a rubber gasket. An air and steam discharge tap, pressure gauge and safety valve are fitted in the lid (Figure 3.1). Water in the bottom of the autoclave is heated by external gas burners, an electric immersion heater or a steam coil.

Operating instructions
There must be sufficient water inside the chamber. The autoclave is loaded and the lid is fastened down with the discharge tap open. The safety valve is then adjusted to the required temperature and the heat is turned on.

  When the water boils, the steam will issue from the discharge tap and carry the air from the chamber with it. The steam and air should be allowed to escape freely

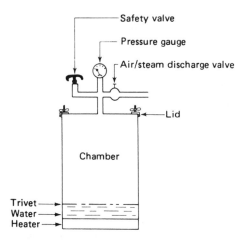

Figure 3.1 'Pressure cooker' laboratory autoclave

until all of the air has been removed. This may be tested by attaching one end of a length of rubber tubing to the discharge tap and inserting the other end into a bucket or similar large container of water. Steam condenses in the water and the air rises as bubbles to the surface; when all of the air has been removed from the chamber, bubbling in the bucket will cease. When this stage has been reached, the air–steam discharge tap is closed and the rubber tubing removed.The steam pressure then rises in the chamber until the desired pressure and temperature are reached and steam issues from the safety valve.

When the load has reached the required temperature (see Testing autoclaves, below), the pressure is held for 15 min.

At the end of the sterilizing period, the heater is turned off and the autoclave allowed to cool.

The air and steam discharge tap is opened very slowly after the pressure gauge has reached zero (atmospheric pressure). If the tap is opened too soon, while the autoclave is still under pressure, any fluid inside (liquid media, etc.) will boil explosively and bottles containing liquids may even burst. The contents are allowed to cool. Depending on the nature of the materials being sterilized, the cooling (or 'run-down') period needed may be several hours for large bottles of agar to cool to 80°C, when they are safe to handle.

Autoclaves with air discharge by gravity displacement

These autoclaves are usually arranged horizontally and are rectangular in shape, thus making the chamber more convenient for loading. A pallet and trolley system can be used.

Figure 3.2 shows in diagrammatic form a jacketed gravity displacement type of autoclave. Similar autoclaves can be constructed without jackets. The door should have a safety device to ensure that it cannot be opened while the chamber is under pressure.

The jacket surrounding the autoclave consists of an outer wall enclosing a narrow space around the chamber, which is filled with steam under pressure to keep the chamber wall warm. The steam enters the jacket from the mains supply, which is at high pressure, through a valve that reduces this pressure to the working level. The working pressure is measured on a separate pressure gauge fitted to the jacket. This jacket also has a separate drain for air and condensate to pass through.

The steam enters the chamber from the same source which supplies steam to the jacket. It is introduced in such a way that it is deflected upwards and fills the chamber from the top downwards, thus forcing the air and condensate to flow out of the drain at the base of the chamber by gravity displacement. The drain is fitted with strainers to prevent blockage by debris. The drain is usually fitted with a thermometer for registering the temperature of the issuing steam. The temperature recorded by the drain thermometer is often lower than that in the chamber. The difference should be found with thermocouple tests (see below). A 'near to steam' trap is also fitted.

The automatic steam trap or 'near-to-steam' trap is designed to ensure that only saturated steam is retained inside the chamber, and that air and condensate, which are at a lower temperature than saturated steam, are automatically discharged. It is called a 'near-to-steam' trap because it opens if the temperature falls to about 2°C below that of saturated steam and closes within 2°C or nearer to the saturated steam temperature. The trap operates by the expansion and contraction of a metal

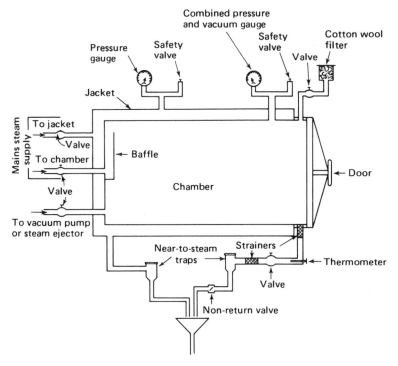

Figure 3.2 Gravity displacement autoclave

bellows, which open and close a valve. The drain discharges into a tundish in such a way that there is a complete air-break between the drain and the dish. This ensures that no contaminated water can flow back from the waste-pipe into the chamber.

Operation of a gravity displacement autoclave
In the autoclave is jacketed, the jacket must first be brought to the operating temperature. The chamber is loaded, the door is closed and the steam-valve is opened, allowing steam to enter the top of the chamber. Air and condensate flow out through the drain at the bottom (Figure 3.2). When the drain thermometer reaches the required temperature a further period must be allowed for the load to reach the temperature. This should be determined initially and periodically for each autoclave as described below. Unless this is done the load is unlikely to be sterilized. The autoclave cycle is then continued for the holding time. When it is completed the steam valves are closed and the autoclave allowed to cool until the temperature dial reads less than 80°C. Not until then is the autoclave safe to open. It should first be 'cracked' or opened very slightly and left in that position for several minutes to allow steam to escape and the load to cool further (see Safety of the operator, below).

Tests for autoclave efficiency; determining time cycles

Steam sterilizers should be tested when purchased and at regular intervals thereafter with thermocouples.

Measurement of temperatures is by thermocouples built into the autoclave so that sensors may be placed anywhere in the load. Those recording the temperature in the drain are misleading; the drain temperature may reach 121°C while that in the load is less than 100°C.

Autoclaves should be tested under the 'worst load' conditions. In most laboratories this would be a container full of 5 ml screw-capped bottles. This should be placed in the centre of the autoclave, and if space is available other loaded containers placed around it. A thermocouple should be placed in a bottle in the middle of this load. Other thermocouples may be distributed in other parts of the autoclave. The sterilization cycle is then started and timed. The time taken for the temperature in the chamber drain to reach 121°C is noted and then the time taken for the thermocouple in the load to register that temperature. This is when the sterilization time begins. After not less than 15 min the steam may be turned off when the cooling time begins. Thus there are four periods:

(a) *warming-up*, until chamber drain thermometer reaches 121°C;
(b) *steam penetration into load*, until centre of load reaches 121°C;
(c) *sterilization*, during which the load is maintained at 121°C, usually 15 min;
(d) *cooling down time*, until the temperature in the load falls to 80°C, when it may be removed.

As it is not practicable to use a thermocouple for each load (unless a sensor is built into the chamber), it is important to note these times for normal operation. The period of exposure *after* the temperature in the drain reaches 121°C is therefore (b)+(c) min. Automatic cycle autoclaves, which depend on the temperature in the drain should be modified accordingly.

Although thermocouples connected to recorders or digital read-out devices are the ideal there are other kinds of indicators which can be placed in the load. They contain chemicals in tubes or on paper strips which change colour or appearance when a suitable time/temperature relationship is achieved.

In some autoclaves the door cannot be opened until the temperature in the drain falls to 80°C. This does not imply that the temperature in the load has also fallen to a safe level. That in large, sealed bottles may still be over 100°C, when the contents will be at a high pressure. Sudden cooling may cause the bottles to explode. The autoclave should not be opened until the temperature in the load has fallen to 80°C or below. This may take a very long time, and in some autoclaves there are locks which permit the door to be opened only fractionally to cool the load further before it is finally released. In others, complicated cooling arrangements with air blasts and cold water sprays are used.

Material suspected of containing the agents of spongiform encephalopathies should be autoclaved at 134°C for 18 min or given six consecutive 3 minute cycles at that temperature.

For further information on autoclave problems see PHLS (1978, 1981), Gardner and Peel (1986) and Kennedy (1988).

Protection of the operator

Serious accidents, including burns and scalds to the face and hands have occurred when autoclaves have been opened, even when the temperature gauge read below 80°C and the door lock has allowed the door to be opened. Liquids in bottles may still be over 100°C and under considerable pressure. The bottles may explode on contact with air at room temperature.

When autoclaves are being unloaded operators should wear full-face visors of the kind that cover the skin under the chin and throat. They should also wear thermal-protective gloves (see Kennedy, 1988).

Steam at 100°C ('Tyndallization')

This process, named after the Irish physicist cum bacteriologist John Tyndall, employs a Koch or Arnold steamer, which is a metal box in the bottom of which water is boiled by a gas burner, electric heater or steam coil. The articles to be processed rest on a perforated rack just above the water level. The lid is conical so that the condensation runs down the sides instead of dripping on the contents. A small hole in the top of the lid allows air and steam to escape. This method is used to sterilize culture media that might be spoiled by exposure to higher temperatures, e.g. media containing easily hydrolysed carbohydrates or gelatin. These are steamed for 30–45 min on each of three successive days. On the first occasion, vegetative bacteria are killed; any spores that survive will germinate in the nutrient medium overnight, producing vegetative forms that are killed by the second or third steaming.

Filtration

Bacteria can be removed from liquids by passing them through filters with very small pores that trap bacteria but, in general, not mycoplasmas or viruses. The method is used for sterilizing serum for laboratory use, antibiotic solutions and special culture media that would be damaged by heat. It is also used for separating the soluble products of bacterial growth (e.g. toxins) in fluid culture media.

Types of filters

The following types of filters are mainly of historic interest:

(1) earthenware candles such as the Berkefeld and Mandler type, which are made from Kieselghur, and the Chamberland filter made from unglazed porcelain;
(2) asbestos pads (Seitz filters);
(3) sintered-glass filters, which are made of finely ground glass, fused sufficiently to make the glass granules adhere to one another.

The disadvantages of earthenware candles and asbestos pads are that they are absorptive and have a comparatively slow filtration rate.

Sintered-glass filters have the advantage that little absorption takes place, but the rate of filtration is comparatively slow.

Membrane filters
These filters are made from cellulose esters (cellulose acetate, cellulose nitrate, collodion, etc.). They have many advantages over the early types of bacteriological filters because they are much less absorptive and have a high filtration rate. A range of pore sizes is available. Bacterial filters have a pore size of less than 0.75 μm. The membranes can be sterilized by autoclaving.

For use, the membrane is mounted on a perforated platform, usually made of stainless steel, which is sealed together between the upper and lower funnels. Filtration is achieved by applying either a positive pressure to the entrance side of the filter or a negative pressure to the exit side.

Small filter units for filtering small volumes of fluid (e.g. 1–5 ml) are available. The fluid passes through the filter by gravitational force in a centrifuge or is forced through a small filter from a syringe.

Membrane filters and filter holders to suit different purposes are obtainable and several manufacturers publish useful booklets or leaflets about their products.

Chemical disinfection

Some disinfectants present health hazards (see Table 3.1). It is advisable to wear eye and hand protection when making dilutions.

Many different chemicals may be used and they are collectively described as disinfectants or biocides. The former term is used in this book. Some are ordinary reagents, others are special formulations, marketed under trade names. Microbiologists and laboratory managers are often under pressure from salesmen to buy products for which extravagant claims are made. There are usually marked differences between the activity of some disinfectants when tested under optimal conditions by the Rideal–Walker or similarly discredited techniques and when they are used in practice. The effects of time, temperature, pH, and the chemical and physical nature of the article to be disinfected and of the organic matter present are often not fully appreciated.

Types and laboratory uses of disinfectants

There is an approximate spectrum of susceptibility of microorganisms to disinfectants. The most susceptible are vegetative bacteria, fungi and lipid-containing viruses. Mycobacteria and non lipid-containing viruses are less susceptible and spores are generally resistant.

Some consideration should be given to the disinfectant's toxicity and any harmful effects that they may have on the skin, eyes and respiratory tract.

Only those disinfectants which have a laboratory application are considered here. For further information and other applications see Russell *et al.* (1982), Ayliffe *et al.* (1984) and Gardner and Peel (1986).

The most commonly used disinfectants in laboratory work are clear phenolics and hypochlorites. Aldehydes have a more limited application, and alcohol and alcohol mixtures are less popular but deserve greater attention. Iodophors and quaternary ammonium compounds (QAC) are more popular in the USA than in the UK, while mercurial compounds are the least used. The properties of these disinfectants are summarized below and in Table 3.1. Other substances, such as ethylene oxide and propiolactone, are used commercially in the preparation of sterile equipment for hospital and laboratory use but are not used for decontaminating laboratory waste and the other activities mentioned above.

Clear phenolics
These compounds are effective against vegetative bacteria (including mycobacteria) and fungi. They are inactive against spores and non lipid-containing viruses.

Table 3.1 Properties of some disinfectants

	Active against					Inactivated by					Toxicity		
	Fungi	G+	Bacteria G−	Myco-bacteria	Spores	Protein	Natural materials	Man-made materials	Hard water	Detergent	Skin	Eyes	Lungs
Phenolics	+++	+++	+++	++	−	+	++	++	+	C	+	+	−
Hypochlorites	+	+++	+++	++	++	+++	+	++	+	C	+	++	+++
Alcohols	−	+++	+++	+++	+++	+++a	+	+	+	+	−	++	+++
Formaldehyde	+++	+++	+++	+++	+++a	+	+	+	+	−	+	+	
Glutaral-dehyde	+++	+++	+++	++	+++b	NA	+	+	+	−	+	++	+
Iodophors	+++	+++	+++	++	+	+++	+	+	+	A	+	++	−
QAC	+	+++	++	−	−	+++	+++	+++	+++	A(C)	+	+	−

+++ good   G Gram
++ fair   C Cationic
+ slight   A Anionic
− nil
a above 40°C
b above 20°C
From Collins (1988)

Most phenolics are active in the presence of considerable amounts of protein but are inactivated to some extent by rubber, wood and plastics. They are not compatible with cationic detergents. Laboratory uses include discard jars and disinfection of surfaces. Clear phenolics should be used at the highest concentration recommended by the manufacturers for 'dirty situations', i.e. where they will encounter relatively large amounts of organic matter. This is usually 2–5%, as oppposed to 1% for 'clean' situations where they will not encounter much protein. Dilutions should be prepared daily and diluted phenolics should not be stored for laboratory use for more than 24 h, although many diluted clear phenolics may be effective for more than 7 days.

Skin and eyes should be protected.

Hypochlorites
The activity is due to chlorine, which is very effective against vegetative bacteria (including mycobacteria), spores and fungi. Hypochlorites are considerably inactivated by protein and to some extent by natural non-protein material and plastics and they are not compatible with cationic detergents. Their uses include discard jars and surface disinfection but as they corrode some metals care is necessary. They should not be used on the metal parts of centrifuges and other machines which are subjected to stress when in use.

The hypochlorites sold for industrial and laboratory use in the UK contain 100 000 ppm available chlorine. They should be diluted as follows:

Reasonable clean surfaces	1:100	giving	1 000	ppm
Pipette and discard jars	1:40		2 500	
Blood spillage	1:10		10 000	

Some household hypochlorites (e.g. those used for babies' feeding bottles) contain 10 000 ppm and should be diluted accordingly. Household 'bleaches' in the UK and USA contain 50 000 ppm available chlorine and dilutions of 1:20 and 1:5 are appropriate. Solid preparations including those that contain hypochlorites, used for domestic disinfection, and chlorinated isocyanurates, used in swimming pools, may have laboratory applications.

Hypochlorites decay rapidly in use, although the products as supplied are stable. Diluted solutions should be replaced after 24 h. The colouring matter added to some commercial hypochlorites is intended to identify them: it is not an indicator of activity.

Hypochlorites may cause irritation of skin, eyes and lungs.

Aldehydes
Formaldehyde (gas) and glutaraldehyde (liquid) are good disinfectants. They are active against vegetative bacteria (including mycobacteria), spores and fungi. They are active in the presence of protein and are not very much inactivated by natural or man-made materials, or detergents.

Formaldehyde is not very active at temperatures below 20°C and requires a relative humidity of at least 70%. It is not supplied as a gas, but as a solid polymer, paraformaldehyde, and a liquid, formalin, which contains 37–40% of formaldehyde. Both forms are heated or mixed with potassium permanganate and water to liberate the gas, which is used for disinfecting enclosed spaces such as safety cabinets and rooms. Formalin, diluted 1:10 to give a solution containing 4% formaldehyde, is used for disinfecting surfaces and, in some circumstances, cultures. Solid, formaldehyde-releasing compounds are now on the market and these may have laboratory applications but they have not yet been evaluated for this purpose. Formaldehyde is used mainly for decontaminating safety cabinets (p.19) and rooms.

Glutaraldehyde usually needs an activator, such as sodium bicarbonate, which is supplied with the bulk liquid. Most activators contain a dye so that the user can be sure that the disinfectant has been activated. Effectiveness and stability after activation varies with product and the manufacturers' literature should be consulted. Kelsey *et al.* (1974) found that the sporicidal activity of one brand was halved in 7 days, and Coates (1980) suggested that if activated glutaraldehydes are used up to this age then the exposure time should be doubled. They may be used in discard jars (but are expensive) and are particularly useful for disinfecting metal surfaces as they do not cause corrosion.

Aldehydes are toxic. Formaldehyde is particularly unpleasant as it affects the eyes and causes respiratory distress. Special precautions are required (see below). Glutaraldehyde is less harmful, but contact with skin and eyes should be avoided.

Alcohol and alcohol mixtures
Ethanol and propanol, at concentrations of about 70–80% in water are effective, albeit slowly, against vegetative bacteria. They are not effective against spores or fungi. They are not especially inactivated by protein and other material or detergents.

Effectiveness is enhanced by the addition of formaldehyde, e.g. a mixture of 10% formalin in 70% alcohol, or hypochlorite to give 2000 ppm of available chlorine (USPHS, 1978; Coates and Death, 1978).

Alcohols and alcohol mixtures are useful for disinfecting surfaces and, alcohol–hypochlorite mixtures excepted, for balancing centrifuge buckets.

They are relatively harmless to skin but may cause eye irritation.

Quaternary ammonium compounds
These are cationic detergents known as QACs or quats, and are effective against vegetative bacteria and some fungi but not against mycobacteria or spores. They are inactivated by protein and by a variety of natural and plastic materials and by anionic detergents and soap. Their laboratory uses are therefore limited but they have the distinct advantages of being stable and of not corroding metals. They are usually employed at 1 – 2% dilution for cleaning surfaces and are very popular in food hygiene laboratories because of their detergent nature.

QACs are not toxic and are harmless to the skin and eyes.

Iodophors
Like chlorine compounds these iodines are effective against vegetative bacteria (including mycobacteria), spores, fungi, and both lipid-containing and non lipid-containing viruses. They are rapidly inactivated by protein, and to a certain extent by natural and plastic substances and are not compatible with anionic detergents. For use in discard jars and for disinfecting surfaces they should be diluted to give 75–150 ppm iodine (USPHS, 1974), but for hand-washing or as a sporicide, diluted in 50% alcohol to give 1600 ppm iodine (USPHS, 1978). As sold, iodophors usually contain a detergent and they have a built-in indicator: they are active as long as they remain brown or yellow. They stain the skin and surfaces but stains may be removed with sodium thiosulphate solution.

Iodophors are relatively harmless to skin but some eye irritation may be experienced.

Mercurial compounds

Activity against vegetative bacteria is poor and mercurials are not effective against spores. They do have an action on viruses at concentrations of 1:500 to 1:1000 and a limited use, as saturated solutions, for safely making microscopic preparations of mycobacteria.

Their limited usefulness and highly poisonous nature make mercurials unsuitable for general laboratory use.

Precautions in the use of disinfectants

As indicated above, some disinfectants have undesirable effects on the skin, eyes and respiratory tract. Disposable gloves and safety spectacles, goggles or a visor should be worn by anyone who is handling strong disinfectants, e.g. when pouring from stock and preparing dilutions for use.

The laboratory testing of disinfectants

The following are the main tests that have been devised for assessing the efficiency of disinfectants.

Manufacturers' tests

These tests are used to control the quality of batches during production. The Rideal–Walker (BSI, 1934 but under revision) and Chick–Martin (BSI, 1938) tests were originally developed for testing the efficiency of phenolic-type disinfectants against a pure phenol standard. The Rideal–Walker test compares the performance of a disinfectant with that of phenol, and the results are calculated as a phenol coefficient, which is given as a number following the letters RW. Higher numbers indicate better disinfectant performances. This test was introduced during the early years of this century when phenol was much used as a disinfectant. A few years after the Rideal–Walker test had been described, Chick and Martin published their phenol coefficient test method. It is similar in design to the Rideal–Walker test in many ways but is done in the presence of organic material. As most disinfectants are inactivated by organic material, the Chick–Martin coefficient (CM) is therefore normally a lower number than the RW coefficient.

The Rideal–Walker and Chick–Martin tests are manufacturers' tests for phenolic disinfectants but the more recently introduced non-phenolic types of disinfectants have very different properties and vary greatly, and they are unsuitable for either the Rideal–Walker or the Chick–Martin test.

Other tests include AOAC (1960) and the Kelsey–Sykes (Kelsey and Maurer, 1974) tests. Much has been written about these tests which need not be reviewed here and the reader is referred to the work by Croshaw (1981).

There now seems to be no justification for doing any of them in the clinical or research microbiology laboratory. They are best left to the reference and other specialized establishments. Results of 'one-off' tests may not be reliable except in very skilled hands. Tests may be wrong in principle, when like is not compared with like and – a common mistake – neutralization procedures may be inadequate or incorrectly applied (Russell, 1981).

There is also another good reason why certain of these tests should not be done in clinical and teaching laboratories. Most of them retain *Salmonella typhi* as the test organism, although there is a great deal of scientific and commonsense objection to the use of this organism (Croshaw, 1981). It is claimed that the strains of typhoid bacilli specified, e.g. *S.typhi* NCTC 786 is not virulent because it has no Vi antigen, an assumption now questioned by bacteriologists.

A more serious objection, noted by ourselves, is that clinical and teaching laboratories have used strains of *S.typhi*, recently isolated from cases of typhoid fever, for these tests. This observation, made earlier to an official working party, contributed to the decision to ban the use of *S.typhi* for testing disinfectants in clinical laboratories in the UK (DHSS, 1978)

To meet a growing need for more realistic tests for hospital use, the Public Health Laboratory Service (PHLS) has been responsible for the development of tests that are employed to determine a practical use-dilution for a particular brand of disinfectant which is to be used under hospital or laboratory conditions, e.g. for decontaminating dirty instruments, cleaning hospital walls and operating theatre trolleys and in the bacteriological laboratory for such uses as in laboratory discard jars. These tests are:

(1) the capacity use-dilution test (Kelsey and Maurer, 1974)
(2) the stability test (Maurer, 1969) and
(3) the 'in-use' test (Maurer, 1972) described in detail below.

The capacity and stability tests are carried out mainly in manufacturers' and reference laboratories. The 'in-use' test should be employed by the users of disinfectants to check the efficiency of the disinfectants actually in use in their own hospitals or laboratories. This test is easy to perform and is strongly recommended for routine checks.

The Kelsey–Sykes capacity test

The Kelsey–Sykes capacity test is used to estimate the concentration of disinfectants that can be recommended for use in hospitals under 'clean' and 'dirty' conditions, that is, in the absence or in the presence of organic material. It is suitable for testing all types of disinfectants.

The term 'clean condition' implies a situation in which organic material is absent, such as food preparation surfaces and trolley tops, which are first washed clean with hot water and detergent before the application of a disinfectant. The term 'dirty condition' is applied to those situations where organic material is present, e.g. the surfaces of bedpans and laboratory discard jars, and higher concentrations of disinfectants are needed to neutralize the inactivating effect of the organic material and still ensure disinfection.

Basically, the test consists in adding a known volume of a suspension of a test organism at set intervals to a standard volume of the particular concentration of the disinfectant under test, with or without added organic matter. The organic matter added is usually yeast. At regular time intervals after the addition of each of three volumes of the test suspension, a standard amount is removed from the reaction mixture and added to five tubes of recovery medium.

During the progress of the test, the disinfectant is increasingly diluted, but it is the particular concentration at the start which is said to pass or fail the test.

A starting concentration which, after the second addition of organism suspension, yields no growth in at least two out of the five tubes of recovery medium is said to pass the test and can be recommended for use under 'clean' conditions, or under 'dirty' conditions when yeast was included in the test.

The test is not intended for use in hospital laboratories, but is for official laboratories or for manufacturers, as a guide for recommending the concentrations of disinfectants to be used in hospitals, catering establishments, etc.

The stability test

This is a simplified test designed to check the stability and long-term effectiveness of disinfectants in concentrations recommended by the manufacturers for use in 'clean' and 'dirty' situations. The test is used to supplement the information given by the Kelsey–Sykes capacity test and is suitable only for specialist laboratories. A study of the paper by Maurer (1969) shows evidence that bacteria can not only survive but also multiply in some disinfectant solutions.

The 'in-use' test

Samples of liquid disinfectants are taken from such sources as laboratory discard jars, floor-mop buckets, mop wringings, disinfectant liquids in which cleaning materials or lavatory brushes are stored, disinfectants in Central Sterile Supply Departments, used instrument containers and stock solutions of diluted disinfectants. The object is to determine whether the fluids contain living bacteria, and in what numbers. The test is described here in detail for use in any laboratory, because meaningful results can be obtained only in the light of local circumstances.

Method (as described by Maurer, 1972)
Stage 1   A 1-ml volume of disinfectant solution in use is taken from each pot or bucket with a separate sterile pipette.
Stage 2   The 1-ml sample is added to 9 ml of diluent in a sterile universal container or a 25-ml screw-capped bottle. The diluent should be selected according to the group to which the disinfectant belongs (Table 3.2).
Stage 3   The bottle of diluent is returned to the laboratory within 1 h of the addition of the disinfectant. A separate sterile pasteur pipette or 50-drop pipette is used to withdraw a small volume of the disinfectant/diluent and to place ten drops, separately, on the surface of each of two well-dried nutrient agar plates.
Stage 4   The two plates are incubated as follows:

(1) One plate is incubated for 3 days at 32 or 37°C, whichever is most convenient. The optimum temperature for most pathogenic bacteria is 37°C but those which have been damaged by disinfectants often recover more readily at 32°C.
(2) The second plate is incubated for 7 days at room temperature.

Stage 5 After incubation, the plates are examined and bacterial growth is recorded.

Table 3.2 'In-use' test neutralizing diluents

Diluent	Disinfectant group
Nutrient broth	Alcohols
	Aldehydes
	Hypochlorites
	Phenolics
Nutrient broth + Tween 80, 3% w/v	Hypochlorites + detergent
	Iodophors
	Phenolics + detergent
	QACs

'In-use' test results

The growth of bacterial colonies on one or both of a pair of plates is evidence of the survival of bacteria in the particular pot from which the sample is taken. One or two colonies on a plate may be ignored. A disinfectant is not a sterilant and the presence of a few live bacteria in a pot is to be expected.

However, the growth of five or more colonies on one plate should arouse suspicion that all is not well. The relationship between the number of colonies on the plate and the number of live bacteria in the pot can easily be calculated, as the disinfectant sample is diluted 1 in 10 and the 50-drop pipette delivers 50 drops/ml. When five colonies are grown from ten drops of disinfectant/diluent, then five live bacteria were present in one drop of disinfectant and 250 live bacteria were present in 1 ml of disinfectant.

Precautions when testing disinfectants

In spite of the misgivings expressed above and by others about the use of *S.typhi* in the testing of disinfectants it is apparent that the manufacturers wish to cling to the practice. They cannot be blamed entirely for this, as they are in business to sell disinfectants and their customers demand 'phenol coefficients', even when they do not understand them. Too little has been done in the past to educate the purchaser.

A Code of Practice embodying appropriate safety precautions was drawn up by the British Association of Chemical Specialities (BACS, 1981). This Code was timely and necessary, because of misapplication, in certain areas, of the (Howie) Code of Practice to non-clinical and industrial laboratories. The Howie Code, as mentioned above, forbids the use of *S.typhi* for testing disinfectants in clinical laboratories. It also places *S.typhi* in what is now Risk/Hazard Group 3. In the UK Containment Level 3, special accommodation and conditions for containment are specified for handling microorganisms in this group and these include biological safety cabinets. This caused some confusion because the Howie Code did not specify which organisms were likely to cause infection by the airborne route and therefore require handling in safety cabinets, and which, like *S.typhi* infect (usually) by the alimentary route and may be handled on the open bench but away from the mainstream of laboratory activities. The BACS Code of Practice, which has official approval, further clarifies this issue and it is this Code, not the Howie

Code which should be observed in non clinical laboratories where disinfectants are tested.

The use of disinfectants in hospitals

It is advisable that hospitals work out a detailed code of practice on the use of disinfectants and antiseptics which applies to their own particular circumstances. Guides for this purpose have been published by the Public Health Laboratory Service (Ayliffe *et al.*, 1984).

For further information about disinfection and sterilization see the books by Russell *et al.* (1982) and Gardner and Peel (1984).

Treatment and disposal of infected materials

It is a cardinal rule that no infected material shall leave the laboratory.

In the context of the treatment and disposal of contaminated laboratory waste (including that contained in re-usable articles), this simple precept offers no problems to properly-equipped and well-managed laboratories.

It is clearly the responsibility of the laboratory management to ensure that no waste containing viable microorganisms leaves the laboratory premises. The only possible exception would be when it is to be incinerated under the direct supervision of a member of the laboratory staff. Unfortunately, placing this responsibility firmly on the laboratory management may not solve the problem. In some establishments there are very sketchy ideas on freeing material from living organisms. There is a touching faith in the ability of disinfectants at varying concentrations and indefinite ages to kill microbes submerged in or even placed near to them.

We have always believed that disinfection is a first line defence, and for discarded bench equipment it is a temporary measure, to be followed as soon as possible by autoclaving or incineration. Disinfectants should not be used as the sole method of treating bacterial cultures, even if they are completely submerged and all air bubbles are removed.

Containers for discarded infected material

In the laboratory there should be four important types of receptacles for discarded infected materials:

(1) discard bins or bags to receive cultures and specimens;
(2) discard jars to receive slides, pasteur pipettes and small disposable items;
(3) pipette jars for graduated (recoverable) pipettes; and
(4) colour-coded plastic bags for combustibles such as specimen boxes and wrappers which might be contaminated

Recommended colours in the UK are:
Yellow   For incineration.
Light blue or transparent with blue inscription   For autoclaving (but may subsequently be incinerated).
Black   Normal household waste: local authority refuse collection.

White or clear plastic   Soiled linen (e.g. laboratory overalls).

Other colour codes encountered in UK hospitals, but not normally affecting laboratories, include red or red band on white for foul or infected linen.

Discard bins and bags

Discard bins should have solid bottoms which should not leak; otherwise contaminated materials may escape. To overcome steam penetration problems these containers should be shallow, not more than 8 in deep and as wide as will fit loosely into the autoclave. They should never be completely filled. Suitable plastic (polypropylene) containers are available commercially although not specifically designed for this purpose.

Plastic bags are popular but only those made for this purpose should be used. They should be supported in buckets or discard bins. Even the most reliable may burst if roughly handled.

For safe transmission to the preparation room the bins may require lids and the bags may be fastened with wire ties.

Discard bins and bags should be colour coded, i.e. marked in a distinctive way so that all workers recognize them as containing infected material.

Discard jars

The jars or pots of disinfectant that sit on the laboratory bench and into which used slides, pasteur pipettes and other rubbish are dumped have a long history of neglect and abuse. In all too many laboratories these jars are filled infrequently with unknown dilutions of disinfectants, are overloaded with protein and articles that float and are infrequently emptied. The contents are rarely disinfected.

Choice of container
Old jam jars and instant coffee jars are not suitable. Glass jars are easily broken and broken glass is an unnecessary laboratory hazard especially if it is likely to be contaminated. Discard jars should be robust and autoclavable and the most serviceable articles are 1-litre polypropylene beakers or screw-capped polypropylene jars. These are deep enough to hold submerged most of the things that are likely to be discarded, are quite unbreakable, and survive many autoclave cycles. They go dark brown in time, but this does not affect their use. Screw-capped polypropylene jars are better because they can be capped after use, inverted to ensure that the contents are all wetted by the disinfectants and air bubbles which might protect objects from the fluid are removed.

Correct dilution
A 1-litre discard jar should hold 750 ml of diluted disinfectant and leave space for displacement without overflow or the risk of spillage when it is moved. A mark should be made at 750 ml on each jar, preferably with paint (grease pencil and felt-pen marks are less permanent). The correct volume of neat disinfectant to be added to water to make up this volume for 'dirty situations' can be calculated from the manufacturers' instructions. This volume is then marked on a small measuring jug, e.g. of enamelled iron, or a plastic dispenser is locked to deliver it from a bulk container. The disinfectant is added to the beaker and water added to the 750-ml mark.

Sensible use

Laboratory supervisors should ensure that inappropriate articles are not placed in discard jars. There is a reasonable limit to the amount of paper or tissues that such a jar will hold, and articles that float are unsuitable for disinfectant jars, unless these can be capped and inverted from time to time to wet all the contents.

Large volumes of liquids should never be added to dilute disinfectants. Discard jars, containing the usual volume of neat disinfectant can be provided for fluids such as centrifuge supernatants, which should be poured in through a funnel that fits into the top of the beaker. This prevents splashing and aerosol dispersal. At the end of the day water can be added to the 750 ml mark and the mixture left overnight. Material containing large amounts of protein should not be added to disinfectants but should be autoclaved or incinerated.

Regular emptying

No material should be left in disinfectant in discard jars for more than 24 h, or surviving bacteria may grow. All discard jars should therefore be emptied once daily, but whether this is at the end of the day or the following morning is a matter for local choice. Even jars that have received little or nothing during that time should be emptied.

'Dry discard jars'

Instead of jars containing disinfectants there is a place in some laboratories for the plastic containers used in hospitals for discarded disposable syringes and their needles. These will accommodate pasteur pipettes, slides, etc. and can be autoclaved or incinerated.

Pipette jars

Jars for recoverable pipettes should be made of polypropylene or rubber. These are safer than glass. The jars should be tall enough to allow pipettes to be completely submerged without the disinfectant overflowing. A compatible detergent should be added to the disinfectant to facilitate cleaning the pipettes at a later stage. Tall jars are inconvenient for short people, who tend to place them on the floor. This is hazardous. The square based rubber jars may be inclined in a box or on a rack which is convenient and safer.

Plastic bags for combustibles

These should be colour coded (yellow in the UK, see p.44).

Treatment and disposal procedures

There are three practical methods for the treatment of contaminated, discarded laboratory materials and waste:

(1) autoclaving,
(2) incineration,
(3) chemical disinfection.

The choice is determined by the nature of the material: if it is disposable, or recoverable, and if the latter if it is affected by heat. With certain exceptions none of the methods excludes the others. It will be seen from Figure 3.3 that incineration alone is advised only if the incinerator is under the control of the laboratory staff (DHSS, 1978; Collins 1988). Disinfection alone is advisable only for graduated, recoverable pipettes.

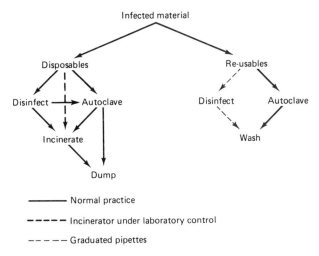

Figure 3.3 Flow chart for the treatment of infected material

Organization of treatment

The design features of a preparation room for dealing with discarded laboratory materials should include autoclaves, a sluice, a waste disposal unit plumbed to the public sewer, deep sinks, glassware washing machines, drying ovens, sterilizing ovens and large benches.

These should be arranged to preclude any possible mixing of contaminated and decontaminated materials. The designers should therefore work to a flow, or critical pathway chart provided by a professional microbiologist.

Such a chart is shown in Figure 3.4. The contaminated materials arrive in coloured coded containers onto a bench or into an area designated and used for that purpose only. They are then sorted according to their colour codes and despatched to the incinerator or loaded into the autoclave. Nothing bypasses this area. After autoclaving the containers are taken to the sorting bench where the contents are separated into:

(1) waste for incineration, which is put into colour coded bags,
(2) waste for the rubbish tip, which is also put into different colour coded bags,
(3) waste suitable for the sluice or waste disposal unit,
(4) recoverable material which is passed to the next section or room for washing and re-sterilizing.

This section or room should have a separate autoclave. Contaminated waste and materials for re-use or re-issue should not be processed in the same autoclave.

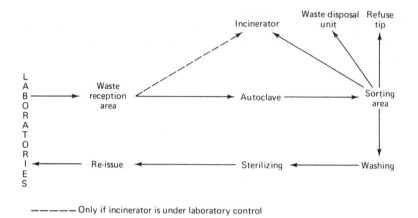

------ Only if incinerator is under laboratory control

Figure 3.4 Design of preparation (utility) rooms. Flow chart for the disposal of infected laboratory waste and re-usable materials

Procedures for various items

Before being autoclaved the lids of discard bins should be removed and then included in the autoclave load in such a way that they do not interfere with steam penetration. Plastic bags should have the ties removed and the bags opened fully in the bins or buckets that support them.

Contaminated glassware

After autoclaving, culture media may be poured away or scraped out and the tubes and bottles, etc., washed by hand or mechanically with a suitable detergent. The washing liquid or powder used will depend on the hardness of the water supply and the method of washing. The advice of several laboratory detergent manufacturers should be sought.

Busy laboratories require glassware washing machines. Before purchasing one of these machines it is best to consider several and to ask other laboratories which models they have found satisfactory. Not all laboratory glassware washing machines are as good as the manufacturers claim. A prerequisite is a good supply of distilled or deionized water.

If hand-washing is practised, double sinks, for washing and then rinsing, are necessary, plus plastic or stainless steel bowls for final rinsing in distilled or deionized water. Distilled water from stripper stills, off the steam line, is rarely satisfactory for bacteriological work.

Rubber liners should be removed from screw-caps and the liners and caps washed separately and re-assembled. Colanders or sieves made of polypropylene are useful for this procedure.

New glassware, except that made of borosilicate or similar material, may require neutralization. When fluids are autoclaved in new soda-glass tubes or bottles, alkali may be released and alter the pH. Soaking for several hours in 2–3% hydrochloric acid is usually sufficient, but it is advisable to test a sample by filling with neutral water plus a few drops of suitable indicator and autoclaving.

Discard jars
After standing overnight to allow the disinfectant to act, the contents of the jars should be poured carefully through a polypropylene colander and flushed down the sluice sink. The colander and its contents are then placed in a discard bin and autoclaved. Rubber gloves should be worn for these operations.

The empty discard jars should be autoclaved before they are returned to the laboratory for further use. There may be residual contamination.

Re-usable pipettes
After total immersion in disinfectant (e.g. hypochlorite, 2500 ppm available chlorine: an anionic detergent may be added) overnight the pipettes should be removed with gloved hands.

Before the pipettes are washed, the cotton wool plugs must be removed. This can be carried out by inserting the tip into a piece of rubber tubing attached to a water tap. Difficult plugs can be removed with a small crochet hook. Several excellent pipette washing machines are manufactured that rely on water pressure and/or a siphoning action, but the final rinse should be in distilled or deionized water.

24-h urines
Although rare in microbiology laboratories other pathology departments may send them for disposal on the grounds that they may contain pathogens. Ideally they should be processed in the department concerned as follows.

Sufficient disinfectant, e.g. hypochlorite, should be added to the urine to give the use-dilution. After standing overnight the urine should be poured carefully down the sink or sluice to join similar material in the public sewer. The containers, which are usually plastic, may then be placed in colour coded bags for incineration.

Incineration

The problems with this method of disposal are in ensuring that the waste actually reaches the incinerator, and that if it does it is effectively sterilized; and that none escapes, either as unburned material or up the flue. Incinerators are rarely under the control of laboratory staff. Sometimes they are not even under the control of the staff of the hospital or institution, but are some distance away and contaminated and infectious material has to be sent to them on the public highway. It may never arrive or may not be incinerated.

Incinerators are not always efficient. Nor are incinerator operators. Unburned material may be found among the ash, and from its appearance it may be deduced that it may not have been heated enough to kill microorganisms. We have recovered unconsumed animal debris, including fur and feathers and entrails, from a laboratory incinerator. The up-draught of air may carry microorganisms up the flue and into the atmosphere if the load is too large or badly distributed. Darlow (personal communication) has recovered organisms on culture plates exposed over a flue. He recommends that incinerators used for infected materials should have afterburners so that they consume their own smoke and render their effluent harmless. This is endorsed in the Code of Practice (DHSS, 1978) which also requires adequate supervision of the incineration of infected laboratory waste. There are British Standards (BSI 1573a, b).

In view of the problems and uncertainties associated with incineration it seems advisable to use this method only for material which has been autoclaved or disinfected and which, for aesthetic or other reasons, cannot be placed on a rubbish tip. There must, of course, be exceptions, and polythene 24-h urine containers are an example. Care must be taken when plastics are incinerated. A highly toxic smoke may be produced and it is usually recommended that not more than about 20% of any load is composed of plastic materials.

References

AOAC (1960) *Official Methods of Analysis of the Association of Analytical Chemists*, Washington DC

Ayliffe, G. A. J., Coates, D. and Hoffman, P. N. (1984) *Chemical Disinfectants in Hospitals*, Public Health Laboratory Service, London

BACS (1981) *Code of Practice for Handling Salmonella typhi NCTC 786*, British Association for Chemical Specialities, London

BSI (1934) *BS 541: Technique for Determining the Rideal–Walker Coefficient of Disinfectants*, British Standards Institution, London

BSI (1938) *BS 808: Modified Technique for the Chick–Martin Test for Disinfectants*, British Standards Institution, London

BSI (1961) *BS 3421: Specification for the Performance of Electrically Heated Sterilizing Ovens*, British Standards Institution, London

BSI (1973a) *BS 3107: Specification for small incinerators*, British Standards Institution, London

BSI (1973b) *BS 3316: Specification for large incinerators*, British Standards Institution, London

Coates, D. A. (1980) Laboratory use of disinfectants and the Howie Code. *Gazette, Institute of Medical Laboratory Sciences*, **24**, 555–556

Coates, D. A. and Death, J. E. (1978) Sporicidal activities of mixtures of alcohol and hypochlorites. *Journal of Clinical Pathology*, **31**, 148–152

Collins, C. H. (1988) *Laboratory-acquired Infections*, 2nd edn, Butterworths, London

Croshaw, B. (1981) Disinfectant testing, with particular reference to the Rideal–Walker and Kelsey–Sykes tests. In *Disinfectants, their Use and Evaluation of Effectiveness* (eds C. H. Collins, M. C. Allwood, S. F. Bloomfield and A. Fox) Society for Applied Bacteriology Technical Series No. 16, Academic Press, London, pp. 1–14

DHSS (1978) *Code of Practice for the Prevention of Infection in Clinical Laboratories*, Department of Health and Social Security, HMSO, London

DHSS (1980) *Sterilizers*, Health Technical Memorandum No. 10, HMSO, London

Gardner, J. F. and Peel, M. M. (1986) *Introduction to Sterilization and Disinfection*, Churchill Livingstone, London

Kelsey, J. C. and Maurer, I. M. (1974) An improved (1974) Kelsey–Sykes test for disinfectants. *Pharmaceutical Journal*, **213**, 528–536

Kelsey, J. C., McKinnon, I. H. and Maurer, I. M. (1974) Sporicidal activity of hospital disinfectants. *Journal of Clinical Pathology*, **27**, 632-638

Kennedy, D. A. (1988) Equipment-related hazards. In *Safety in Clinical and Biomedical Laboratories* (ed. C. H. Collins), Chapman & Hall, London, pp. 11–46

Maurer, I. M. (1969) A test for stability and long-term effectiveness in disinfectants. *Pharmaceutical Journal*, **203**, 529–534

Maurer, I. M. (1972) The management of laboratory discard jars. In *Safety in Microbiology* (eds D. A. Shapton and R. G. Board) Society for Applied Bacteriology Technical Series No.6, Academic Press, London, pp. 53–59

PHLS (1978) Autoclaving practice in microbiological laboratories. A report of a survey. *Journal of Clinical Pathology*, **31**, 418–422

PHLS (1981) Specification for laboratory autoclaves. *Journal of Hospital Infection*, **2**, 377–384

Russell, A. D. (1981) Neutralization procedures in the evaluation of bacterial activity. In *Disinfectants. Their Use and Evaluation of Effectiveness* (eds C. H. Collins, M. C. Allwood, S. F. Bloomfield and A. Fox), Society for Applied Bacteriology Technical Series No. 16. Academic Press, London, pp. 45–57

Russell, A. D., Hugo, W. B. and Ayliffe, G. A. J. (1982) *Principles and Practice of Disinfection, Preservation and Sterilization*, Blackwell Scientific, London

USPHS (1974) *Laboratory Safety at the Center for Disease Control*, Government Printing Office, Washington DC

USPHS (1978) *National Institutes of Health Laboratory Safety Monograph*. Government Printing Office, Washington DC

Chapter 4

Culture media

Most laboratories now purchase their culture media in dehydrated form or ready for use.

Commercially prepared media, which are subjected to strict internal quality control, may be purchased from three large companies that have international supply organizations (BBL, Difco and Oxoid), or from smaller companies that may supply certain regions only. Many culture media have generic names and are supplied by the majority of companies, but it should not be assumed that they are identical in performance because this may be influenced by the unique characteristics of the individual manufacturer's ingredients.

Some culture media have proprietary names and are therefore obtainable only from particular companies. Although similarly formulated media may be purchased elsewhere it must be stressed that such 'equivalent' products should not be assumed to be identical.

Only two commercially available media are rigorously controlled to meet outside regulatory specifications: Standard Plate Agar (American Public Health Association; APHA 1985) and Mueller–Hinton agar (National Committee for Clinical Laboratory Standards, 1979). Culture media meeting these standards should give a uniform performance irrespective of manufacturer or batch number.

The information given in this chapter is intended to guide users in their choice of media for particular organisms and purposes. Further guidance may be found in the publications of the various manufacturers. In particular, the *Oxoid Manual*, new editions of which appear regularly, is very useful.

We do not give the formulas for the preparation of most of the commercially available media in this book, but formulas are given for a number of media which are not on sale or which we have preferred to make ourselves.

Ingredients of some culture media and reagents (e.g. azides, barbitones, cyclohexamide, selenites, tellurites) may be hazardous to health. It is advisable to wear dust masks (e.g. 3M) when handling these. See manufacturers' handbooks and data sheets.

General purpose media

These include a wide variety of protein digest meat infusion broths and agars, containing meat particles, meat or yeast extracts. Robertson's cooked meat medium is included as a typical example. They are formulated to support the growth of the less fastidious organisms without specific selection or inhibition.

Media for fastidious organisms

These are broths and agars which are enriched with certain nutrients and may be used alone or supplemented with various other ingredients. They include blood agar bases and formulations containing brain, heart and veal extracts and mixtures of soya and meat digests from different proteolytic enzyme activities. Brain Heart Infusion, Columbia and GC Agar Base are typical examples.

Media for the selective isolation of bacteria

Not all of the media mentioned in this chapter are referred to in other parts of the book. This is because the editors and contributors have no experience of them. They are included to indicate that readers have a wider choice.

We have followed the convention of using capital initials for the names of proprietary media, most of which are described in the *Oxoid Manual*.

Aeromonas
Blood agar plus ampicillin; Aeromonas medium.

Bacillus cereus
B. Cereus Selective agar.

Bacteroides
Blood agar plus menadione–antibiotic supplements.

Bordetella
Bordet Gengou or Charcoal Agar Base plus antibiotic supplements.

Brucella
Blood agar or proprietary Brucella media plus antibiotic supplements.

Campylobacter
Columbia or Campylobacter agar plus supplements.

Clostridia
Wilkins–Chalgren agar plus antibiotics; Reinforced Clostridial agar; Tryptose-Sulphite Cycloserine agar, TSC; Oleandomycin Polymyxin Sulphadiazine Perfringens agar, OPSP; Iron Sulphite agar; Clostrisel; Clostridium Difficile agar; Willis and Hobbs medium (p.77).

Coliform bacilli
Broths   Brilliant Green Bile; Brilliant Green Lactose Bile Broth (BGLBB); lauryl sulphate; MacConkey; Minerals Modified Glutamate; Lactose Broth; EC broth; EE broth. *Agars*   Cystine Lactose Electrolyte-Deficient (CLED); deoxycholate; Endo; Eosin Methylene Blue (EMB); Violet Red Bile (VRB); Brilliant Green.

Corynebacteria
Hoyle; other commercial tellurites; Loeffler.

Enterobacteria
See under Coliform bacilli, Salmonella, Shigella.

Gardnerella
Gardnerella agar and supplements.

Haemophilus
Blood agar plus Fildes or commercial supplements.

Lactobacilli
L-S Differential; MRS; Raka–Ray; Tomato Juice; Rogosa.

Legionella
Blood agar plus supplements; proprietary media plus supplements.

Leptospira
EMJH medium and EMJH selective media (p.69).

Listeria
MacBride's; Listeria Selective Agar (Oxford); Listeria Enrichment Broth; Listeria enrichment medium (p. 72).

Mycobacteria
Lowenstein–Jensen (p. 72); Acid Egg, Kirchner; Middlebrook 7H9, 7H10, 7H11.

Neisseria
GC base plus supplements; New York City.

Pediococci
Raka–Ray; Wort agar.

Pseudomonas
Commercial selective media plus cetrimide and/or antibiotics; King' medium A (p.71).

Salmonellas

Enrichment   Selenite; Mueller–Kauffman Tetrathionate; Brilliant Green; Rappaport–Vassiliadis (RV).

Plating   Bismuth sulphite; Deoxycholate–citrate (many formulations: SS, DCA, XLD, etc.); Hektoen; Mannitol Lysine Crystal Violet Brilliant Green Agar; brilliant green agars (various formulations).

Shigellas
Deoxycholate – citrate (many formulations) as for salmonella; Hektoen.

Staphylococci
Baird-Parker; Mannitol–Salt; milk salt agar (p.73); phenolphthalein polymyxin

phosphate (PPPA) (p.75); Tellurite Polymyxin Egg Yolk (TPEY); Kranep; Staphylococcus 110; Giolotti–Cantoni.

Streptococci
Blood agar plus selective supplements; Azide media; BAGG; Barnes'; Edwards'; Mead's; Slanetz and Bartley's; KF Streptococcus; COBA; Islam's (p.71).

Vibrios
Alkaline peptone water (p.62); Thiosulphate Citrate Bile Sucrose (TCBS); Glucose salt Teepol (p.70); Salt colistin (p.75).

Yersinia enterocolitica
Yersinia Selective medium plus supplements.

Growth-promoting or inhibiting supplements and additives

Single substances or mixtures of growth promoters and antibiotics, packed ready to add to culture media, are supplied by several companies. The most comprehensive selection is offered by Oxoid:

Purpose	Supplement/additive
Anaerobes	Cysteine–HCl–sodium dithiothreitol–Vitamin K
Bacteroides	Haemin–menadione–sodium pyruvate–nalidixic acid
Bacillus cereus	Polymyxin
Bordetella	Cephalexin selective supplement
Brucella	Polymyxin–bacitracin–cycloserine–nalidixic acid–nystatin–vancomycin
Campylobacter	Sodium pyruvate–metabisulphite–ferrous sulphate. Vancomycin–polymyxin–trimethoprim. Bacitracin–cycloheximide––colistin–cephazolin–novobiocin
Clostridium difficile	Cycloserine–cefoxitin
Corynebacteria	Potassium tellurite (3.5%)
Haemophilus	Haemoglobin. Filde's extract
Lactobacilli	Lactic acid
Legionella	Cysteine HCl–ferric pyrophosphate–sodium selenite. Cefamandole–polymyxin–anisomycin medium; Polymyxin–anisomycin–vancomycin medium. Colistin–vancomycin–trimethoprim–amphotericin (CVTA)
Mycobacteria	Bovine albumin. Oleic acid–albumin–dextrose–citrate (OADC)
Neisseria	Haemoglobin. Yeast autolysate. Vancomycin–colistin–nystatin (VCN). Proprietary mixtures of growth factors (Vitox, Isovitalex). Vancomycin–colistin–nystatin–trimethoprim (VCNT) Vancomycin–Colistin–Amphotericin–trimethoprim (VCAT) Lincomycin–colistin–amphotericin–trimethoprim (LCAT)
Pasteurella	Haemoglobin
Salmonellas from sewage	Sulphacetamide–sodium mandelate

Streptococci (mitis,   Potassium tellurite (0.8%)
etc.)   Nalidixic acid polymyxin–neomycin
Yersinia   Cefsulodin–Irgasan–novobiocin (CIN)

Media for dermatophytes and pathogenic yeasts

General media
Czapek–Dox agar; wort agar; Sabouraud agar (various formulations).

Dermatophytes
Littman ox gall agar; malt agar with and without chloramphenicol and cyclohexi-mide. Dermasel.

Yeasts
BiGGY (Nickerson) medium; corn meal agar.

Transport media

Semisolid media with and without charcoal and antibiotics, intended to keep delicate microbes alive during transit to the laboratory are sold under the following names:
   Amies; Cary–Blair; Stuart; Transgrow; Transport; Pertussis.

Media for food microbiology

Most general media are useful, especially glucose (dextrose) tryptone media.

Bacterial counts
Standard (plate) count agar; Standard methods agar; milk agar; yeast extract milk agar. Tryptose Glucose Extract Agar.

Counting yeasts and moulds
Buffered yeast agar (p.65). OGYE agar; WL agar; Lysine agar; glucose salt agar; potato dextrose agar; Malt extract agar; Rose Bengal chloramphenicol agar Dichloran Glycerol agar; Dichloran Rose Bengal chloramphenicol agar.

Acidophilic yeasts
Malt agar.

Beer spoilage
Universal Beer agar; Raka–Ray agar.

Brochothrix
Gardner's STAA (p.70).

Coliform bacilli
See p.53.

Clostridia
See p.53.

Enterobacteria
See coliform bacilli.

Flat sour organisms
Glucose tryptone agar.

Hydrogen swell organisms
Glucose tryptone agar.

Lactobacilli
See p.54.

Lipolytic organisms
Tributyrin agar.

Moulds and yeasts in dairy products and utensils
Malt agar plus chloramphenicol and cycloheximide; Buffered yeast agar (p.65);
Potato dextrose agar.

Moulds on meat
Potato glucose agar.

Staphylococci
See p.54.

Streptococci
See p.55.

Media for water bacteriology

Bacterial counts
Standard methods agar; standard plate count agar.

Coliform bacilli
See p.53.

Clostridium perfringens
See p.53.

Enterobacteria
See coliform bacilli, p.53.

Salmonellas
See p.54.

Membrane filtration media

Companies that sell these media place a suffix before the names of their products: BBL, *M*-; Difco, m-; Gibco, MF-; Merck, Membrane Filtration; Oxoid, M-. Membrane filtration media at present available include the following.

Counting bacteria
Glucose tryptone broths; standard methods broth; tryptone soya broths.

Coliform bacilli
Brilliant green broths; Endo media; eosin methylene blue broth; enrichment broth; lauryl sulphate media; MacConkey broths; resuscitation broths.

Flat sour organisms
Glucose tryptone broths.

Salmonellas, including S.typhi
Bismuth sulphite broth; tetrathionate broth.

Staphylococci
Staphylococcus 110 broth.

Streptococci
Azide broths; enterococcus media.

Moulds
Czapek–Dox broth.

Antibiotic sensitivity test media

Mueller Hinton medium is the best known. Most culture media manufacturers offer this or their own variations under trade names.

Sterility test media

For products not containing preservatives – any non-selective broth medium. For products containing preservatives – Clausen medium; thioglycollate media (various formulations).

Identification media

These are used for various 'biochemical' tests. Most are available commercially; page numbers are given for those that are not. See also Paper strip (p.97) and 'Kit' bacteriology (p.97).

Aesculin hydrolysis
Aesculin medium, Edwards' medium.

Arginine hydrolysis
Arginine broth, combined arginine–ornithine media.

Casein hydrolysis
Skim milk agar (p.76).

Carbohydrate utilization (most available commercially)
These are usually a peptone broth base containing an indicator (phenol red or Andrade) to which the fermentable substance is added at 0.5–2% (1% for most bacteria). A Durham's tube is included if detection of gas is desired.

These simple media are not suitable for all purposes, however, and the following modifications are advisable for special purposes.

Baird-Parker carbohydrate media
For staphylococci and micrococci (p.66).

Basal synthetic medium
Add carbohydrate for aerobic spore bearers and pseudomonads (p.63).

GC serum-free sugar medium
For fastidious organisms (p.67).

Gillies medium
Contains glucose, mannitol, sucrose and salicin for screening enterobacteria.

Kohn two-tube medium
This is a modification of Gillies medium.

Krumwiede double sugar agar
Contains lactose and sucrose.

MRS medium
Plus carbohydrates for lactobacilli (p.74).

Robinson's serum water sugars
Preferred for pathogenic corynebacteria (p.66).

Russell's double sugar agar
Contains glucose and lactose.

Triple sugar iron agar (TSI)
Contains glucose, lactose and sucrose and also indicator for H_2S production.

Citrate utilization
Koser; Simmons.

Decarboxylase tests
Falkow and Moeller lysine media, basal media to which arginine, lysine and ornithine must be added (see also p.67).

Esculin hydrolysis
See Aesculin hydrolysis.

DNase
DNase agars with or without indicator.

Gelatin liquefaction
Nutrient gelatin, gelatin agar, various formulations (p.70). Charcoal gelatin discs
for use with nutrient broths.

Gluconate utilization
Gluconate broth (p.70).

Hippurate hydrolysis
Hippurate broth (p.71).

Hugh and Leifson test
Add glucose to commercial base (p. 71).

Hydrogen sulphide production
Lead acetate agar (p.72) and various formulations of iron agar, iron plus carbohydrate and iron plus lysine medium. Kligler Agar; Kohn Agar.

Indole production
Any commercial peptone or tryptone water medium.

Lecithinase
Egg yolk agar and broth (p.68). Willis and Hobbs medium (p.77). Egg yolk salt
agar (p.68).

Malonate utilization
Malonate and phenylalanine – malonate broths (p.73).

Milk reactions
Litmus milk, bromocresol purple milk, Crossley milk medium.

Methyl red (MR) and Voges Proskauer (VP)
Glucose phosphate broth, MRVP media.

Motility
Semisolid media. See also Craigie method (p.104).

Nitratase test
Nitrate broths, indole–nitrate broths and agars (p.74).

Nagler reaction
Egg yolk agar (p.68). Willis and Hobbs medium (p.77).

OF (Ox-ferm: Hugh and Leifson) test
Add glucose to commercial base (see also p.71).

ONPG test
ONPG broth (p.74).

Organic acid utilization
Add 1–2% organic acid to commercial organic acid base medium (see also p.63).

Oxygen preference
Use sloppy agar (p.76).

PPA (phenylalanine deamination)
Phenylalanine agar.

Phosphatase test
Phenolphthalein phosphate agar (p.75).

Pyocyanin production
King's medium A (p.71).

Starch hydrolysis
Starch agar (p.76).

Sulphatase test
Phenolphthalein sulphate agar and broth (p.75).

Urease test
Usually sold as a base, with sterile (filtered) urea separately.

Tyrosine decomposition
Tyrosine agar (p.77).

Xanthine decomposition
Xanthine agar (p.77).

Supplements and additives for identification media

These are usually supplied weighed or measured and ready to add to the commercial base media. They include:

Horse, sheep blood and serum
Carbohydrates (dulcitol, glucose, lactose, maltose, mannitol, salicin, sucrose, xylose)
Egg yolk emulsion for lecithinase test media
Lead acetate for hydrogen sulphide detection
Ox bile for bile solubility test and bile inhibition tests
Potassium lactate for lysine agar
Sodium nitrate (30%) for nitrate broths

Most commercial media companies supply some or all of these.

Laboratory prepared media

Acid broth

Used for high acid thermally processed foods

Proteose peptone	5 g
Yeast extract	5 g
Glucose	5 g
Dipotassium phosphate K$_2$HPO$_4$	4 g
Distilled water to	1000 ml

Dissolve and dispense in 12–15 ml amounts, sterilize at 121°C for 15 min. Final pH should be 5.0.

Alkaline peptone water

This is used for isolating *V.cholerae*. Adjust the pH of peptone water to pH 8.6 with N NaOH.

Antifungal drug sensitivity testing

(1)	Glucose monohydrate	20 g
	Casamino acids	20 g
	Sodium glycerophosphate	5 g
	Yeast extract 5%	2 ml
	Distilled water	1000 ml

Heat at 50–60°C to dissolve and distribute in 100-ml lots.

(2)	Agar	3 g
	Distilled water	1000 ml

Dissolve by autoclaving and distribute in 100-ml lots.

For sensitivity to amphotericin and imidazole add 2 ml of 1% cytosine to medium (1). For the assay of 5-fluorocytosine mix 100 ml of (1) and (2) and distribute in 10-ml lots for tests on slopes or 20-ml lots for tests in petri dishes.

Antifungal drug assay medium

Yeast Nitrogen Base	3.5 g
Glucose	10.0 g
KH$_2$PO$_4$	1.5 g
Na$_2$HPO$_4$.2H$_2$O	1.0 g
Distilled water	1000 ml

Dissolve, dispense in 110-ml lots. Add 1.2 g of agar to each bottle and autoclave.

Ammonium salt sugars

See under Carbohydrate media (p.65).

Arginine broth

This is useful in identifying some streptococci and Gram-negative rods.

Tryptone	5 g
Yeast extract	5 g

Dipotassium hydrogen phosphate	2 g
L-Arginine monohydrochloride	3 g
Glucose	0.5 g
Water	1000 ml

Dissolve by heating, adjust to pH 7.0; dispense in 5–10-ml amounts and autoclave at 115°C for 10 min.

For arginine breakdown by lactobacilli, use MRS broth in which the ammonium citrate is replaced with 0.3% arginine hydrochloride.

Baird-Parker's sugar media

See under Carbohydrate media (p.65).

Basal synthetic media

These are used when investigations are being made into the ability of bacteria to use various carbon sources or, with the addition of a suitable carbon source, of growth factor or vitamin requirements.

(a)   For carbohydrate utilization

Ammonium dihydrogen phosphate	1.0 g
Potassium chloride	0.2 g
Magnesium sulphate	0.2 g
Agar	10.0 g
Water	1000 ml

Dissolve by heating, add 4 ml of 0.2% bromothymol blue and a final 1% of sterile (filtered) carbohydrate solution (0.1% aesculin; 0.2% starch; these are sterilized by steaming).

To investigate amino acid sources, add 1% glucose and 0.1 g of amino acid.

(b)   For organic acid utilization

Magnesium sulphate	1.0 g
Sodium chloride	1.0 g
Diammonium hydrogen phosphate	1.0 g
Potassium dihydrogen phosphate	0.5 g
Agar	12.0 g
Water	1000 ml

Dissolve salts by heating in 200 ml of water. Dissolve agar in the remaining water. Mix and adjust to pH 6.8. Bottle in 100-ml amounts. Melt, add 0.2 g of organic acid (sodium salts of acetic, benzoic, citric, oxalic, propionic, pyruvic, succinic, tartaric acids; calcium salt of malic acid; mucic acid as free acid), re-adjust to pH 6.8 if necessary and add 0.4 ml of 0.2% phenol red. Dispense and give a final steaming to sterilize.

To investigate amino acid sources, add 1% glucose and the indicator and 0.1 g of the amino acid.

Blood agar media

Blood agar

This is nutrient agar or one of the commercial dehydrated Blood Agar Bases to which sterile horse blood (sheep blood in the USA and some other countries) is added. It is used to detect haemolytic organisms and to encourage the growth of organisms that grow poorly or not at all on nutrient agar. Some specially enriched commercial blood agar bases are available for diagnostic and sensitivity tests.

Melt 100 ml of nutrient agar or prepare 100 ml of Blood Agar Base. Cool to 48°C in a water-bath. Add 10 ml of sterile defibrinated or oxalated horse blood and mix gently so as to avoid bubbles. Pour into petri dishes by raising the lid only enough to introduce the neck of the bottle. This amount should give seven poured plates.

In pathology laboratories, to conserve blood and to see more easily the clear zone around colonies, a very thin layer of blood agar is poured on top of a thin layer of ordinary nutrient agar. Sometimes saline agar, a 1% agar containing 0.9% sodium chloride, at pH 7.4 is used.

Chloral hydrate agar
Used in the purification of cultures of anthrax bacilli. Add 5% chloral hydrate to blood agar while still fluid.

Chocolate agar
Some fastidious bacteria grow best on a blood agar in which the cells have been disrupted and the haemoglobin altered by heat (horse blood is better than sheep blood).

Make blood agar as described above but replace the bottle in the water-bath, raise the temperature of the water-bath slowly to 80°C and leave for 5 min. Cool slightly and pour into petri dishes. The medium should be homogeneous and chocolate coloured, not flaky.

Chocolate cystine agar
For the cultivation of *Neisseria* and *Haemophilus*.

Solution A	
Cystine	0.1 g
Nutrient broth	50 ml
Solution B	
p-Aminobenzoic acid	0.1 g
Nutrient broth	50 ml

Make up when required and steam for 1 h.

To 950 ml of melted blood agar base add 25-ml amounts of each of solutions A and B. Bottle and sterilize at 115°C for 10 min. For use, melt, cool to 50°C, add 10% horse blood and heat at 70–80°C until reddish brown in colour. Pour plates and use as soon as possible.

Colistin–nalidixic acid agar
To isolate Gram-positive cocci from material heavily contaminated with *Pseudomonas, Proteus, Klebsiella*, etc., add 10 µg/ml of colistin and 15 µg/ml of nalidixic acid to blood agar base before adding the blood.

Crystal violet blood agar
The dye inhibits many organisms but not streptococci. Add 1 ml of a 0.001% solution of crystal violet in water to 500 ml of blood agar while still melted.

Gentamicin blood agar
Add 500 µg of gentamicin to 100 ml of blood agar.

Lysed blood agar
This is used for sulphonamide sensitivity tests. Lysed horse blood (not sheep) is obtainable commercially (see above) or made by alternate freezing and thawing of whole blood. The medium is made as described under Blood Agar.

Neomycin blood agar
Add 75 μg of neomycin sulphate to 100 ml of blood agar.

PABA blood agar
To neutralize sulphonamides in an inoculum, add 10 mg% of *p*-aminobenzoic acid to the medium before sterilizing it.

Buffered yeast agar

This is useful for the detection of and counting yeasts and moulds in the food industry, utensils, washes, etc.

Yeast extract	5.0 g
Dextrose	20.0 g
Ammonium sulphate	0.72 g
Ammonium dihydrogen phosphate	0.26 g
Agar	12.0 g
Water	1000 ml

Dissolve by heating and adjust to pH 5.5, bottle and sterilize at 115°C for 10 min. If required for bottle counts, increase agar to 20 g/litre.

To adjust this medium to a more acidic pH, cool to 50°C and add 1% citric acid monohydrate and 10% lactic acid to each 100 ml as follows (Davis, 1931, 1982):

pH	1% citric acid (ml)	10% lactic acid (ml)
4.75	1.26	0.125
4.50	2.24	0.20
4.25	3.92	0.30
4.00	6.16	0.45
3.75	9.52	0.70
3.50	14.56	1.17

Carbohydrate fermentation and oxidation tests

Formerly known as 'peptone water sugars', some of these media are now more nutritious and do not require the addition of serum except for very fastidious organisms. Liquid, semisolid and solid basal media are available commercially and contain a suitable indicator, e.g. Andrade (pH 5–8), phenol red (pH 6.8–8.4) or bromothymol blue (pH 5.2–6.8).

Some sterile carbohydrate solutions (glucose, maltose, sucrose, lactose, dulcitol, mannitol, salicin) are available commercially. Prepare others as a 10% solution in water and sterilize by filtration. Add 10 ml to each 100 ml of the reconstituted sterile basal medium and tube aseptically. Do not heat sugar carbohydrate media. Durham's (fermentation) tubes need to be added to the glucose tubes only; gas from other substrates is not diagnostically significant.

These media are not suitable for some organisms; variations are given below.

Ammonium salt 'sugars'

Pseudomonads and spore bearers produce alkali from peptone water and the indicator may not change colour. Use this basal synthetic medium instead.

Ammonium dihydrogen phosphate	1.0 g
Potassium chloride	0.2 g
Magnesium sulphate	0.2 g
Agar	10.0 g
Water	1000 ml

Dissolve by heating, add 4 ml of 0.2% bromothymol blue and a final 1% of sterile (filtered) carbohydrate solution (0.1% aesculin; 0.2% starch; these are sterilized by steaming).

'Anaerobic sugars'

Peptone water or serum peptone water sugar media need to have a low oxygen tension for testing the reactions of anaerobes, even under anaerobic conditions. Add a clean wire nail to each tube before sterilization. Do not add indicator until after growth is seen.

Baird-Parker's carbohydrate medium

For the sugar reactions of staphylococci and micrococci use Baird-Parker (1966) formula.

Yeast extract	1 g
Ammonium dihydrogen phosphate	1 g
Potassium chloride	0.2 g
Magnesium sulphate	0.2 g
Agar	12 g
Water	1000 ml

Steam to dissolve and adjust to pH 7.0. Add 20 ml of 2% bromocresol purple and bottle in 95-ml amounts. Sterilize at 115°C for 10 min. For use, melt and add 5 ml of 10% sterile carbohydrate.

Lactobacilli fermentation medium

For sugar reactions of lactobacilli make MRS base (p.74) without Lab-Lemco and glucose and adjust to pH 6.2–6.5. Bottle in 100-ml amounts and autoclave at 115°C for 15 min. Melt 100 ml, add 10 ml of 10% sterile (filtered) carbohydrate solution and 2 ml of 0.2% chlorophenol red. Dispense aseptically.

Robinsons's serum water sugars

Some organisms will not grow in peptone water sugars. Robinson's serum water medium is superior to that of Hiss.

Peptone	5 g
Disodium hydrogen phosphate	1 g
Water	1000 ml

Steam for 15 min, adjust to pH 7.4 and add 250 ml of horse serum. Steam for 20 min, add 10 ml of Andrade indicator and 1% of appropriate sugar (0.4% starch).

Unheated serum contains diastase and may contain a small amount of fermentable carbohydrate, so a buffer is desirable, particularly in starch fermentation tests.

Carbon assimilation media for yeasts

Ammonium sulphate	5 g
Magnesium sulphate	0.5 g
Potassium dihydrogen phosphate	1.0 g
Agar	1.5 g
Water	1000 ml

Dissolve by heating. Autoclave in 15-ml amounts at 115°C for 15 min.

Decarboxylase medium

(Moeller, 1955)

Peptone (Evans)	5 g
Lab – Lemco	5 g
Pyridoxine HCl	5 mg
Glucose	0.5 g
Distilled water	990 ml

Dissolve and adjust to pH 6.0. Add 5 ml of 1:500 bromocresol purple and 2.5 ml of 1:500 cresol red. Make up to 1000 ml and divide into three batches. To one add 2 g% DL-lysine and adjust to pH 6.0 with 0.1 N NaOH. To another batch add 2 g% DL-ornithine and adjust to pH 6.0 with 0.1 N NaOH. The third is the control medium. Dispense in 1.1-ml amounts in narrow tubes so that the column of medium is about 2 cm. Seal with a few millimetres of liquid paraffin, plug and steam for 1 h.

Medium to demonstrate arginine decarboxylase or dihydrolase is made in the same way.

Decarboxylase media

(Falkow, 1958)
This formula is preferred for those organisms which require a less anaerobic medium than Moeller's medium.

Peptone	5 g
Yeast extract	3 g
Glucose	1 g
Water	1000 ml

Dissolve by heating, adjust to pH 6.7 and add 10 ml of 0.2% bromocresol purple. Bottle in 100-ml amounts and autoclave at 115°C for 10 min.

To 100 ml add 0.5 g of the appropriate amino acid (lysine, ornithine, arginine) and tube. Ordinary-sized tubes or bottles can be used. Dispense also one batch without amino acid as control. Sterilize by steaming.

GC serum-free sugars

(Flynn and Waitkins, 1972)

Oxoid GC base	33 g
Distilled water	900 ml
Phenol red (0.2%)	9 ml
C-H supplement	18 ml

Bring to boil to dissolve. Adjust to pH 7.6 with N NaOH, distribute in 90 ml lots and autoclave at 115°C for 10 min. Store in cold room.

C-H supplement

L-glutamine	1 g
Ferric nitrate	0.05 g
Sterile distilled water	100 ml

For use melt four bottles of base, cool to 56°C and add 10 ml of each filter-sterilized (10%) sugar solutions and add sterile indicator solution before autoclaving.

Donovan's medium (1966)

Originally devised to differentiate klebsiellas, this is useful for screening lactose fermenting Gram-negative rods.

Tryptone	10 g
Sodium chloride	5 g
Triphenyltetrazolium chloride	0.5 g
Ferrous ammonium sulphate	0.2 g
1% aqueous bromothymol blue	3 ml
Inositol	10 g
Agar	3 g
Water	1000 ml

Dissolve by heating and adjust to pH 7.2. Distribute in 4-ml amounts and autoclave at 115°C for 15 min. Allow to set as butts.

Dorset's egg medium

Wash six fresh eggs in soap and water and break into a sterilized basin. Beat with a sterile fork and strain through sterile cotton gauze into a sterile measuring cylinder. To 3 parts of egg add 1 part of nutrient broth. Mix, tube or bottle and inspissate in a sloped position for 45 min at 80–85°C. Glycerol (5%) may be added if desired. (Griffith's egg medium uses saline instead of broth and is useful for storing stock cultures.)

Egg yolk agar

(Lowbury and Lilly, 1955)

This is selective for *Cl. perfringens* and shows the Nagler and pearly layer reactions.

Nutrient agar or blood agar, base, melted	100 ml
Fildes' extract	5 ml

Steam for 20 min to remove chloroform from Fildes' extract, cool to 55°C, and add 10 ml of egg yolk suspension, 100–125 μg/ml of neomycin and pour plates.

Egg yolk salt broth and agar

For lecithinase tests

Nutrient broth or blood agar base	10 ml
Egg yolk emulsion	1 ml
NaCl (5%)	0.2 ml

Elek's medium (modified)

(Davies, 1974)
Used for plate toxigenicity test on diphtheria bacilli.

Solution A

Evans peptone	20 g
Maltose	3 g
Lactic acid	0.7 g
Distilled water	500 ml

Heat to dissolve. Add 3.5 ml of 10 N NaOH. Mix well, heat to boiling, filter, adjust to pH 7-8.

Solution B

Agar	15 g
Sodium chloride	5 g
Distilled water	500 ml

Heat to dissolve. Adjust to pH 7.8.

Mix solutions A and B and bottle in 15-ml amounts. Autoclave at 115°C for 10 min.

EMJ medium for leptospires

The formula is given by Waitkins (1985) but the complete medium and enrichment is obtainable from Difco.

Selective EMJH/5FU medium

(Waitkins, 1985)
Dissolve 1 g of 5-fluorouracil in approx. 50 ml of distilled water containing 1–2 ml of N NaOH by gentle heat (do not exceed 56°C). Adjust to pH 7.4–7.6 with N HCl, make up to 100 ml with distilled water and sterilize through a membrane filter (0.45 μm). Dispense in 1-ml lots and store at −4°C.

Thaw 1 ml at 56°C and add it to 100 ml of the EMJH medium (final concentration of 5FU is 100 μg/ml).

Ellner's medium

This is used to persuade *Cl. perfringens* to form spores.

Peptone	10 g
Yeast extract	3 g
Starch, soluble	3 g
Magnesium sulphate (MgSO$_4$.7H$_2$O)	0.1 g
Disodium hydrogen phosphate (Na$_2$HPO$_4$.7H$_2$O)	50 g
Potassium dihydrogen phosphate	1.5 g
Water	1000 ml

Dissolve by heating, adjust to pH 7.8 and tube. Before use, steam to drive off oxygen, cool rapidly and inoculate heavily at the bottom of the tube with a pasteur pipette without introducing air.

Gardner's (1966) STAA medium

For the isolation of *Brochothrix* from meat.

Peptone	20 g
Yeast extract (*O*)	2 g
Glycerol	15 g
K₂HPO₄	1 g
MgSO₄.7H₂O	1 g
Agar	13 g
Distilled water	1000 ml

Heat to dissolve, adjust to pH 6.0 and sterilize at 121°C for 15 min. Tube in suitable amounts. Before use melt, cool to 50°C and add these inhibitory agents to give the final concentrations stated: streptomycin sulphate to 500 µg/ml; cycloheximide to 50 µg/ml; thallous acetate to 50 µg/ml.

Gelatin agar

Melt nutrient agar and add 5% bacteriological gelatin.

Gluconate broth

For differentiation among enterobacteria.

Yeast extract	1.0 g
Peptone	1.5 g
Dipotassium hydrogen phosphate	1.0 g
Potassium gluconate	40.0 g
(or sodium gluconate)	37.25 g
Water	1000 ml

Dissolve by heating, adjust to pH 7.0, filter if necessary, dispense in 5–10-ml amounts and autoclave at 115°C for 10 min.

Glucose salt Teepol broth

For the isolation of *Vibrio parahaemolyticus* from water and foods.

Beef extract	3 g
Tryptone	10 g
NaCl	30 g
Glucose	5 g
Methyl violet	0.002 g
Teepol 610	4 ml
Distilled water	1000 ml

Dissolve by heat, adjust to pH 9.4 and autoclave at 115°C for 10 min.

Glucose phosphate medium

For MR and VP tests. Add 0.5% each of glucose and K₂HPO₄ to commercial peptone broth.

Griffith's egg medium

For storing stock cultures, see Dorset's egg medium.

Hippurate broth

Dissolve 10 g of sodium hippurate in 1000 ml of nutrient broth. Steam to dissolve and dispense in small amounts.

Hugh and Leifson's medium

Peptone	2.0 g
Sodium chloride	5.0 g
Dipotassium hydrogen phosphate	0.3 g
Agar	3.0 g
Water	1000 ml

Heat to dissolve, adjust to pH 7.1 and add 15 ml of 0.2% bromothymol blue and sterile (filtered) glucose to give a final 1% concentration. Dispense aseptically in narrow (less than 1 cm) tubes in 8–10 ml amounts.
Baird-Parker (1966) modification for staphylococci and micrococci:

Tryptone	10 g
Yeast extract	1 g
Glucose	10 g
Agar	2 g
Water	1000 ml

Steam to dissolve, adjust to pH 7.2. Add 20 ml of 0.2% bromocresol purple, tube in 10-ml amounts in narrow (12 mm) tubes and sterilize at 115°C for 10 min.

Islam's (1977) medium

For Group B streptococci

Proteose peptone No. 3 (Difco)	23 g
Soluble starch	5 g
$NaH_2PO_4.2H_2O$	1.482 g
Na_2HPO_4	5.749 g
Agar	10 g
Water	1000 ml

Steam to dissolve, adjust to pH 7.4 and sterilize at 115°C for 10 min. For use melt, cool to 55°C and add 5% inactivated horse serum.

King's A broth (modified)

(Drake, 1966)

Peptone	20 g
Ethanol	25 ml
Potassium sulphate	10 ml
Magnesium chloride	1.4 g
Cetrimide	0.5 g
Distilled water	1000 ml

Steam to dissolve, distribute in screw-capped bottles and sterilize at 115°C for 10 min.

Kirchner's medium

For the culture and differentiation of mycobacteria.

Na$_2$HPO$_4$.12H$_2$O	19.0 g
KH$_2$PO$_4$	2.5 g
MgSO$_4$.7H$_2$O	0.6 g
Trisodium citrate	2.5 g
Asparagine	5.0 g
Glycerol	20.0 ml
Phenol red 0.4% aq.	3.0 ml
Distilled water	1000 ml

Steam to dissolve. The pH should be 7.4–7.6. Bottle in 9-ml amounts and autoclave at 115°C for 10 min. To each bottle 1 ml of horse serum and antibiotics as desired.

Lead acetate agar

For detection of H$_2$S.

Nutrient agar	1000 ml
Lead acetate	0.2 g
Sodium thiosulphate	0.08 g

Steam to dissolve, distribute for stab culture and autoclave at 110°C for 10 min.

Listeria enrichment medium (Lovell *et al.*, 1987)

Prepare the following solutions in distilled water:
  (1) Acriflavine   0.15%
  (2) Nalidixic acid 0.4%
  (3) Cycloheximide 0.5%
Sterilize by filtration and store at −4°C. Add 1 ml of each to 100 ml of tryptone broth.

Loeffler medium: inspissated serum

Mix 3 parts of horse serum with 1 part of glucose broth, tube and inspissate in slopes.

Lowbury and Lilly medium

See Egg yolk agar.

Lowenstein–Jensen medium

This is used for the isolation of mycobacteria.

KH$_2$PO$_4$ (AnalaR)	4.0 g
MgSO$_4$.7H$_2$O (AnalaR)	0.4 g
Magnesium citrate	1.0 g
Asparagine	6.0 g
Glycerol (AnalaR)	20.0 ml
Distilled water to	1000 ml

Dissolve in this order, steam for 2 h.

Clean fresh eggs with soap and water and break them into a sterile graduated cylinder. Take 1600 ml of egg fluid, mix in a screw-capped jar or polypropylene container with some large glass beads to break the yolks, filter through gauze and add to 1 litre of salt mixture. Add 50 ml of 1% aqueous malachite green, dispense and inspissate.

A 4-ml volume of pyruvic acid (neutralized with 3 N NaOH solution) can replace glycerol.

Malonate broth

$(NH_4)_2SO_4$	2 g
K_2HPO_4	0.6 g
KH_2PO_4	0.4 g
Sodium malonate	3 g
Yeast extract	1 g
Distilled water	1000 ml
Bromothymol blue (0.2%)	12.5 ml

Heat to dissolve, add indicator and autoclave at 115°C for 10 min.

Malt agar modifications

To isolate acidophilic yeasts, cool the sterilized medium to 50°C, add 1 ml of sterile 10% lactic acid and pour plates at once.

For dermatophytes or pathogenic yeasts add 10 ml of 5% cycloheximide in acetone and 10 ml of 0.5% chloramphenicol in alcohol. Dispense in 10-ml amounts and autoclave at 115°C for 10 min.

To suppress bacterial growth and isolate moulds, add only the chloramphenicol.

Mead's medium (1963)

This medium allows *S. faecalis* to be recognized by its ability to ferment sorbitol, decompose tyrosine and reduce triphenyltetrazolium chloride. Cultures are incubated at 45°C.

Peptone	10 g
Yeastrel	1 g
Sorbitol	2 g
Tyrosine	5 g
Agar	12 g
Water	1000 ml

Dissolve by autoclaving at 115°C for 10 min and adjust to pH 6.2. Add 0.1 g of triphenyltetrazolium chloride and 1 g of thallous acetate. When these have dissolved, add a further 4 g of tyrosine to give a suspension. Cool to 50°C, mix to give an even suspension of tyrosine and pour plates. Layered plates, the lowest layer of the same medium without tyrosine, are best.

Milk salt agar

(Gilbert *et al.*, 1969)
This is used for the recovery of *S.aureus* from food samples.

Peptone	5 g
Lab Lemco	3 g
Sodium chloride	6.5 g

Agar   15 g
Sterilize at 121°C for 15 min. Add 10% skim milk and pour plates.

MRS medium

(de Man *et al.*, 1960)
Lab Lemco	10 g
Yeast extract	5 g
Peptone	10 g
Dipotassium hydrogen phosphate	2 g
Sodium acetate	5 g
Magnesium sulphate (MgSO$_4$.7H$_2$O)	0.2 g
Manganese sulphate (MnSO$_4$.4H$_2$O)	0.05 g
Tween 80	1 ml
Triammonium citrate	2 g
Water	1000 ml

Dissolve by heating, bottle and autoclave. For use, add 2% sterile (filtered) glucose or other carbohydrate solution.

N medium

(Collins, 1962)
This is used in the differentiation of mycobacteria.
NaCl	1.0 g
MgSO$_4$.7H$_2$O	0.2 g
KH$_2$PO$_4$	0.5 g
Na$_2$HPO$_4$.12H$_2$O	3.0 g
(NH$_4$)$_2$SO$_4$	10.0 g
Glucose	10.0 g
Water	1000 ml

Dissolve in that order, adjust to pH 6.8–7.0 and distribute. Autoclave at 115°C for 10 min. This medium can be modified by the addition of 1.2% agar and 0.04% bromocresol purple.

Nitrate broth

Dissolve 1 g of potassium nitrate in 1000 ml of nutrient broth. Steam to dissolve and bottle in small amounts.

Nutrient gelatin

Choose a high-quality gelatin and add 12–15% to a nutrient, Lemco or yeast extract broth. Heat to dissolve at 115°C for 10 min.

ONPG broth

o-Nitrophenyl-D-galactopyranoside	6 g
0.001 M Na$_2$HPO$_4$	1000 ml

Tube in 2-ml amounts. Do not heat.

Phenolphthalein phosphate agar

Phenolphthalein phosphate agar for the phosphatase test is made by adding 1 ml of a 1% solution of phenolphthalein phosphate to 100 ml of nutrient agar or commercial blood agar base and pouring plates.

Phenolphthalein phosphate polymyxin agar (PPPA)

For isolating *Staphylococcus aureus* from foods.
    Add 125 units of polymyxin to 1 litre of phenolphthalein phosphate agar.

Phenolphthalein sulphate agar

This medium for the arylsulphatase test is made by dissolving 0.64 g of potassium phenolphthalein sulphate in 100 ml of water (0.01 M) Dispense one of the Middlebrook Broth media in 2.7-ml amounts and add 0.3 ml of the phenolphthalein sulphate solution to each tube.

Phenylalanine agar

DL-phenylalanine	2 g
Yeast extract	3 g
Na₂HPO	1 g
NaCl	5 g
Agar	20 g
Distilled water	1000 ml

Dissolve by heat; sterilize at 115°C for 10 min and tube as slopes.

Robertson's cooked meat medium

This is a useful general purpose medium and is used for the cultivation of anaerobes.
    Mince 500 g of fresh lean beef and simmer for 1 h in 500 ml of boiling water containing 2 ml of 1 N NaOH. This neutralizes the acid in the medium; filter, dry the meat in a cloth and when dry put it into 28-ml screw-capped bottles so that there is about 25 mm of meat in each. Add 10–12 ml of infusion or Lemco broth to each tube, cap and autoclave at 115°C for 10 min.

Salt agar

For cultivation of plague bacilli. Add 3% sodium chloride to nutrient agar.

Salt colistin broth

For staphylococci

Yeast extract	3 g
Tryptone	10 g
NaCl	20 g
Distilled water	1000 ml

    Steam to dissolve; sterilize at 115°C for 10 min; add 10mg/l of colistin sulphate and dispense aseptically in 10-ml lots.

Purple milk

Supersedes litmus milk

Skim milk	1000 ml
Bromocresol purple, 0.2%	10 ml

Dispense in 5 or 10-ml lots and autoclave at 110°C for 10 min.

Salt meat broth

Advantage is taken of the salt-tolerance (halophily) of *S.aureus* in investigating outbreaks of food poisoning due to *S. aureus*. Many other organisms are inhibited by high salt concentrations.

Prepare Robertson's cooked meat medium as above but add 10% sodium chloride to the liquid before tubing.

Semisolid oxygen preference medium

For identifying mycobacteria. Add 0.1% agar to one of the fluid Middlebrook liquid media. Autoclave at 115°C for 10 min. Dispense in 12-ml amounts in narrow bottles. Before use add 0.5 ml of oleic acid albumin dextrose catalase (OADC) supplement.

Skim milk agar

Reconstitute skim milk powder and mix equal volumes of this and double-strength nutrient agar melted and at 55°C. Mix and pour plates.

'Sloppy agar' Craigie tube

Broth containing 0.1–0.2% agar will permit the migration of motile organisms but will not allow convection currents. This permits motile organisms to be separated from non-motile bacteria in the Craigie tube.

Dispense sloppy agar in 12-ml amounts in screw-capped bottles. To each bottle add a piece of glass tubing of dimensions 50 × 5 or 6 mm. There must be sufficient clearance between the top of the tube and the surface of the medium so that a meniscus bridge does not form and the only connection between the fluid and the inner tube is through the bottom.

Starch agar

Prepare a 10% solution of soluble starch in water and steam for 1 h. Add 20 ml of this solution to 100 ml of melted nutrient agar and pour plates.

Sucrose agar

To demonstrate levan or dextran formation, add 5% sucrose to 100 ml of melted nutrient agar. Steam for 30 min, cool to 55°C, add 5 ml of sterile horse serum and pour plates.

Tellurite blood agar media for corynebacteria

The addition of 0.04% potassium tellurite (*caution*) to blood agar inhibits many other organisms and permits differentiation of *gravis, mitis* and *intermedius* types of *C.diphtheriae* and other corynebacteria. The base medium should be an infusion, not a digest medium, and Hoyle's base gives good results. For recognition and differentiation at 18–24 h, Brain–Veal Infusion agar is excellent. Some workers prefer lysed to fresh blood. Horse blood may be lysed by alternate freezing and thawing or by the addition of 5 ml of 2% saponin solution to 100 ml of blood.

Potassium tellurite is stored as a 2% solution.

Tributyrin agar

Fat-splitting organisms cause spoilage in butter, etc. Most of these bacteria split glyceryl tributyrate (tributyrin).

Peptone	5 g
Yeast extract	3 g
Tributyrin	10 g
Agar	12 g
Water	1000 ml

Heat to dissolve, adjust to pH 7.5, dispense and autoclave at 115°C for 15 min. Colonies of lipolytic organisms clear the medium.

Tyrosine agar: xanthine agar

To 100 ml of melted nutrient agar, add 5 g of tyrosine or 4 g of xanthine and steam for 30 min. Mix well to suspend amino acid and pour plates. See also Mead's medium.

Willis and Hobbs (1959) medium

This excellent medium for isolating and identifying clostridia sometimes needs minor modification to give the best results for particular species. Nagler reaction, 'pearly layer' proteolysis and lactose fermentation can be observed.

Nutrient broth	1000 ml
Lactose	12 g
Agar	12 g
Water	1000 ml

Heat to dissolve, adjust to pH 7.0, add 3.5 ml of 1% neutral red and 1 g of sodium thioglycollate, bottle in 100-ml portions and autoclave at 115°C for 10 min.

To each 100 ml of melted base add 4 ml of egg yolk emulsion and 15 ml of sterile whole milk. Pour plates.

Neomycin may be added to inhibit non-spore-bearing anaerobes and aerobic spore bearers but should be used with caution. It is desirable to experiment with varying concentrations up to 100 μg/ml.

Yeast extract milk agar

This is the same as commercial Milk agar.

Diluents

Saline and Ringer solutions

Physiological saline, which has the same osmotic pressure as microorganisms, is a 0.85% solution of sodium chloride in water. Ringer solution, which is ionically balanced so that the toxic effects of the metallic ions neutralize one another, is better than the saline for making bacterial suspensions. It is used at one-quarter the original strength.

Sodium chloride (AnalaR)	2.15 g
Potassium chloride (AnalaR)	0.075 g
Calcium chloride, anhydrous (AnalaR)	0.12 g
Sodium thiosulphate pentahydrate	0.5 g
Distilled water	1000 ml

The pH should be 6.6. Tablets of sodium chloride and Ringer are useful.

Calgon Ringer

Alginate wool, used in surface swab counts, dissolves in this.

Sodium chloride	2.15 g
Potassium chloride	0.075 g
Calcium chloride	0.12 g
Sodium hydrogen carbonate	0.05 g
Sodium hexametaphosphate	10.00 g
Water	1000 ml

The pH is 7.0. It is conveniently purchased as tablets.

Thiosulphate Ringer

This may be preferred for dilutions of rinses where there may be residual chlorine. Conveniently manufactured in tablet form.

Peptone water diluent (maximum recovery diluent)

The safest and least lethal of diluents is 0.1% peptone water.

Phosphate buffered saline

This is a useful general diluent at pH 7.3.

Sodium chloride	8.0 g
Potassium dihydrogen phosphate	0.34 g
Dipotassium hydrogen phosphate	1.21 g
Water	1000 ml

Indicators

Andrade

Dissolve 0.5 g of acid fuchsin in 100 ml of water. Add 1 N NaOH solution until the colour changes to yellow. The indicator is colourless at pH 7.2, pink in acidic solutions and yellow in alkaline solutions.

Bromocresol purple

Dibromo-*o*-cresolsulphonphthalein. Dissolve 0.1 g in 9.2 ml of 0.02 N NaOH and dilute to 250 ml with water. The indicator (0.4% solution) is yellow at pH 5.8 and blue at 6.8.

Bromothymol blue

Dibromothymolsulphonphthalein. Dissolve 0.1 g in 8 ml of 0.02 N NaOH and make up to 250 ml with water. This indicator (0.4% solution) is yellow at pH 6.0 and blue at pH 7.6.

Congo red

This indicator (0.1% solution in water) is blue at pH 3.0 and red at pH 5.2.

Methyl red

4'-Dimethylaminoazobenzene-4-sulphonate. Dissolve 0.1 g in 300 ml of ethanol and add 200 ml of water. This solution is red at pH 4.4 and yellow at pH 6.2.

Phenolphthalein

Dissolve 1 g in 100 ml of 95% ethanol. The solution is colourless at pH 8.3 and red at pH 10.

Phenol red

Phenolsulphonphthalein. Dissolve 0.1 g in 14.1 ml of 0.02 N NaOH and dilute to 250 ml with water. This indicator (0.4% solution) is yellow at pH 6.8 and red at pH 8.4.

Thymol blue

Thymolsulphonphthalein. Dissolve 0.1 g in 10.75 ml of 0.02 N NaOH and dilute to 250 ml with water. It has an acid range, red at pH 1.2 and yellow at 2.8, and an alkaline range, yellow at pH 8.0 and blue at pH 9.6.

Preparation of culture media

Utensils

Copper or zinc containers must not be used. Small amounts of these metals will dissolve in the culture media and are bactericidal. Large amounts of media can be made in the stainless steel buckets that are used in the dairy and food trades. Smaller amounts can be made in resistance-glass laboratory flasks. Large stainless steel funnels can be used for filtration, with folded filter-papers of the heavy Chardin type. Filtration and adjustment of pH should be completed before agar is added. The present-day agars give a clear gel but if it is necessary to filter agar medium, cut a sheet of heavy filter-paper to fit a large Büchner funnel and place wet filter-paper pulp on it to a depth of 5 mm. Place the funnel on a filter flask and suck the pulp dry. Heat the whole apparatus in a steamer in order to prevent gelling of the medium in the filter.

Inhibitory substances

Metals such as zinc and copper (see above) and other inhibitory substances such as fatty acids may be present in peptones, agar, carbohydrates and other chemicals which are not specifically made for culture media. It is inadvisable, for reasons of cost and convenience, to use any material that is not specified as being of 'bacteriological quality' and purchasing officers should be severely discouraged from seeking, for reasons of 'economy', suppliers of media ingredients, including chemicals and dyes, who are not known to bacteriologists.

Distilled water from commercial bulk supplies, 'factory' or 'battery' quality may contain substances that inhibit bacterial growth, e.g. oily materials from stripper stills attached to steam lines. Manesty-type distilled water, glass-distilled water and deionized water should be used, although it must be realized that some trace elements necessary for bacterial growth are in fact supplied by the distilled water, by glassware, or by recognized impurities in analytical quality reagents.

Glucose, when autoclaved with salts such as phosphate, may yield inhibitory substances. It is best to add carbohydrates as sterile solutions after the medium has been sterilized.

Some cotton wools, when sterilized in a hot-air oven, yield fatty materials. Sometimes these materials can be seen as condensates on glass tubes. Again, cotton wool from a reliable source should be used.

Excessive heating and holding at high temperatures may destroy growth factors and gelling capacity and cause darkening and pH drift of culture media.

Adjustment of pH

The pH of reconstituted and laboratory-prepared media should be checked with the meter, but for practical purposes devices such as the BDH Lovibond Comparator are good enough.

Pipette 10 ml of the medium into each of two 152 × 16 mm test-tubes. To one add 0.5 ml of 0.04% phenol red solution. Place the tubes in a Lovibond Comparator with the blank tube (without indicator) behind the phenol red colour disc, which must be used with the appropriate screen. Rotate the disc until the colours seen through the apertures match. The pH of the medium can be read in the scale aperture. Rotate the disc until the required pH figure is seen, add 0.05 N sodium hydroxide solution or hydrochloric acid to the medium plus indicator tube and mix until the required pH is obtained.

From the amount added, calculate the volume of 1 or 5 N alkali or acid that must be added to the bulk of the medium. After adding, check the pH again.

Example   10 ml of medium at pH 6.4 requires 0.6 ml of 0.05 N sodium hydroxide solution to give the required pH of 7.2.

Then the bulk medium will require

$$\frac{0.6 \times 100}{20 \times 10} \text{ ml of 1 N NaOH/litre} = 3.0 \text{ ml}$$

Discs of other pH ranges are available.

All final readings of pH must be made with the medium at room temperature because hot medium will give a false reaction with some indicators.

Identification and storage

Most laboratories employ a colour code to identify media that look alike. Coloured beads may be placed in bottles of bulk medium. Coloured cotton wool or coloured aluminium or polypropylene caps (see p.23) can be used for tubed media, and the caps of bottled media can be dabbed with coloured enamel paint. This is particularly useful for carbohydrate fermentation test media.

The labelling machines used to place prices on goods in supermarkets are very useful for labelling and dating tubes, bottles and petri dishes of culture media.

Bulk bottled media in screw-capped bottles can be stored at room temperature.

Dispensing culture media

Liquid media can be dispensed with the aid of a large funnel held in a retort stand and fitted with a short piece of rubber tubing and a spring clip. The automatic commercial fillers (p.22) are more convenient.

To make agar slopes, dispense the medium in 3–7 ml amounts according to the size of the tube required. After sterilization and before the medium sets, slope the tubes individually on the bench by leaning them against a length of glass or metal rod 6 or 7 mm in diameter. When making large batches, slope the tubes or bottles on wire racks, e.g. old refrigerator shelves, sloped at a convenient angle.

When pouring plates, raise the lid only far enough to permit the mouth of the tube or bottle to enter. Pour about 12–15 ml in each plate. Dry the plates slightly open in an incubator, and store them medium side-up in a refrigerator, protected from moisture loss. An automatic plate pouring machine is useful in laboratories where many plates are used. This reduces the number of plates contaminated with airborne organisms to almost nil.

Contamination of sterile media during tubing and pouring may be a serious problem in some buildings. It can be minimized by using clean air cabinets of the simple, horizontal outflow variety, or (more expensive) Class II microbiology safety cabinets.

Quality control and performance tests

Manufacturers of culture media and their ingredients impose stringent controls on the quality and performance of their products. Reconstitution and handling of these media in the laboratory (e.g. overheating) may reduce its efficiency. Laboratory-prepared media are rarely tested.

Some kind of quality control should therefore be imposed on all materials used. The national system of quality control (in the UK) and performance tests (in the USA) operate satisfactorily only in clinical laboratories and problems arise mainly in the smaller industrial and food control establishments. A full scale system is expensive in staff and materials and entails the maintenance of a large number of stock cultures. Two simpler systems are described here, one for isolation and selective media, the other for identification media.

Isolation and selective media: efficiency of plating technique (EOP)

New batches of culture media may vary considerably and should be tested in the following way.

Prepare serial tenfold dilutions, for example. 10^{-2} to 10^{-7} of cultures of various organisms which will grow on or be inhibited by the medium. For example, when testing DCA use *S. sonnei, S. typhi, S. typhimurium,* several other salmonellas and *E. coli.* Use at least four well-dried plates of the test medium for each organism and at least two plates each of a known satisfactory control medium, and of a non-selective medium, e.g. nutrient agar. Do Miles and Misra drop counts with the serial dilutions of the organisms on these plates so that each plate is used for several dilutions (Figure 4.1) (see p.131).

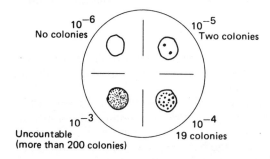

Figure 4.1 Miles and Misra count

Count the colonies, tabulate the results and compare the performances of the various media. These may suggest that in a new batch of a special medium it is necessary to alter the proportion of certain ingredients.

Stock cultures, however, will frequently grow on media which will not support the growth of 'damaged' organisms from natural materials. It is best, therefore, to use dilutions or suspensions of such material.

With some culture media, these drop counts do not give satisfactory results because of reduced surface tension. An alternative procedure is recommended.

Make cardboard masks to fit over the tops of petri dishes and square 25 × 25 mm in the centre of each. Place a mask over a test plate and drop one drop of the suspension through the square on the medium. Spread the drop over the area limited by the mask. This method has the additional advantage of making colony counts easier to perform but more plates must, of course, be tested.

In addition to these EOP tests, ordinary plating methods should be used to compare colony size and appearance.

Identification media

Maintain stock cultures (p.90) of organisms known to give positive or negative reactions with the identification tests in routine use. The same organism may often be used for a number of tests. When a new batch of any identification medium is anticipated, subculture the appropriate stock strain so that a young, active culture is available. Test the new medium with this culture.

References

APHA (1985) *Standard Methods for the Examination of Dairy Products*, 14th edn, American Public Health Association, Washington DC

Baird-Parker, A. C. (1966) Methods for classifying staphylococci and micrococci. In *Identification Methods for Microbiologists* Part A (eds B. M. Gibbs and F. A. Skinner), Society for Applied Bacteriology Technical Series No. 1. Academic Press, London, pp. 59–64

Collins, C. H. (1962) The classification of 'anonymous' acid fast bacilli from human source. *Tubercle*, **43**, 293–298

Davis, J. G. (1931) Standardisation of media in the acid ranges with special reference to the use of citric acid and buffer mixtures for yeast and mould media. *Journal of Dairy Research*, **3**, 133–144

Davis, J. G. (1982) Unhopped beer wort as a medium for yeast and mould counts. *Laboratory Practice*, **31**, 219

Davies, J. R. (1974) Elek's test for the toxigenicity of *C. diphtheriae*. In *Laboratory Methods, 1*, Public Health Laboratory Service Monograph No. 5. HMSO, London.

De Man, J. C., Rogosa, M. and Sharpe, M. E. (1960) A medium for the cultivation of lactobacilli. *Journal of Applied Bacteriology*, **23**, 130–133

Donovan, T. J. (1966) A *Klebsiella* screening medium. *Journal of Medical Laboratory Technology*, **23**, 194–196

Drake, C. H. (1966) Evaluation of culture media for the isolation and enumeration of *Pseudomonas aeruginosa*. *Health Laboratory Service*, **3**, 10–14

Ellner, P. D. (1956) A medium promoting rapid quantitative sporulation in *Clostridium perfringens*. *Journal of Bacteriology*, **71**, 495–496

Falkow, S. (1958) Activity of lysine decarboxylase as an aid in the identification of salmonellae and shigellae. *American Journal of Clinical Pathology*, **29**, 598–560

Flynn, J. and Waitkins, S. A. (1972) A serum-free medium for testing the fermentation reactions in *Neisseria*. *Journal of Clinical Pathology*, **25**, 525–527

Gardner, G. A. (1966) A selective medium for the enumeration of *Microbacterium thermosphactum* in meat and meat products. *Journal of Applied Bacteriology*, **29**, 455–460

Gilbert, R. J., Kendall, M. and Hobbs, B. C. (1969) Media for the isolation and enumeration of coagulase positive staphylococci. In *Isolation Methods for Microbiologists* (eds D. A. Shapton and G. W. Gould), Society for Applied Bacteriology Technical Series No. 3. Academic Press, London, pp. 9–14

Islam, A. K. M. S. (1977) Rapid recognition of Group B streptococci. *Lancet*, **i**, 356–357

Lovell, J., Francis, D. W. and Hunt, W. H. (1987) *Listeria monocytogenes* in raw milk: detection, incidence and pathogenicity. *Journal of Food Protection*, **50**, 188–197

Lowbury, G. J. L. and Lilly, H. A. (1955) A selective plate medium for *Cl. welchii*. *Journal of Pathology and Bacteriology*, **70**, 105–107

Mead, G. C. (1963) A medium for the isolation of *Streptococcus faecalis sensu stricto*. *Nature (London)*, **197**, 1323–1324

Moeller, V. (1955) Simplified tests for some amino acid decarboxylases and for the arginine dihydrolase system. *Acta Pathologica et Microbiologica Scandinavica*, **36**, 158–172

National Committee for Clinical Laboratory Standards (1979) Approved Standard ASM-2. Performance standards for antimicrobic disc susceptibility tests, 2nd edn, Villanova, PA

Petts, D. N. (1984) Colistin–oxolinic acid–blood agar: a new selective medium for streptococci. *Journal of Clinical Microbiology*, **19**, 4–7

Waitkins, S. (1985) Leptospiras and Leptospirosis. In *Isolation and Identification of Micro-organisms of Medical and Veterinary Importance* (eds C. H. Collins and J. M. Grange), Society for Applied Bacteriology Technical Series No. 21, Academic Press, London, pp. 251–296

Willis, A. T. and Hobbs, G. (1959) A medium for the identification of clostridia producing opalescence in egg yolk emulsions. *Journal of Pathology and Bacteriology*, **77**, 299–300

Chapter 5

Cultural methods

General bacterial flora

To observe the general bacterial flora of a sample of any material, use non-selective media. No one medium nor any one temperature will support the growth of all possible organisms. It is best to use several different media, each favouring a different group of organisms, and to incubate at various temperatures aerobically and anaerobically. The pH of the medium used should approximate that of the material under examination.

Digest media preferably with the addition of blood, glucose tryptone agar, skim milk agar and Rogosa agar are suitable media. Selective and differential media and enrichment additives are described in Chapter 4 and under methods for the isolation and identification of bacteria and fungi.

Methods for estimating bacterial numbers are given in Chapter 9, and the dilution methods employed can be used to make inocula for the following techniques. Three methods are given here for plating material. They depend on separating units (see p.127) of bacteria so that each will grow into a separate colony. These colonies can be subcultured or picked for further examination.

Pour plate method

Melt several tubes containing 15 ml of medium. Place in water-bath at 45–50°C to cool. Emulsify the material to be examined in 0.1% peptone water and prepare 1:10, 1:100 and 1:1000 dilutions in the same solution (see p.130). To 15 ml of melted medium at 45°C add 1 ml of a dilution, mix by rotating the tube between the palms of the hands and pour into a petri dish. Make replicate pour plates of each dilution for incubation at suitable temperatures. Allow the medium to set, invert the plate and incubate. This method gives a better distribution of colonies than that used for the plate count (p.130), but cannot be used for that purpose because some of the inoculum, diluted in the agar, remains in the test-tube.

Spreader method

Dry plates of a suitable medium. Make dilutions of the emulsified material as described above. Place about 0.05 ml (1 drop) of dilution in the centre of a plate and spread it over the medium by pushing the glass spreader backward and forward

84

while rotating the plate. Replace the petri dish lid and leave for 1–2 h to dry before inverting and incubating as above.

Looping-out method

Make dilutions of material as described above. Usually the neat emulsions and at 1:10 dilution will suffice. Place one loopful of material on the medium near the rim of the plate and spread it over the segment (Figure 5.1). Flame the loop and spread from area *A* over area *B* with parallel streaks, taking care not to let the streaks overlap. Flame the loop and repeat with area *C*, and so on. Each looping-out dilutes the inoculum. Invert the plates and incubate.

When looping-out methods are used, two loops are useful; one is cooling while the other is being used.

These surface plate methods are said to give better isolation of anaerobes than pour plates.

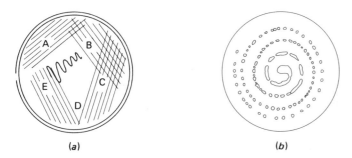

(a)   (b)

Figure 5.1 (a) 'Spreading' or 'looping-out' in a plate; (b) rotary plating

Spiral or rotary plating

This mechanical method is invaluable for plating large numbers of specimens or cultures and for doing sensitivity tests by Stokes method (p.160). Hand-operated (with electrically-driven turntable) and fully-automated models are available. With the latter, exact volumes may be plated. A laser colony counter is on the market.

To use the hand-operated machine place the open plate on the turntable. Touch the rotating plate with the charged loop at the edge of the medium and draw it slowly towards the circumference. A spiral of bacterial growth will be obtained after incubation which will show discrete colonies (Figure 5.1).

Multipoint inoculators

These are convenient tools when many replicate cultures are needed. They are fully or semi-automatic and spot-inoculate large numbers of petri dish cultures each with 10–20 cultures. They are particularly useful for combined antibiotic susceptibility and minimum inhibitory concentration tests as well as for the quality control of culture media.

Shake tube cultures

These are useful for observing colony formation in deep agar cultures, especially of anaerobic or microaerophilic organisms.

Dispense media, glucose agar or thioglycollate agar, in 15–20-ml amounts in bottles or tubes 20–25 mm in diameter. Melt and cool to approximately 45°C. Add about 0.1 ml of inoculum to one tube, mix by rotating between the palms, remove one loopful to inoculate a second tube, and so on. Allow to set and incubate. Submerged colonies will develop and will be distributed as follows: obligate aerobes grow only at the top of the medium, obligate anaerobes only near the bottom; microaerophiles grow near but not at the top; facultative organisms grow uniformly throughout the medium.

Stab cultures

These can be used to observe motility, gas production and gelatin liquefaction. The medium is dispensed in 5-ml amounts in tubes or bottles of 12.5 mm diameter and inoculated by stabbing the wire down the centre of the agar.

Subcultures

For further examination, pure cultures are prepared from the various types of colonies in mixed cultures.

Place the plate culture under a low-power binocular microscope and select the colony to be examined. With a flamed straight wire (*not* a loop) touch the colony. There is no need to dig or scrape; too vigorous handling may result in contamination with adjacent colonies or with those beneath the selected colony.

With the charged wire, touch the medium on another plate, flame the wire and then use a loop to loop-out the culture to obtain individual colonies. This should give a pure culture but the procedure may need repeating.

To inoculate tubes or slopes, follow the same procedure. Hold the charged wire in one hand, the tube in the other in a nearly horizontal position. Remove the cap or plug of the tube by grasping it with the little finger of the hand holding the wire. Pass the mouth of the tube through a bunsen flame to kill any organism that might fall into the culture and inoculate the medium by drawing the wire along the surface of a slope or touching the surface of a liquid. Replace the cap or plug and flame the wire. Several tubes can be inoculated in succession without recharging the wire.

Anaerobic culture

Anaerobic jars

Modern anaerobic jars are made of metal or transparent polycarbonate, are vented by Schrader valves and use 'cold' catalysts. They are therefore safer than earlier models; internal temperatures and pressures are lower. More catalyst is used and the escape of small particles of catalyst, which can ignite hydrogen–air mixtures is prevented. Nevertheless the catalyst should be kept dry, dried after use and replaced frequently.

For ordinary clinical laboratory work the commercial sachets are a convenient source of hydrogen and hydrogen–carbon dioxide mixtures.

Place the plates 'upside down', i.e. with the media at the bottom in the vessel. If they are incubated the right way up, the medium sometimes falls into the lid when the pressure is reduced. Dry the catalyst capsule by flaming and replace the lid.

Some workers prefer to use pure hydrogen or a mixture of 90% hydrogen and 10% carbon dioxide. A non-explosive mixture of 10% hydrogen, 10% carbon dioxide and 80% nitrogen is safer (see Kennedy, 1988) but may be difficult to obtain in some countries. There is a common misconception about storing these cylinders horizontally: they may safely be stored vertically.

For the evacuation-replacement jar technique with a mixture containing 90% or more hydrogen attach a vacuum pump to the jar across a manometer and remove air to about −300 mmHg. Replace the vacuum with the gas mixture, disconnect both tubes and allow the secondary vacuum to develop as the catalyst does its work. Leave for 10 min and reconnect the manometer to check that there is a secondary vacuum of about −100 mmHg. This indicates that the seals of the jar are gas tight and that the catalyst is working properly. Replace the partial vacuum with gas to atmospheric pressure.

NB If the gas mixture contains only 10% hydrogen the jar must be evacuated to at least −610 mmHg so that enough air is removed for the catalyst and hydrogen to remove the remainder.

Indicators of anaerobiosis

'Redox' indicators based on methylene blue and resazurin are available commercially. There is a simple test to demonstrate that anaerobiosis is achieved during incubation: inoculate a blood agar plate with *Clostridium perfringens* and place on it a 5 μg metronidazole disc. This bacterium is not very exacting in its requirement for anaerobiosis and will grow in suboptimal conditions. A zone of inhibition around the growth indicates that conditions were sufficiently reduced for the drug to work and that anaerobiosis is adequate. There is little to commend the use of *Pseudomonas aeruginosa* as a 'negative' control. Even if a 'positive' control such as *Cl. tetani* fails to grow it does not necessarily suggest inadequate anaerobic conditions.

Anaerobic cabinets

These are now commonly used in diagnostic laboratories. They may be used simply as anaerobic incubators with the added advantage that cultures do not have to be removed for examination, or as anaerobic work stations for inoculating pre-reduced media. Modern cabinets are very sophisticated, allowing 'bare hand' working, humidity control and atmospheric detoxification.

Culture under carbon dioxide

Some microaerophiles grow best when the oxygen tension is reduced. Capnophiles require carbon dioxide. For some purposes, candle jars are adequate. Place the cultures in a biscuit or dried-milk tin with a lighted candle or night-light. Replace the lid. The light will go out when most of the oxygen has been removed and replaced with carbon dioxide.

Alternatively, place 25 ml of 0.1 N HCl in a 100-ml beaker in the container and add a few marble chips before replacing the lid.

Commercial carbon dioxide sachets, hydrogen–carbon dioxide sachets or carbon dioxide from a cylinder may be used with an anaerobic jar. Remove air from the jar down to −76 mmHg and replace with carbon dioxide from a football bladder. This is the best method for capnophiles, which require 10–20% carbon dioxide to initiate growth, but if extensive work of this nature is carried out, a carbon dioxide incubator is invaluable. A nitrogen–carbon dioxide mixture can be used instead of pure carbon dioxide.

Incubation of cultures

Incubators are discussed in Chapter 2. Instead of using an incubator for culturing at 20–22°C, plates and tubes may be left on the bench, but a cupboard is better as draughts can be avoided. A maximum and minimum thermometer is necessary. Ambient temperatures often fluctuate widely when windows are open or heating is switched off at night.

Psychrophiles are usually incubated at 4–7°C. A domestic refrigerator can be used provided that it is not opened too often. Temperatures of 20–25°C are widely used for food bacteria and 35–37°C for medical bacteria. Thermophiles require 55–60°C.

Petri dish cultures are incubated upside down, i.e. medium uppermost (except in anaerobic vessels), otherwise condensation water collects on the surface of the medium and prevents the formation of isolated colonies.

If it is necessary to incubate petri dish cultures for several days, they must be sealed in order to prevent the medium drying up. Plastic bags or plastic food containers can be used.

Colony and cultural appearances

Much time and labour may be saved in diagnostic bacteriology if workers familiarize themselves with colonial appearances. These appearances and the terms used to describe them are illustrated in Figure 5.2.

Examine colonies and growths on slopes with a hand-lens or plate microscope. Observe pigment formation both on top of and under the surface of the colony and note if pigment diffuses into the medium. In broth culture, surface growth may occur in the form of a pellicle. Some cultures develop characteristic odours.

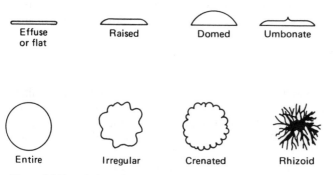

Figure 5.2 Description of colonies and cultures

Auxanograms

These are used to investigate the nutritional requirements of bacteria and fungi, for example, which amino acid or vitamin substance is required by 'exacting strains' or which carbohydrates the organisms can utilize as carbon sources.

Make a minimal inoculum suspension of the organism to be tested in phosphate-buffered saline. Centrifuge, re-suspend in fresh phosphate-buffered saline. Wash again in the same way and add 1 ml to several 15-ml amounts of basal medium (p.63) with indicator. Pour into plates. On each plate place a filter-paper disc of 5-mm diameter and on the filter-paper place one loopful of a 5% solution of the appropriate carbohydrate. Incubate and observe growth. Commercially-prepared discs are available.

When testing for amino acid or vitamin requirements in this way, use the basal medium plus glucose. Several discs, each containing a different amino acid, can be used on a single plate.

API make an auxanogram kit.

Stock cultures

Most laboratories need stock cultures of 'standard strains' of microorganisms for testing culture media, controlling certain tests and so that workers may become familiar with cultural appearances and other properties. These must be subcultured periodically to keep them alive, with the consequent risk of contamination, and more importantly in laboratories engaged on assay work, the involuntary selection of mutants. It is because of mutations in bacterial cultures that the so-called 'standard strains' or 'type cultures' so often vary in behaviour from one laboratory to another.

To overcome these difficulties, Type Culture Collections are maintained by government laboratories. In these establishments, the cultures are preserved in a freeze-dried state, and in the course of maintenance of the stocks they are tested to ensure that they conform to the official descriptions of the standard or type strains.

Cultures of microorganisms may be obtained from: the National Collection of Type Cultures (NCTC), Central Public Health Laboratory, Colindale Avenue, London NW9 5HT; the National Collection of Industrial and Marine Bacteria (NCIMB), Torry Research Station, PO Box 13, Abbey Road, Aberdeen AB9 8DG; the American Type Culture Collection (ATCC), 12301 Parklawn Drive, Rockville, Maryland 20852; the National Collection of Dairy Organisms (NCDO), National Institute for Research in Dairying, Shinfield near Reading, Berkshire; the National Collection of Pathogenic Fungi (NCPF), PHLS Mycology Reference Laboratory, Central Public Health Laboratory, Colindale Avenue, London NW9 5HT the Commonwealth Mycological Collection (CMC), Ferry Lane, Kew, Surrey; the National Collection of Yeast Cultures (NCYC), Food Research Institute, Colney Lane, Norwich, Norfolk NR4 7UA. A fee is usually charged. These institutions also keep lists of private and commercial collections. Strains that are to be used occasionally, or rarely, should be obtained when required and not maintained in the laboratory. Strains that are in constant use may be kept for short periods by serial subculture, for longer periods by one of the drying methods and indefinitely by freeze-drying. When stocks of dried or freeze-dried cultures need renewing, however, it is advisable to start again with a culture from one of the National Collection rather than perpetuate one's own stock.

Serial subculture

The principle is to subculture as infrequently as possible in order to keep the culture alive and to arrest growth as early as possible in the logarithmic phase so as to avoid the appearance and possible selection of mutants. Useful media for keeping stock cultures of bacteria are Griffith's egg, fastidious anaerobe medium, Robertson's cooked meat and litmus milk. Fungi can be maintained on Sabouraud medium. These media must be in screw-capped bottles. It is difficult to keep cultures of some of the very fastidious organisms, for example gonococci, and it is advisable to store them above liquid nitrogen.

Inoculate two tubes, incubate until growth is just obvious and active, for example, 12–18 h for enterobacteria, other Gram-negative rods and micrococci, 2 or 3 days for lactobacilli. Cool the cultures and store them in a refrigerator for 1–2 months.

Keep one of these cultures, A, as the stock, and use the other, B, for laboratory purposes.

$$\text{Stock} \quad A_1 \rightleftharpoons A_2 \rightleftharpoons A_3 \rightleftharpoons A_4$$
$$\text{Use} \quad\quad \text{use } B \ \text{use } B \ \text{use } B$$

Drying cultures

Two methods are satisfactory. To dry by Stamp's method, make a small amount of 10% gelatin in nutrient broth and add 0.25% of ascorbic acid. Add a thick suspension of the organisms before the medium gels (to obtain a final density of about 10^{10} organisms/ml). Immerse paper sheets in wax, drain and allow to set. Drop the suspension on the waxed surface. Dry in a vacuum desiccator over phosphorus pentoxide. Store the discs that form in screw-capped bottles in a refrigerator. Petri dishes smeared with a 20% solution of silicone in light petroleum and dried in a hot-air oven can be used instead of waxed paper.

Rayson's method is slightly different. Use 10% serum broth or gelatin broth and drop on discs of about 10 mm diameter cut from Cellophane and sterilized in petri dishes. Dry in a vacuum desiccator over phosphorus pentoxide. Store in screw-capped bottles in a refrigerator.

To reconstitute, remove one disc with sterile forceps into a tube of broth and incubate.

Freezing

Make thick suspensions of bacteria in 0.5 ml of sterile tap water or skim milk in screw-capped bottles and place in a deep freeze at −40°C or, better, −70°C. Many organisms remain viable for long periods in this way.

Suspensions may also be stored at the vapour phase above liquid nitrogen.

Freeze-drying

Suspensions of bacteria in a mixture of 1 part of 30% glucose in nutrient broth and 3 parts of commercial horse serum are frozen and then dried by evaporation under vacuum. This preserves the bacteria indefinitely and is known as lyophilization. Yeasts, cryptococci, nocardias and streptomycetes can be lyophilized by making a thick suspension in sterile skimmed milk fortified with 5% sucrose. This simple protective medium can be sterilized by autoclaving in convenient portions, e.g. 3 ml in bijou bottles.

Several machines are available. The simplest consists of a metal chamber in which solid carbon dioxide and ethanol are mixed. The bacterial suspension is placed in 0.2-ml amounts in ampoules or small tubes together with a label and these are held in the freezing mixture, which is at about $-78°C$, until they freeze. They are then attached to the manifold of a vacuum pump capable of reducing the pressure to less than 0.01 mmHg. Between the pump and the manifold there is a metal container, usually part of the freezing container, in which a mixture of solid carbon dioxide and phosphorus pentoxide condenses the moisture withdrawn from the tubes. The tubes or ampoules are then sealed with a small blowpipe.

Simultaneous freezing and drying is carried out in centrifugal freeze-driers in which bubbling of the suspension is prevented by centrifugal action.

For further information on the preservation of bacteria by freeze-drying and other methods see Lapage and Redway (1975).

References

Kennedy, D. A. (1988) Equipment-related hazards. In *Safety in Clinical and Biomedical Laboratories*, (ed. C. H. Collins), Chapman & Hall, London, pp.11–46

Lapage, S. P. and Redway, M. (1975) *The Preservation of Bacteria and other Organisms*, Public Health Laboratory Service Monograph No. 7, HMSO, London

Chapter 6

Staining methods

The morphology of bacteria may be observed in wet, unstained preparations using light, darkfield or phase contrast microscopy but these are not permanent. It is usual to stain thin films of organisms prepared as follows.

Place a very small drop of saline or water in the centre of a 76 × 25 mm glass microslide. Remove a small amount of bacterial growth with an inoculating wire or loop, emulsify the organisms in the liquid, spreading it to occupy about 1 or 2 cm². Allow to dry in air or by waving high over a bunsen burner, taking care that the slide becomes no warmer than can be borne on the back of the hand. Pass the slide, film side down, once only through the bunsen flame to 'fix' it. This coagulates bacterial protein and makes the film less likely to float off during staining. It may not kill all the organisms, however, and the film should still be regarded as a source of infection. After staining, drain, blot with fresh, clean filter paper and dry by gentle heat over a bunsen.

Most laboratories now buy stains in solution ready for use, but some formulas are given below.

In the USA only stains certified by the Biological Stains Commission should be used. In the UK there is no comparable system and stains prepared by one of the specialist organizations should be used or purchased in solution ready for use.

A large number of stains, mostly basic aniline dyes, have been used and described. For practical purposes one or two simple stains, the Gram method and an acid-fast stain, are all that are required, plus one stain for fungi. A few other stains are described here including methods for staining capsules and spores.

Methylene blue stain

Stain for 1 min with a 0.5% aqueous solution of methylene blue or Loeffler's methylene blue: mix 30 ml of saturated aqueous methylene blue solution with 100 ml of 0.1% potassium hydroxide. Prolonged storage, with occasional shaking, yields 'polychrome' methylene blue for McFadyean's reaction for *B. anthracis* in blood films.

Fuchsin stain

Stain for 30 s with the following: dissolve 1 g basic fuchsin in 100 ml 95% ethanol. Stand for 24 h. Filter and add 900 ml water. *Or* use carbol fuchsin (see 'Acid-fast stain') diluted 1:10 with phenol saline.

Gram stain

There are many modifications of this. This is Jensen's version:

(1) Dissolve 0.5 g methyl violet in 100 ml distilled water.
(2) Dissolve 2 g potassium iodide in 20 ml distilled water. Add 1 g of finely ground iodine and stand overnight. Make up to 300 ml when dissolved.
(3) Dissolve 1 g of safranin or 1 g of neutral red in 100 ml of distilled water.

Stain with methyl violet solution for 20 s. Wash off and replace with iodine solution. Leave for 1 min. Wash off iodine solution with 95% alcohol or acetone, leaving on for a few seconds only. Wash with water. Counterstain with fuchsin or safranin for 30 s.

Some practice is required with this stain to achieve the correct degree of decolorization. Acetone decolorizes much more rapidly than alcohol.

The Hucker method is commonly used in US laboratories.

(1) Dissolve 2 g of crystal violet in 20 ml of 95% ethanol. Dissolve 0.8 g of ammonium oxalate in 80 ml of distilled water. Mix these two solutions, stand for 24 h and then filter.
(2) Dissolve 2 g of potassium iodide and 1 g of iodine in 300 ml of distilled water using the method described for Jensen's Gram stain.
(3) Grind 0.25 g of safranin in a mortar with 10 ml of 95% ethanol. Wash into a flask and make up to 100 ml with distilled water.

Stain with the crystal violet solution for 1 min. Wash with tap water. Stain with the iodine solution for 1 min. Decolorize with 95% ethanol until no more stain comes away. Wash with tap water. Stain with the safranin solution for 2 min.

Acid-fast stain

The Ziehl–Neelsen method employs carbol fuchsin, acid–alcohol and a blue or green counterstain. Colour-blind workers should use picric acid counterstain.

(1) Basic fuchsin   5 g
     Crystalline phenol (*caution*)   25 g
     95% alcohol   50 ml
     Distilled water   500 ml
     Dissolve the fuchsin and phenol in the alcohol over a warm water-bath, then add the water. Filter before use.
(2) 95% ethyl alcohol   970 ml
     Conc. hydrochloric acid   30 ml
(3) 0.5% methylene blue or malachite green, or 0.75% picric acid in distilled water.
     Pour carbol–fuchsin on the slide and heat carefully until steam rises. Stain for 3–5 min but do not allow to dry. Wash well with water, decolorize with acid–alcohol for 10–20 s, changing twice and counterstain for a few minutes with methylene blue or malachite green.
     Some workers prefer the original decolorizing procedure with 20% sulphuric acid followed by 95% alcohol.

When staining nocardias or leprosy bacilli use 1% sulphuric acid and no alcohol.

Cold staining methods are popular with some workers. For the Muller–Chermack method add 1 drop of Tergitol 7 (sodium heptadecyl sulphate) to 25 ml or carbol fuchsin and proceed as for Ziehl–Neelsen stain.

For Kinyoun's method make the carbol–fuchsin as follows:

Basic fuchsin	4 g
Melted phenol (*caution*)	8 g
95% ethanol	20 ml
Distilled water	100 ml

Dissolve the fuchsin in the alcohol. Shake gently while adding the water. Then add the phenol. Proceed as for Ziehl–Neelsen stain but do not heat. Decolorize with 1% sulphuric acid in water.

Fluorescent stain for acid-fast bacilli

There are several of these. We have found this auramine–phenol stain adequate.

(1)	Phenol crystals (*caution*)	3 g
	Auramine	0.3 g
	Distilled water	100 ml
(2)	Conc. HCl	0.5 ml
	NaCl	0.5 g
	Ethanol	75 ml

(3) Potassium permanganate 0.1% in distilled water.

Prepare and fix films in the usual way. Stain with auramine phenol for 4 min. Wash in water. Decolorize with acid alcohol for 4 min. Wash with potassium permanganate solution.

For microscopy use one of the commercial fluorescent microscopes (see Chapter 7).

Staining corynebacteria

To show the barred or beaded appearance and the metachromatic granules of these organisms Laybourn's modification of Albert's stain may be used.

(1) Dissolve 0.2 g malachite green and 0.15 g toluidine blue in a mixture of 100 ml water, 1 ml glacial acetic acid and 2 ml 95% alcohol.

(2) Dissolve 3 g of potassium iodide in 50–100 ml of distilled water. Add 2 g of finely ground iodine and leave overnight. Make up to 300 ml.

Stain with Solution (1) for 4 min. Wash, blot and dry and stain with Solution (2) for 1 min. The granules stain black and the barred cytoplasm light and dark green.

Spore staining

Make a thick film, stain for 3 min with hot carbol–fuchsin (Ziehl–Neelsen). Wash, flood with 30% aqueous ferric chloride for 2 min. Decolorize with 5% sodium sulphite solution. Wash and counterstain with 1% malachite green.

Flemming's technique substitutes nigrosin for the counterstain. Spread nigrosin over the film with the edge of another slide.

Spores are stained red. Cells are green, or with Flemming's method are transparent on a grey background.

Capsule staining

Place a small drop of india ink on a slide. Mix into it a small loopful of bacterial culture or suspension. Place a cover-glass over the drop avoiding air bubbles and press firmly between blotting paper. Examine with high-power lens. (Dispose of blotting paper in disinfectant.)

For dry preparations mix one loopful of indian ink with one loopful of a suspension of organisms in 5% glucose solution at one end of a slide. Spread the mixture with the end of another slide, allow to dry and pour a few drops of methyl alcohol over the film to fix. Stain for a few seconds with methyl violet (Jensen's Gram solution No. 1). The organisms appear stained blue with capsules showing as haloes.

Churchman's method uses Wright's stain (used for staining blood films). Allow stain almost to dry on the slide and wash well with water. Capsules stain pink, bacteria blue.

Lactophenol – cotton blue

Phenol, melted (*caution*)	20 g
Lactic acid	20 ml
Glycerol	40 ml
Water	20 ml

Warm the water and add the melted phenol, followed by the lactic acid and glycerol.

Add 0.05% cotton blue (methyl blue) to lactophenol.

Dimethyl sulphoxide for clearing tissue for fungi

Mix 40 ml of dimethyl sulphoxide (*caution*) with 60 ml of water. Then dissolve 10 g of potassium hydroxide pellets in the mixture.

Chapter 7

Identification methods

Accurate identification requires pure cultures and although time may be saved by subculture directly from the primary plate to identification media or kits it is advisable to plate the colonies selected to check purity.

Identification procedures should be checked with cultures of known organisms that give positive and negative results. Such cultures are available from type culture collections.

Many conventional methods for the identification of microorganisms are being replaced by paper strip and disc methods and by batteries of tests in 'kits'. These new methods offer certain advantages, e.g. the saving of time and labour but before an arbitrary choice is made between conventional tests and the newer methods, and between the several products available, the following factors should be considered.

(1) There are no 'Universal Kits'. Some organisms, e.g. *Corynebacterium diphtheriae*, nocardias, mycobacteria, cannot be identified by these rapid methods.

(2) Caution is required in interpretation. A kit result may suggest unlikely organisms, e.g. *Yersinia pestis*, *Brucella melitensis*, *Pseudomonas mallei* and ill-considered reporting of these could cause havoc.

(3) Conventional, paper strip and disc methods enable the user to make his own choice of tests. Kit methods do not; they give 10–20 test results whether the user needs them or not.

(4) Identification of organisms by conventional, paper strip or disc methods, usually requires the judgement of the user. With kit methods the results are interpreted by numeric charts or computers.

(5) If final identification depends on serology then a few screening tests are usually adequate. If identification depends on biochemical tests then the more tests used the more reliable will be the results. Kits may then be the methods of choice, especially with unusual organisms and in epidemiological investigations when the kit manufacturers' services are particularly useful.

(6) Some techniques may be more hazardous than others. Those that involve pipetting increase the risk of dispersing aerosols and of environmental contamination. Those employing syringe and needle work increase the risk of self-inoculation.

(7) Some methods are much more expensive per identification than others.

It follows that choices should be made according to the nature of the investigations, the professional knowledge and skill of the workers and the size of the laboratory budget. The relative merits of the individual products mentioned below are not assessed here but assistance may be obtained from the survey of an Advisory Group (Bennett and Joynson, 1986) in which ten commercial kits were tested with over 1000 bacterial strains. Kit manufacturers will supply other references on request.

As kit tests are unlikely to replace all conventional tests in all laboratories the latter are also described below.

Paper strip and touch stick tests

The strip is placed in a tube containing a small amount of culture, or a loopful of the organisms is rubbed on to the paper. The culture is touched with the stick.

The following paper strips and touch sticks are available:

Beta lactamase	Mast, Medical Wire, Oxoid
Gram reaction	Medical Wire
Indole	Medical Wire
Oxidase	Medical Wire, Oxoid
ONPG	Oxoid
PPA	Medical Wire, Oxoid
Pyridoxyl peptidase	EY Laboratories
Urease	Medical Wire
Butylase	Lab M

Antibiotic discs

Gram-negative anaerobes	Mast, Oxoid
Carbohydrates, ONPG, bacitracin optochin, 0129 SPS for peptostreptococci	Oxoid
X and V factors	Mast, Oxoid

Tablets

Enzyme substrates for biochemistry	LabM
Antibiotics	Mast

Kit identification systems

These range from rows of small cells containing dry substrates arranged in plastic strips, through discs dispensed into special trays or dishes, to sets of substrates in plastic tubes or test-tubes.

API systems
API systems are available for: Enterobacteria (20E, Rapid, 20E, 10S); Non-enteric Gram-negative rods (20NE); Staphylococci (20/STAPH); Streptococci (20/STREP); Anaerobes (20A); Heterotrophic aerobes (20B); Yeasts (20C/AUX); Fermentation, oxidation, assimilation (50CH); Enzyme tests (ZYM); Coryneform (CORYNE); Neisseria (QUADFERM).

Suspensions of the organisms are added to cupules in plastic galleries. These are incubated overnight. Some tests then require reagents to be added. Results are read from codes and a Profile Register. There is a computer identification service.

Enterotubes (Roche)
These are designed for the identification of enterobacteria. A plastic tube contains the substrates for 14 tests. A wire needle passes through each substrate, from end to end. After uncapping, the wire is touched on the culture and then drawn through

the tube and discarded. The tubes are recapped and incubated. One reagent must be added, using a hypodermic needle and syringe. The results are read from a set of codes and keys (ENCISE). There is also a computer consultation service.

Biotest
Multiwell plates contain substrates for biochemical and susceptibility tests. Incubation time 16–24 h. There is a code book.

Mast ID
A range of carbohydrates and biochemicals is provided in agars for use in petri dishes with multipoint inoculation. They may be used singly or in combination. Results are read by eye or with the Mastascan computerized system. The principles are similar to those of conventional tests but allow the user more flexibility. One set is recommended for enteric bacteria.

Rapid ID (Innovator Diagnostic Systems)
These identify streptococci, enterococci, haemophilus and neisserias. Suspensions are added to cups in a plastic gallery and reagents are added after incubation. They are designed for use with a multipoint inoculator and allow user flexibility. There is a coded manual and a computer service.

Automated identification systems
There are several of these, e.g. Sceptor (Becton Dickinson), Hewlett Packard, Bactomatic.

Kits for individual organisms
Apart from those that detect certain properties and permit the identification of a large range of organisms there are kits that are intended to simplify the identification of a particular species or genus, e.g. staphylococci, streptococci, meningococci, haemophilus, salmonellas.

There are also kits for detecting the toxins of certain organisms in food, e.g. of *Vibrio cholerae*, *Clostridium perfringens*, *Escherichia coli*, *Staphylococcus aureus*. References are made to these in the relevant sections of this book.

Oxi-Ferm (Roche)
This system permits the identification of a number of Gram-negative organisms.

Flow Laboratories (Diagnostic Research)
These systems use tubes and petri dishes of a special design. The *r/b* set is for oxidase negative organisms. Four tubes allow enterobacteria to be identified. Results are interpreted from a computer code book after overnight incubation. The *N/F* set screens for some *Pseudomonas* species in 24 h and identifies other non-enterobacteria in a further 24 h. Results are interpreted by a 'logic' scheme or by computerized numerical coding. Two other sets are for identifying yeasts: Uni-Yeast-Tek identifies *Candida albicans* and *C. stellatoidea*; the *C/N* single tube identifies *Cryptococcus neoformans*. There is also a single tube, *Uni-OF*, for the Hugh and Leifson test and a 5-h system for enterics (TTE–RAS).

Microbact (Disposable Products)
This provides three systems, for enterobacteria, enterobacteria and other Gram-

negative rods and anaerobic bacteria. Suspensions of organisms are added to wells in prepared trays. Some wells must be covered with mineral oil. Some reagents have to be added after incubation. The results are read from charts.

The Micro ID system (General Diagnostics)
This identifies 33 species of enterobacteria with 15 tests. Suspensions of the organisms are added to impregnated paper in plastic trays and incubated for 5 h. Results are interpreted with the aid of a coded manual.

Minitek systems (Becton Dickinson)
These allow the identification of enterobacteria, neisseria, staphylococci, anaerobes and yeasts. Discs, impregnated with substrates, are dispensed into specially prepared plates, inoculated with suspensions of the organisms and incubated overnight. Results are read from colour charts.

Enteroset (Fisher)
This is for identifying enterobacteria and consists of porous paper strips containing substrates sandwiched between plastic films. Suspensions of organisms are absorbed on the strips and results are read from charts after overnight incubation.

Auxotab (Wilson Diagnostics)
This consists of cards carrying reagent filled capillary chambers to which the bacterial suspensions are added. Enterobacteria are identified after 7-h incubation.

AutoMicrobic (Vitek)
This is an automated system incorporating a dilution dispenser and filling module for dispensing the suspensions into prepared wells in trays. A computer module provides identifications after 8-h incubation.

Conventional tests

Aesculin hydrolysis

Inoculate aesculin medium or Edwards' medium and incubate overnight. Organisms that hydrolyse aesculin blacken the medium.

Ammonia test

Incubate culture in nutrient or peptone broth for 5 days. Wet a small piece of filter-paper with Nessler reagent and place it in the upper part of the culture tube. Warm the tube in a water-bath at 50–60°C. The filter-paper turns brown or black if ammonia is present.

Arginine hydrolysis

Incubate the culture in arginine broth for 24–48 h and add a few drops of Nessler reagent. A brown colour indicates hydrolysis. (But for lactobacilli, see p.62.)

Carbohydrate fermentation and oxidation

Liquid, semisolid and solid media are described on p.65. They contain a ferment-able carbohydrate, alcohol or glucoside and an indicator to show the production of acid. To demonstrate gas production, a small inverted tube (Durham's tube; gas tube) is placed in the fluid media. In solid media (stab tube p.86), gas production is obvious from the bubbles and disruption of the medium. Normally, gas formation is recorded only in the glucose tube and any that occurs in tubes of other carbohy-drates results from the fermentation of glucose formed during the first part of the reaction.

For most bacteria, use peptone or broth-based media.

For *Lactobacillus* spp., use MRS base (p.74) without glucose or meat extract and adjust to pH 6.2–6.5. Add chlorphenol red indicator.

For *Bacillus* spp. and *Pseudomonas* spp., use ammonium salt 'sugar' media (p.66). These organisms produce ammonia from peptones and this may mask acid production.

For *Neisseria* spp. and other fastidious organisms use solid or semisolid media. Neisserias do not like liquid media. Enrich with Fildes' extract (5%) or with rabbit serum (10%) or use the GC serum-free medium of Flynn and Waitkins. Horse serum may give false results as it contains fermentable carbohydrates. Robinson's buffered serum sugar medium (p.66) overcomes this problem and is the medium of choice for *Corynebacterium* spp.

Treat all anaerobes as fastidious organisms. The indicator may be decolorized during incubation so add more after incubation. A sterile iron nail added to liquid media may improve the anaerobic conditions.

Tests may be carried out on solid medium containing indicator. Inoculate the medium heavily, spreading it all over the surface and place carbohydrate discs (p.97) on the surface. Acid production is indicated by a change of colour in the medium around the disc. We do not recommend placing more than four discs on one plate.

Casein hydrolysis

Use skim milk agar (p.76) and observe clearing around colonies of casein-hydrolysing organisms. The milk should be dialysed so that acid production by lactose fermenters does not interfere. To detect false clearing due to this, pour a 10% solution of mercuric chloride in 20% hydrochloric acid over this medium. If the cleared area disappears, casein was not hydrolysed.

Catalase test

(1) Emulsify some of the culture in 0.5 ml of a 1% solution of Tween 80 in a screw-capped bottle. Add 0.5 ml of 20-vol hydrogen peroxide (*caution*) and replace the cap. Effervescence indicates the presence of catalase. Do not do this test on an uncovered slide as the effervescence creates aerosols. Cultures on low-carbohydrated medium give the most reliable results.

(2) Add a mixture of equal volumes of 1% Tween 80 and 20-vol hydrogen peroxide to the growth on an agar slope. Observe effervescence after 5 min.

(3) Test mycobacteria by Wayne's method (p.354).

(4) Test minute colonies which grow on nutrient agar by the blue slide test. Place

one drop of a mixture of equal parts of methylene blue stain and 20-vol hydrogen peroxide on a slide. Place a cover-slip over the colonies to be tested and press down firmly to make an impression smear. Remove the cover-slip and place it on the methylene blue–peroxide mixture. Clear bubbles, appearing within 30 s, indicate catalase activity.

Citrate utilization

Inoculate solid medium (Simmons) or fluid medium (Koser) with a straight wire. Heavy inocula may give false-positive results. Incubate at 30-35°C and observe growth.

Coagulase test

Staphylococcus aureus and a few other organisms coagulate plasma (see p.300).

Decarboxylase tests

Falkow's method can be used for most Gram-negative rods but Moeller's method gives the best results with *Klebsiella* spp. and *Enterobacter* spp. Commercial Falkow media allow only lysine decarboxylase tests but commercial Moeller media permit lysine, ornithine and arginine tests to be used.

Inoculate Falkow medium (p.67) and incubate for 24 h at 37°C. The indicator (bromocresol purple) changes from blue to yellow due to fermentation of glucose. If it remains yellow, the test is negative; if it then changes to purple, the test is positive.

When using Moeller medium (p.67), include a control tube that contains no amino acid. After inoculation, seal with liquid paraffin to ensure anaerobic conditions and incubate at 37°C for 3–5 days. There are two indicators, bromothymol blue and cresol red. The colour changes to yellow if glucose is fermented. Decarboxylation is indicated by a purple colour. The control tube should remain yellow.

DNase test

Streak or spot the organisms heavily on DNase agar. Incubate overnight and flood the plate with 1 N hydrochloric acid, which precipitates unchanged nucleic acid. A clear halo around the inoculum indicates a positive reaction.

Eijkman test

Inoculate suspected *E. coli* into brilliant green broth with a gas tube. Incubate at 44 ± 0.2°C for 24 h. *E. coli* is one of the few organisms that produce gas at this temperature.

Gelatin liquefaction

(1) Inoculate nutrient gelatin stabs, incubate at room temperature for 7 days and observe digestion. For organisms that grow only at temperatures when gelatin is fluid, include an uninoculated control and after incubation place both tubes

in a refrigerator overnight. The control tube should solidify. This method is not entirely reliable.

(2) Inoculate nutrient broth. Place a denatured gelatin charcoal disc in the medium and incubate. Gelatin liquefaction is indicated by the release of charcoal granules, which fall to the bottom of the tube.

(3) For a rapid test, use 1 ml of broth of 0.01 M calcium chloride in saline. Inoculate heavily (whole growth from slope or petri dish). Add a gelatin charcoal disc and place in a water-bath at 37°C. Examine at 15-min intervals for 3 h.

(4) Inoculate gelatin agar medium. Incubate overnight at 37°C and then flood the plate with saturated ammonium sulphate solution. Haloes appear around colonies of organisms producing gelatinase.

Gluconate oxidation

Inoculate gluconate broth (p.70) and incubate for 48 h. Add an equal volume of Benedict's reagent and place in a boiling water-bath for 10 min. An orange or brown precipitate indicates gluconate oxidation.

Hippurate hydrolysis

Inoculate hippurate broth, incubate overnight and add excess of 5% ferric chloride. A brown precipitate indicates hydrolysis.

Hugh and Leifson (HL) test. Oxidation fermentation test

This is also known as the 'oxferm' test. Some organisms metabolize glucose oxidatively, i.e. oxygen is the ultimate hydrogen acceptor and culture must therefore be aerobic. Others ferment glucose, when the hydrogen acceptor is another substance: this is independent of oxygen and cultures may be aerobic or anaerobic.

Heat two tubes of medium (p.71) in boiling water for 10 min to drive off oxygen, cool and inoculate; incubate one aerobically and the other either anaerobically or seal the surface of the medium with 2 cm of melted Vaseline or agar to give anaerobic conditions.

Oxidative metabolism: acid in aerobic tube only.

Fermentative metabolism: acid in both tubes.

For testing staphylococci and micrococci, use the Baird-Parker modification (p.71).

Hydrogen sulphide production

(1) Inoculate a tube of nutrient broth. Place a strip of filter-paper impregnated with a lead acetate indicator paper in the top of the tube, holding it in place with a cotton wool plug. Incubate and examine for blackening of the paper.

(2) Inoculate one of the iron or lead acetate media. Incubate and observe blackening.

TSI (Triple Sugar Iron Agar) and similar media do not give satisfactory results with sucrose-fermenting organisms. The indicator paper (lead acetate) is the most sensitive method and the ferrous chloride media the least sensitive method, but the latter is probably the method of choice for identifying salmonellas and for differentiation in other groups where the amount of hydrogen sulphide produced varies with the species.

Indole formation

Grow the organisms for 2–5 days in peptone or tryptone broth.
(1) *Ehrlich's method*   Dissolve 4 g of *p*-dimethylaminobenzaldehyde in a mixture of 80 ml of concentrated hydrochloric acid and 380 ml of ethanol (do not use industrial spirit; this gives a brown instead of a yellow solution). To the broth culture add a few drops of xylene and shake gently. Add a few drops of the reagent. A rose pink colour indicates indole.
(2) *Kovac's method*   Dissolve 5 g of *p*-dimethylaminobenzaldehyde in a mixture of 75 ml of amyl alcohol and 25 ml of concentrated sulphuric acid. Add a few drops of this reagent to the broth culture. A rose pink colour indicates indole.
(3) *Spot test*   This is described by Miller and Wright (1982). Dissolve 1 g of *p*-dimethylamino cinnamaldehyde (DMAC) (*caution*) in 100 ml of 10% hydrochloric acid. Moisten a filter paper with it and smear on a colony from blood or nutrient agar. A blue green colour is positive, pink negative. This test may be applied directly to colonies but is unreliable in the presence of carbohydrates, e.g. on MacConkey or CLED media.

Lecithinase activity

(1) Inoculate egg yolk agar (p.68) or Willis and Hobbs medium (p.77) and incubate for 3 days. Lecithinase-producing colonies are surrounded by zones of opacity.
(2) Inoculate egg yolk salt broth and incubate for 3 days. Lecithinase producers make the broth opalescent. Some organisms (e.g. *B. cereus*) give a thick turbidity in 13 h.

Levan production

Inoculate nutrient agar containing 5% of sucrose. Levan producers give large mucoid colonies after incubation for 24–48 h.

Lipolytic activity

Inoculate tributyrin agar. Incubate for 48 h at 25–30°C. A clear zone develops around colonies of fat-splitting organisms. These media can be used for counting lipolytic bacteria in dairy products.

Malonate test

Inoculate one of the malonate broth media and incubate overnight. Growth and a deep blue colour indicate malonate utilization.

Methyl red test

Inoculate glucose phosphate broth (p.70) and incubate for 5 days at 30°C. Add a few drops of methyl red solution, prepared by dissolving 0.1 g of the dye in 300 ml of ethanol and making up to 500 ml with distilled water. A red colour indicates that the pH has been reduced to 4.5 or less. A yellow colour indicates a negative reaction.

Motility

Hanging drop method

Place a very small drop of liquid bacterial culture in the centre of a 16-mm square No. 1 cover-glass, with the aid of a small (2-mm) inoculating loop. Place a small drop of water at each corner of the cover-glass. Invert over the cover-glass a microslide with a central depression – a 'well-slide'. The cover-glass will adhere to the slide and when the slide is inverted the hanging drop is suspended in the well.

Instead of using 'well-slides' a ring of Vaseline may be made on a slide. This is supplied in collapsible tubes or squeezed from a hypodermic syringe.

Bring the edge of the hanging drop, or the air bubbles in the water seal into focus with the 16-mm lens before turning to the high-power dry objective to observe motility.

Bacterial motility must be distinguished from Brownian movement. There is usually little difficulty with actively motile organisms, but feebly motile bacteria may require prolonged observation of individual cells.

Careful examination of hanging drops may indicate whether a motile organism has polar flagellation – a darting zig-zag movement – or peritrichate flagellation – a less vigorous and more vibratory movement.

To examine anaerobic organisms for motility grow them in a suitable liquid medium. Touch the culture with a capillary tube 60–70-mm long and about 0.5–1-mm bore. Some culture will enter the tube. Seal both ends of the tube in the bunsen flame and mount on Plasticine on the microscope stage. Examine as for hanging drops.

Craigie tube method

Craigie tubes (p.76) can be used instead of hanging drop cultures. Inoculate the medium in the inner tube. Incubate and subculture daily from the outer tube. Only motile organisms can grow through the sloppy agar.

TTC method

Donovan's medium detects motility. This motility agar is semisolid and contains triphenyl tetrazolium chloride (TTC). Motility is shown by a diffuse red line in stab culture. Compartment petri dishes may be used.

Nagler test

This tests for lecithinase activity (see above) but the word has come to mean the half antitoxin plate test for *Cl.perfringens*. Egg yolk agar containing Fildes enrichment or Willis and Hobbs' medium are satisfactory.

Niacin test

This test is used exclusively for typing mycobacteria (see p.352).

Nitratase test (nitrate reduction)

(1) Inoculate nitrate broth medium and incubate overnight.
(2) Make a dense suspension of the test organisms in 0.01 M sodium nitrate in M/45 phosphate buffer of pH 7.0 and incubate at 37°C for 24 h
(3) Grow the organisms in a suitable broth. Add a few drops of 1% sodium nitrate solution and incubate for 4 h.

Acidify with a few drops of 1 N hydrochloric acid and add 0.5 ml each of a 0.2% solution of sulphanilamide and 0.1% N-naphthylethylenediamine hydrochloride (*caution*) (these two reagents should be kept in a refrigerator and prepared fresh monthly). A pink colour denotes nitratase activity. But as some organisms further reduce nitrite, if no colour is produced add a very small amount of zinc dust. Any nitrate present will be reduced to nitrite and produce a pink colour, i.e. a pink colour in this part of the test indicates no nitratase activity and no colour indicates that nitrates have been completely reduced.

ONPG test

Lactose is fermented only when β-galactosidase and permease are present. Deficiency of the latter gives late fermentation. True non-lactose fermenters do not possess β-galactosidase.
    Inoculate ONPG broth (p.74) and incubate overnight. If β-galactosidase is present a yellow colour due to *o*-nitrophenol is formed.

Optochin test

Pneumococci are sensitive but streptococci are resistant to optochin (ethylhydro-cupreine hydrochloride).
    Streak the organisms on blood agar and place an optochin disc on the surface. Incubate overnight and examine the zone around the disc.

Oxidase test (cytochrome oxidase test)

(1) Soak small pieces of filter-paper in 1% aqueous tetramethyl-*p*-phenylenediamine dihydrochloride or oxalate (which keeps better). Some filter-papers give a blue colour and these must not be used. Dry or use wet. Scrape some of a fresh young culture with a clean platinum wire or a glass rod (dirty or Nichrome wire gives false positives) and rub on the filter-paper. A blue colour within 10 s is a positive oxidase test. Old cultures are unreliable. Tellurite inhibits oxidase as do fermentable carbohydrates. Organisms which have produced acid from a carbohydrate should be subcultured to a sugar-free medium.
(2) Incubate cultures on nutrient agar slopes for 24–48 h at the optimum temperature for the strain concerned. Add a few drops each of freshly prepared 1% aqueous *p*-aminodimethylaniline oxalate and 1% α-naphthol in ethanol. Allow the mixture to run over the growth. A deep blue colour is a positive reaction.

Oxidative or fermentative metabolism of glucose

See Hugh and Leifson test above.

Phenylalanine test (PPA or PPD test)

Inoculate phenylalanine agar and incubate overnight. Pour a few drops 10% ferric chloride solution over the growth. A green colour indicates deamination of phenylalanine to phenylpyruvic acid. Among the enterobacteria only proteus and providence strains have this property.

Phosphatase test

Some bacteria, e.g. *S. aureus*, can split ester phosphates. Inoculate a plate of phenolphthalein phosphate agar (p.75) and incubate overnight. Expose the culture to ammonia vapour (*caution*). Colonies of phosphatase producers turn pink.

Proteolysis

Inoculate cooked meat medium (p.75) and incubate for 7–10 days. Proteolysis is indicated by blackening of the meat or digestion (volume diminishes). Tyrosine crystals may appear.

Starch hydrolysis

Inoculate starch agar (p.76). Incubate for 3–5 days then flood with dilute iodine solution. Hydrolysis is indicated by clear zones around the growth. Unchanged starch gives a blue colour.

Sulphatase test

Some organisms (e.g. certain species of mycobacteria) can split ester sulphates. Grow the organisms in media (Middlebrook 7H9 for mycobacteria) containing 0.001 M potassium phenolphthalein disulphate for 14 days. Add a few drops of ammonia solution. A pink colour indicates the presence of free phenolphthalein.

Tellurite reduction

Some mycobacteria reduce tellurite to tellurium metal (p.356).

Tween hydrolysis

Some mycobacteria can hydrolyse Tween (polysorbate) 80, releasing fatty acids that change the colour of an indicator. This test is used mostly with mycobacteria (p. 356).

Tyrosine decomposition

Inoculate parallel streaks on a plate of tyrosine agar (p.77) and incubate 3–4 weeks. Examine periodically under a low-power microscope for the disappearance of crystals around the bacterial growth.

Urease test

Inoculate one of the urea media heavily and incubate for 3–12 h. The fluid media give more rapid results if incubated in a water-bath. If urease is present, the urea is split to form ammonia, which changes the colour of the indicator from yellow to pink.

Voges Proskauer Reaction (VP test)

This tests for the formation of acetyl methyl carbinol (acetoin) from glucose. This is oxidized by the reagent to diacetyl, which produces a red colour with guanidine residues in the media.

Inoculate glucose phosphate broth (p.70) and incubate for 5 days at 30°C. A very heavy inoculum and 6 h may suffice. Test by one of the following methods.

(1) Add 3 ml of 5% alcoholic α-naphthol solution and 3 ml of 40% potassium hydroxide solution (Barritt's method).
(2) Add a 'knife point' of creatinine and 5 ml of 40% potassium hydroxide solution (O'Meara's method).
(3) Add 5 ml of a mixture of 1 g of copper sulphate (blue) dissolved in 40 ml of saturated sodium hydroxide solution plus 960 ml of 10% potassium hydroxide solution (APHA method).

A bright pink or eosin red colour appearing in 5 min is a positive reaction. For testing *Bacillus* spp., add 1% sodium chloride to the medium.

Xanthine decomposition

Inoculate parallel streaks on a plate of xanthine agar (p.77) and incubate for 3–4 weeks. Examine periodically under a low-power microscope for the disappearance of xanthine crystals around the bacterial growth.

Agglutination tests

These tests are performed in small test-tubes (75 × 9 mm). Dilutions are made in physiological saline with graduated pipettes controlled by a rubber teat or (for a single-row test) with pasteur pipettes marked with a grease pencil at approximately 0.5 ml. Automatic pipettes or pipettors are useful for doing large numbers of tests.

Standard antigen suspensions and agglutinating sera can be obtained from BBL, Difco, Pasteur Institute and Wellcome. Antigen suspensions are used mainly in the serological diagnosis of enteric fever, which may be caused by several related organisms, and of brucellosis.

Standard agglutinating sera are used to identify unknown organisms.

Testing unknown sera against standard H and O antigen suspensions

To test a single suspension
Prepare a 1:10 dilution of serum by adding 0.2 ml to 1.8 ml of saline. Set up a row of seven small tubes. Add 0.5 ml of saline to tubes 2-7 and 0.5 ml of 1:10 serum to

tubes 1 and 2. Rinse the pipette by sucking in and blowing out saline several times. Mix the contents of tube 2 and transfer 0.5 ml to tube 3. Rinse the pipette. Continue with doubling dilutions but discard 0.5 ml from tube 6 instead of adding it to tube 7. Rinse the pipette between each dilution. The dilutions are now:

Tube no.	1	2	3	4	5	6	7
Dilution	1:10	1:20	1:40	1:80	1:160	1:320	0

Add 0.5 ml of standard suspension to each tube. The last tube, containing no serum, tests the stability of the suspension. The dilutions are now:

Tube no.	1	2	3	4	5	6	7
Dilution	1:20	1:40	1:80	1:160	1:320	1:640	0

To test with several suspensions
In some investigations, e.g. in enteric fevers, a number of suspensions will be used. Set up six large tubes (152 × 16 mm). To tube 1 add 9 ml of saline, to tubes 2-6 add 5 ml of saline and to tube 1 add 1 ml of serum. Mix and transfer 5 ml from tube 1 to tube 2, and continue in this way, rinsing the pipette between each dilution.

Set up a row of seven small tubes (75 × 9 mm) for each suspension to be tested. Transfer 0.5 ml from each large tube to the corresponding small tube. Work from right to left, i.e. weakest to strongest dilution to avoid unnecessary rinsing of the pipette. The final dilutions are now 1:20 to 1:640.

Incubation temperature and times
Incubate tests with O suspensions in a water-bath at 37°C for 4 h, then allow to stand overnight in refrigerator.

Incubate H tests for 2 h in a water-bath at 50–52°C (*not* 55–56°C, as the antibody may be partially destroyed). The level of water in the bath should be adjusted so that only about half of the liquid in the tubes is below the surface of the water. This encourages convection in the tubes, mixing the contents.

Brucella agglutinations
To avoid false negative results due to the prozone phenomenon, double the dilutions for at least three more tubes, i.e. use 1:20 to 1:5120.

Reading agglutinations
Examine each tube separately from right to left, i.e. beginning with the negative control. Wipe dry and use a hand-lens. If the tubes are scratched, dip them in xylene. The titre of the serum is that dilution in which agglutination is easily visible with a low-power magnifier.

Testing unknown organisms against known sera

Preparation of O suspensions
The organism must be in the smooth phase. Grow on agar slopes for 24 h. Wash off in phenol–saline and allow lumps to settle. Remove the suspension and dilute it so that there are approximately 1000×10^6 bacteria/ml by the opacity tube method (p.129). Heat at 60°C for 1 h. If this antigen is to be stored, add 0.25% chloroform.

If a K antigen is suspected, heat the suspension at 100°C in a water-bath for 1 h (but the B antigen is thermostable).

Preparation of H suspensions
Check that the organism is motile and grow it in nutrient broth for 18 h, or in glucose broth for 4–6 h. Do not use glucose broth for the overnight culture as the bacteria grow too rapidly. Add formalin to give a final concentration of 1.0% and leave for 30 min to kill the organisms. Heat at 50–55°C for 30 min (this step may be omitted if the suspension is to be used at once). Dilute to approximately 1000×10^6 organisms/ml by the opacity tube method.

Agglutination tests
Use the same technique as that described under 'Testing unknown sera...' but as most sera are issued with a titre of at least 1:250 (and labelled accordingly) it is not necessary to dilute beyond 1:640. In practice, as standard sera are highly specific and an organism must be tested against several sera, it is usually convenient to screen by adding 1 drop of serum to about 1 ml of suspension in a 75 × 9 mm tube. Only those sera which give agglutination are then taken to titre.

Slide agglutination
This is the normal procedure for screening with *O* sera.
    Place a loopful of saline on a slide and next to it a loopful of serum. With a straight wire, pick a colony and emulsify in the saline. If it is sticky or granular, or autoagglutinates, the test cannot be done. If the suspension is smooth, mix the serum in with the wire. If agglutination occurs it will be rapid and obvious. Dubious slide agglutination should be discounted.
    Only *O* sera should be used. Growth on solid medium is used for slide agglutination and this is not optimum for the formation of flagella. False-negative results may be obtained with *H* sera unless there is fluid on the slope.
    Suspensions of organisms in the R phase are agglutinated by 1:500 aqueous acriflavine solution.

Fluorescent antibody techniques

The principle of fluorescent antibody techniques is that proteins, including serum antibodies, may be labelled with fluorescent dyes by chemical combination without alteration or interference with the biological or immunological properties of the proteins. These proteins may then be seen in microscopical preparations by fluorescence microscopy.
    The preparation is illuminated by ultraviolet or ultraviolet blue light. Any fluorescence emitted by the specimen passes through a barrier filter above the object. This filter transmits only the visible fluorescence emission. Microscopes suitable for this purpose are now readily available, and it is a simple matter to convert standard microscopes for fluorescence work. Fluorescent dyes are used instead of ordinary dyes as they are detectable in much smaller concentrations. They are available in a form which simplifies the conjugation procedure considerably. The fluorochromes in frequent current use are fluorescein isothiocyanate (FITC) and Lissamine rhodamine B (RB200). Of these, FITC is the most commonly used. It gives an apple-green fluorescence.
    Fluorescence labelling is often used in microbiology and immunology along with or instead of traditional serological tests. Immune serum globulin conjugated with

fluorochrome is usually employed to locate the corresponding antigen in microbio-logical investigations.

The fluorescence microscope

The main object of the fluorescence microscope is to transmit as much as possible of the fluorescence emitted by the specimen to the eye. As this fluorescence is seldom of high intensity, great care must be taken that as little of it as possible is lost. For this reason incident light microscopy is to be preferred.

The greatest problem with both dark-ground and bright-field transmitted light is the relatively low brightness of the fluorescence that results from absorption of both the exciting light from the illuminating system and the longer wavelength light emitted by the fluorescing specimen. This is mainly due to the passage of the light through the various optical components of the microscope system and also, particularly in dark-ground equipment, to the necessity for accurate alignment of the lenses.

For incident light microscopy, the light source is above the specimen and both the exciting light and the light transmitted from the fluorescing specimen are passed through the objective. This eliminates any loss of illumination that would have occurred due to pasage through a condenser, and also means that the system is always optimally aligned. Such a system is the Zetopan equipment, which is manufactured by Reichert (Figure 7.1).

Incident light microscopy is clearly a system of bright-ground microscopy. Theoretically, any such system may be used for fluorescence microscopy, provided that the light source is suitable. In practice, however, the fluorescence normally obtained is very poor, particularly if ultraviolet light is used as the stimulating source. The cause of the poor fluorescence is the beam-splitting mirror used in the opaque illuminator. At the very best, the system reflects 50% of incident illumination with visible light wavelengths, and even less with ultraviolet light. As the light coming from the specimen will suffer a further 50% loss as it passes back through the illuminator, the low brightness is easily explained.

This system for fluorescence microscopy was improved by altering the reflecting characteristics of the mirror to achieve improved light recovery with the incident light system, particularly in use for fluorescence microscopy. The original beam-splitting mirror was replaced by a multi-layer interference filter that produces up to 90% reflection of the exciting illumination and transmits a similar amount of light from the specimen. With an ideal beam-splitting mirror, only certain portions of the spectrum are completely reflected, while the remainder are transmitted without loss. Compared with the original methods of obtaining incident light, this system gives an appreciable increase in brightness.

Hypothetically, this system eliminates the need for both the exciter and the barrier filters, because in theory only the shortwave exciting light reaches the specimen, and only the fluorescing light of longer wavelength reaches the eye. In practice, however, such perfection with the beam-splitting dichroic mirror is impossible to achieve as the spectral spacing of the two wavelengths of light is not sufficiently wide. Consequently, it is still necessary to use an exciter filter and a barrier filter; these are built into the system.

The exciter filter transmits only that wavelength of light which will stimulate the fluorochrome used. The difference in this system is that the light source and filters are placed above the specimen, and the exciting light reaches the specimen through

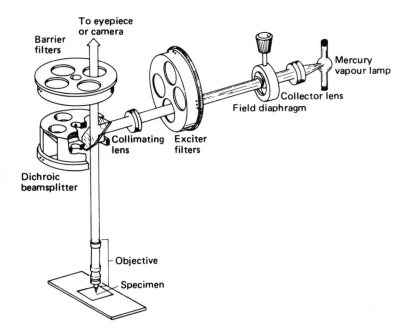

Figure 7.1 Incident light fluorescence: mercury vapour lamp provides rich desirable ultraviolet radiation sources. Collector lens directs light to the exciter filter, which transmits selected wavelengths. Dichroic beam-splitter reflects virtually all of the desirable short wavelength, exciting light down through the objective to the specimen. Fluorochromes in specimen react to excitation wavelengths and emit longer wavelengths of visible fluorescence up through the system to the eyepiece. The barrier filter passes the emitted light from the specimen and blocks out unwanted background illumination. (Reproduced by permission of the British American Optical Co. Ltd)

the objective and not through a condenser. Light of the optimum wavelength is then passed through the dichroic beam splitter, which reflects it with great efficiency towards the specimen. The resulting fluorescence, together with any residual exciting light, again passes through the beam splitter, where most of the exciting light is reflected towards the lamp, while a small proportion passes through towards the eyepiece. The fluorescent light of longer wavelength passes unaltered through the beam-splitter and straight on towards the eyepiece. The barrier filter below the eyepiece removes any remaining exciting light, thus protecting the eye against any possibility of contact with ultraviolet light, and also allowing only the required fluorescent wavelength to be observed.

By altering the dichroic lenses and permutating the exciter and barrier filters, this system can be used for a variety of fluorochromes.

The fluorescent antibody test

Many factors in the preparation of the specimen may affect fluorescence one way or another and of these the commonest is probably variation in pH. All dilutions and washing procedures are carried out in buffer solution. This is available as a lyophilized powder from Difco and Mercia. Difco, BBL, the Pasteur Institute and Wellcome, supply a variety of fluorescent antibody reagents. The catalogues and handbooks of these companies contain much useful information.

In microbiology, two fluorescence methods may be applied, a direct method or the indirect method, which is also known as the sandwich technique. The direct test can be used to identify microbes while the indirect test can be used both for identification and for the examination of patients' sera for antibodies.

The direct test depends upon the attachment of a fluorochrome-conjugated known specific antiserum to a microbe that may be in a pathological exudate, in a culture or in tissue. The fluorescent antibody combines specifically with the microbe, which can then be viewed with the optical system described above.

The indirect test is performed in two stages. In the first, the patient's serum is applied to a known antigen. When antibody is present, it will combine with the antigen but cannot be seen as it is non-fluorescent. In the second stage, fluorescent anti-human globulin is applied, which will attach to any human globulin left in the washed preparation, and antibody being 'sandwiched' between it and the antigen. When the indirect technique is used for identification ordinary specific antisera are used and, after washing, a fluorescent antiserum is applied which contains antibodies to the globulins of the animal species in which the specific antiserum was prepared.

The direct test

A typical example of the direct fluorescent antibody test is that used to identify the pathogenic *Neisseria*.

Method

(1) Take a colony of the organism to be tested and emulsify on a clean slide.
(2) Dry in air without heating and fix by immersion in 3% formalin–saline buffer solution.
(3) Wash in saline buffer for 10 min with two changes of solution immersing the slides in a Coplin jar. Blot gently to dry.
(4) Apply a drop of the conjugated antiserum, which is kept deep-frozen, and spread it over the specimen.
(5) Incubate in a moist petri dish for 30 min, preferably mechanically rotating at 100 rev/min.
(6) Wash in saline buffer for 10 min with three changes of solution. Blot dry, mount in buffered glycerol and apply a clean cover-slip.
(7) Inspect for fluorescence under the fluorescence microscope. If fluorescing cocci are seen, this means that the specific antibody has reacted with the unknown antigen. A positively reacting organism may be either *N. gonorrhoeae* or *N. meningitidis*, as these two species have antigens in common.

Test known positive and negative controls with each batch of investigations.

The indirect test

The most commonly practised form of this test is the absorbed fluorescent treponemal antibody test.

(1) Reconstitute a lyophilized culture of *Treponema pallidum* and make a smear by taking a fairly large loopful of the suspension and spreading it in a small circle

on a thin slide, between 0.8 and 1.0 mm in thickness. PTFE-coated slides are particularly suitable for this purpose. Dry in air and fix in 10% methanol for 5 min. Fixed slides can be stored for several months in a deep-freeze.

(2) Dilute the previously inactivated sera for testing 1/5 in FTA sorbent and allow to absorb for 30 min.
(3) Place one or two drops of each adsorbed serum on a circle of a slide containing the treponemas, and incubate at 37°C for 30 min in a moist chamber.
(4) Wash the slide in at least three changes of fluorescent antibody test buffer solution of pH. 7.2 (Mercia) and gently blot dry.
(5) Spread over each inoculum one or two drops of fluorescein-conjugated anti-human globulin and incubate as before.
(6) Wash in buffer as before. Blot gently and mount in buffered glycerol with a clean cover-slip. Include a positive, weakly positive and negative control with each batch of tests.

Inspect with a fluorescence microscope using a dark-ground condenser. If antibody is present in the tested serum, it will react with the organisms and they will become coated with human globulin. The fluorescent anti-human globulin will then react with the globulin and it will fluoresce. A positive test indicates the presence of treponemal antibodies in the patient's serum. In a completely negative test there is always the possibility that no treponema are present on the slide. This can be easily checked by removing the exciter filter. As a dark-ground condenser is used, spirochaetes are easily seen.

The advantage of the indirect test is that a single conjugated species-specific antiserum can be used for demonstrating the presence of many antigens provided that the specific antisera used are prepared in the same species.

The fluorescent antibody technique can be applied in microbiology to any organism against which it is possible to prepare specific high-titre antisera. Conjugated antisera for many medically important bacteria are available commercially. The technique is capable, in certain circumstances, of greatly shortening the time taken to identify an organism. Microcolonies of organisms that have been growing for only 2.5 h may be identified by cover-slip impression techniques with an homologous conjugated antiserum.

Although the fluorescent antibody technique has proved its worth in the field of microbiology, and is the only technique that combines the specificity of an antigen–antibody reaction with the speed and precision of microscopy, it has its limitations. The technique is employed to detect antigens and not organisms (Nairn, 1975). Pathogenic organisms have antigens not only in common with each other but also in common with harmless commensals of the same or related genera. False-positive reactions are therefore inevitable unless the antisera employed are very highly specific. With the introduction of monoclonal antibodies, however, these techniques will become more widely used. For more information about fluorescent antibody techniques see Wick *et al.*, (1982).

Immersion oils
If it is necessary to use immersion objectives, it is best to avoid the usual immersion oils. These become slightly fluorescent after exposure to air and this process is hastened by intense illumination. AnalaR-grade glycerol is used as the immersion medium but even this slightly impairs the performance of high-power lens systems

as it does not have the ideal refractive index, although lenses made for glycerol immersion can be purchased. It is better to use dry lenses of the highest available magnification. For magnifications higher than × 40, it is also necessary to mount the specimen under a cover-glass. The mounting medium is buffered glycerol.

References

Bennett, C. H. N. and Johnson, D. H. M. (1986) Kit systems for identifying Gram-negative aerobic bacilli: report of the Welsh Standing Specialist Advisory Working Party in Microbiology. *Journal of Clinical Pathology,* **39,** 666–671

Miller, J. M. and Wright, J. W. (1982) Spot indole test: evaluation of four methods. *Journal of Clinical Pathology,* **15,** 589–592

Nairn, R. C. (1974) *Fluorescent Protein Tracing,* Churchill-Livingstone, Edinburgh

Wick, G., Traill, K. N. and Schauenstein, K. (eds) (1982) *Immunofluorescence Technology,* Elsevier, Amsterdam

Chapter 8

Mycological methods

Direct examination

Mount hair, skin or nail fragments on a slide in a drop of 10% potassium hydroxide in 40% dimethyl sulphoxide. Apply a cover-glass and then leave for a few minutes till the preparation 'clears'. Wet films of pus may be made with an equal volume of this mountant or an equal volume of glycerol if a more permanent preparation is required. Mix cerebrospinal fluid deposit with an equal volume of 2.5% nigrosin in 50% glycerol. Glycerol prevents mould growth in the solution, delineates capsules more sharply and prevents preparations, which should be thin, from drying out.

Cultural examination for moulds and yeasts

Primary isolation

Inoculate slopes or plates of chloramphenicol malt agar. If dermatophytes are sought, use a medium containing cycloheximide, either malt agar or one of the proprietary media containing an indicator. Push pieces of hair or skin into the medium at several points. Scrape nail clippings to a powder with a scalpel; do not put them on in lumps. Pieces of skin, hair or nail powder can be picked up more easily if the heated inoculating wire is first pushed into the medium to coat it with agar.

Spread pus, CSF deposit or biopsy material from cases of the deeper mycoses on slopes of glucose peptone agar with chloramphenicol. Cut tissues into small pieces or grind them in a Griffith's tube. Place in the bottom of a petri dish and pour on chloramphenicol malt agar (moulds) or glucose peptone agar (actinomycetes), melted and at 45°C.

Wash mycetoma grains several times with sterile saline to remove most contaminants before culturing.

Incubate cultures at 28°C. Most yeasts grow in 2–3 days, though cryptococci are rather slower. Common moulds, too, will grow up in a few days while cultures for dermatophytes should be kept for 2 weeks though most are identifiable after 1 week. It is useful, then, to examine all mould cultures once or twice a week and ensure adequate aeration.

Cultures from food, soil, plant material, etc.

Inoculate plates of malt agar with chloramphenicol and Czapek–Dox agar. Zygomycetes if present, will outgrow everything else but are markedly restrained by Czapek–Dox. To grow moulds from physiologically dry materials such as jam increase the sucrose content of the Czapek–Dox to 20%. Some yeasts can tolerate low pH and can be isolated on malt agar to which 1% lactic acid has been added after melting and cooling to 50°C. Incubate cultures at room temperature and at 30°C.

To isolate yeasts from highly acidic environments add 1% lactic acid to melted and cooled malt agar before pouring plates, or use buffered yeast agar plus lactic acid.

Sampling air for mould spores

A cheap, simple device for this purpose is the Porton impinger (p.229) in which particles from a measured volume of air are trapped in a liquid medium on which viable counts are done. Slit or SAS samplers (p.229) are also useful. More information is gained by using the Andersen sampler, which sizes airborne particles using the principle that the higher the air speed the smaller the particle which will escape the air stream and be impacted on an agar surface. Plastic petri dishes give lower counts than glass ones but they make the whole machine much easier to handle.

The medium and time of incubation will depend on the organism sought. For general use one run with Czapek–Dox and one with chloramphenicol malt agar will serve; for thermophilic actinomycetes use glucose peptone agar without antibiotics; incubate one set at 40°C, the other at 50°C.

Mould growths on and from foods, etc.

Examine with a lens or low-power microscope *in situ* – to see the arrangement of spores, etc. before distributing the growth. Remove a small piece of mycelium with a wire or needle, the end of which has been bent at right angles to give a short, sharp hook. Use the point of the needle to cut out a piece of the growth near the edge where sporulation is just beginning. Transfer it to a drop of PVA mountant on a slide, apply a cover-slip then examine after 30 min when the stain has penetrated the hyphae.

Identification of moulds

Only well-sporing mould colonies can be readily identified. If spores are not seen reincubate and/or subculture to other media.

The usual methods of examining sporing structures are needle mounts (small fragments of mycelium taken from the colony with a mounted needle) and tape mounts as follows.

Cut a strip about 1.5 cm wide from a roll of 2.5 cm Sellotape, then attach by a narrow end to a mounted needle, forming a flag. Blot the surface of the colony with this near the growing edge, then mount the tape, sticky side down, in a drop of mountant and examine at once.

The centre of a well-sporing colony may show nothing but spores. Make a second preparation nearer to the edge of the colony.

Slide culture

This permits a more critical examination. Cut small blocks of medium (0.5 by 0.5 cm) from a poured plate and transfer them to the centres of sterile slides. Inoculate the edges of each block and apply a cover-slip. Make a moist chamber by cutting two lengths from a plastic drinking straw and gluing them, 2 in apart, to the bottom of a 100-mm square plastic petri dish with a drop of chloroform. Drop a rectangle of filter-paper soaked in 10% glycerol between the supports. Up to three prepared slide cultures may be incubated in one dish. Watch the progress of growth using the low power of the microscope when visible growth has developed. When typical structures are seen, prepare a slide and cover-slip each with a drop of mountant. Lift the cover-slip off the slide culture and mount on the prepared slide, then lift off and discard the agar block: the slide is now mounted on the prepared cover-slip. Examine both preparations after 30 min.

Cultural examination of yeasts

Morphology

To study yeast morphology use corn meal agar: streak the organism on a plate of the medium, cover with a sterile cover-glass and incubate at 30°C. Examine daily through the bottom of the petri dish, using a low-power microscope. Look for cell shape, mycelium and spore formation.

To encourage ascospore formation inoculate sodium acetate agar made by adjusting the pH of 0.5% sodium acetate to 6.5 with acetic acid, then solidifying it with agar. Stain the growth for spores after a week.

Fermentation tests

These differ from those used in bacteriology in that the sugar concentration is 3%. Durham tubes are always used and a 10-ml volume of medium is preferable to encourage fermentation rather than oxidation.

Inoculate these 'sugar' media (with Durham's tubes): glucose, maltose, lactose, raffinose and galactose. Incubate cultures at 25–30°C for at least 7 days. Gas production indicates fermentation (see p.375).

Auxanograms

Fermentation tests alone are insufficient to identify species of yeasts. Make a thick suspension by adding about 5 ml of sterile water to a malt agar slope: it is not necessary to wash the organisms. Add about 0.25 ml of this suspension to 20 ml bottle of auxanogram (carbon assimilation) medium then pour half into each of two 9-cm petri dishes. When set, place discs of carbon or nitrogen sources on the surface of the agar, placing four well apart, near to the edge of the dish and the fifth in the centre. Carbon sources commonly used are glucose (control), maltose, sucrose, lactose, inositol, galactose, raffinose, mannitol and cellobiose.

Repeat the procedure with a single dish of nitrogen assimilation medium and discs of sodium nitrate and asparagine as controls (see p.375).

'Kit' tests

The commercial kits are satisfactory for the identification of most strains encountered in clinical laboratories. Some use methods based on traditional criteria; others rely on novel approaches. In all of them, however, the data bases are drawn from a restricted list of species and occasional misidentifications may occur.

Serological methods

Antibody detection

To detect circulating antibodies a crude antigen is made. The fungus is grown in a well-aerated broth culture. It is harvested by filtration and stored overnight at 20°C before mechanical disruption, extraction and lyophilization. Alternatively, antigens may be made by concentrating the culture filtrate.

The freeze-dried antigens are reconstituted in phosphate buffered saline at a strength indicated by previous titration against positive antisera. Several commercial kits are now available and these obviate the difficulties of standardizing reagents. A selection of kits is shown in Table 8.1.

Table 8.1 Some suppliers of reagents and kits for serology of mycoses

	Candidosis	Cryptococcosis	Aspergillosis	Farmer's lung	Systemic mycoses	Sporotrichosis
For detecting antigen	1,2,5	1,2,5	—	3	—	—
For detecting antibody	1,2,3,4	—	1,2,3,5	—	1,2,3,5	1,2,5
For identification of cultures (exoantigen)	—	—	—	—	1,5	—

1, Immuno-Mycologics, PO Box 1151, Norman, OK 73070, USA
2, Gibco Ltd, PO Box 35, Trident House, Renfrew Road, Paisley PA3 4EF, UK
3, Mercia Brocades Ltd, Broadford Park, Shalford, Guildford, Surrey GU4 8EW, UK
4, Roche Products Ltd, PO Box 8, Welwyn Garden City, Herts AL7 3AY, UK
5, Alpha Laboratories Ltd, 40 Parham Drive, Eastleigh, Hants SO5 4NU, UK

Complement fixation tests

These tests are useful in histoplasmosis and coccidoidomycosis. Both yeast and mycelial phase antigens of histoplasma are used. Those who have facilities and expertise for growing and handling dangerous pathogens can prepare histoplasma mycelial antigen as described above. The cultures must be inoculated in an exhaust protective cabinet and killed by exposure for 3 days to 1:5000 thiomersal before processing. A yeast phase produced on chocolate cystine agar can be used to inoculate glucose peptone broth fortified with 0.1% of cystine. The culture is then stirred or shaken at 37°C for 3–4 days and killed with thiomersal. The washed yeasts are suspended in saline with 1:5000 thiomersal and this whole yeast cell suspension constitutes the antigen.

The preparation of *Coccidioides immitis* antigen is best left to the experts: laboratory infections are almost always fatal.

Agglutination of whole yeast cells

This may be used to detect antibodies as follows. The organisms are grown on glucose peptone agar plates for 24 h at 37°C. Prolonged incubation should be avoided as it encourages mycelial formation. The growth is washed off in saline, centrifuged, washed three times with saline and the packed cells are suspended to about 20% by eye. A Wintrobe haematocrit tube is filled with this suspension and the packed cell volume estimated. The stock solution is then distributed in bottles sufficient to make a 2% v/v suspension by the addition of 10 ml of phosphate-buffered saline with azide. These bottles of stock suspension are kept at −20°C and one is diluted for use as required.

Sera need no inactivation. Volumes of 0.4 ml of phosphate-buffered saline are dispensed into a WHO tray and serial twofold dilutions of serum made as usual. One drop (0.02 ml) of yeast suspension is added to each well and the tray gently shaken to mix and then left at room temperature for 24 h. The deposit is then re-suspended by gentle shaking and the end-point read at +agglutination. Trace agglutinations are ignored.

To detect cryptococcal antibodies a suspension of cells of serotype A is made and the test carried out as for candida agglutination (above) using, however, a medium that discourages florid capsule formation. The following has been found suitable:

Glucose	10 g
Tryptone	10 g
$NaH_2PO_4.2H_2O$	5 g
Agar	15 g
Water	1000 ml

Dissolve by heating, distribute and sterilize at 115°C for 10 min.

Immunodiffusion

Agar base
This is made by dissolving 2 g of agar (Oxoid No. 1) in 100 ml of water by autoclaving. 100-ml of buffer is heated to 50°C and mixed with the agar. The medium is kept in a water-bath at 50°C and discarded if not used within 48 h of preparation.

Buffer

Boric acid (H_2BO_3)	10 g
Powdered borax ($Na_2B_4O_7.10H_2O$)	20 g
EDTA, disodium salt	10 g
Water	to 1000 ml

The pH should be 8.2. This buffer is, of course, double strength; it is diluted with agar for plates and slides or with an equal volume of water for use in electrophoresis tanks.

Plates
A 30-ml volume of agar is measured into a plastic petri dish and a Perspex jig with metal pegs is fitted in place of the lid to produce the pattern of wells shown in Figure 8.1. The large wells are 6 mm in diameter and the smaller wells 2 mm. The distance between the central and peripheral wells is also 6 mm. The test serum occupies the central well, antigens the pairs of large and small wells (shown shaded)

and appropriate control sera the top and bottom wells. Two antigens can thus be tested and a single solution suffices as the relative volumes (60 and 6 µl in the two holes) give high and low concentration gradients in the agar. This arrangement also allows reactions of identity to be obtained between test and control sera, thus eliminating some anomalous reactions, particularly among the aspergilli.

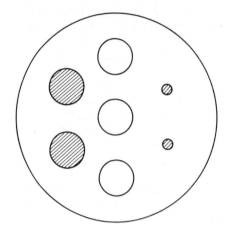

Figure 8.1 Double diffusion tests

Counter immunoelectrophoresis (CIE)

Reagents

Veronal buffer	Barbitone (*caution*)	6.88 g
	Sodium barbitone (*caution*)	15.14 g
	Sodium azide (*caution*)	1.00 g
	Distilled water	1000 ml
Agarose gel	1% agarose	50 ml
	Veronal buffer	50 ml
Coomassie blue	Coomassie brilliant blue	2 g
	Trichloracetic acid, 10%	50 ml
	Distilled water	1000 ml
Destain solution	Methanol	300 ml
	Acetic acid	200 ml
	Distilled water	500 ml

Method
Pre-coat clean glass microscope slides with a very thin layer of 1% agarose gel and heat to 50°C for 30 min.

Pour 5.5 ml of the same gel on the slides and allow to set. Cut wells, 4.5 mm diameter (to hold 20 µl) for sera and 2.5 mm (to hold 8 µl) for the antigens. Allow 5 mm between the neatest edges of respective wells.

Add sera and antigens and include suitable controls.

Electrophorese with antigens at the cathode side of a CIE tank, using veronal buffer diluted 1:1 with water at 4 V/cm for 90 min.

If rapid results are required wash the wet gel for 1 h and examine by dark-ground illumination. This gives preliminary result only and must be confirmed after washing and staining:

(1) Wash with 50% trisodium phosphate for 30 min.
(2) Wash with two changes of saline, 40 min each.
(3) Place in saline overnight.
(4) Wash in distilled water for 30 min.
(5) Cover with blotting paper and dry at 60°C for 1 h.
(6) Stain in Coomassie brilliant blue for 5 min.
(7) Wash in distilled water for 5 min.
(8) Place in destaining solution for 5 min.

Antigen detection

Antigen
Polystyrene latex, coated with rabbit anti-cryptococcus globulin, is agglutinated by cryptococcal capsular polysaccharide. Because of extensive cross-reaction, type A globulin acts as a 'polyvalent' reagent.

Preparation of globulin
Rabbits are immunized with a small capsule strain of cryptococcus until their serum has an agglutination titre of 1 in 500 or better. The animals are then bled and the globulin fraction of their serum isolated using ammonium sulphate or caprylic acid. After dialysis, it is freeze-dried and reconstituted at a concentration of 4% for use.

Sensitization of latex
The stock latex solution is made from Dow or Difco latex 0.81. A 1:20 dilution is normally satisfactory. When this is further diluted 1:100 in a round 12-mm tube, it should have an absorbance of 0.29–0.31 at 650 nm. To this stock an equal volume of the maximally reactive dilution of globulin in glycine–saline is added. This dilution is determined by a chess-board titration using latex sensitized with 1:100, 1:200, 1:400, 1:600 and 1:800 globulin against a known positive human serum or CSF. Alternatively, diluted washings from cryptococcal cells can be used.

The routine reagent can usually be made by mixing equal volumes of 1:250 globulin and 1:10 latex and leaving the mixture at 4°C overnight.

Diluent for cryptococcal latex agglutination
Glycine	7.31 g
Sodium chloride	10 g
1 N sodium hydroxide solution	3.5 ml
Water	to 1000 ml

Leave the solution overnight to equilibrate then check the pH, which should be 8.2.

Use this buffer as it is to dilute the latex; for diluting test and control sera, add 0.1% of bovine albumin. (This buffer is an excellent culture medium; keep the stock frozen at −20°C).

Screening test
Eight drops of diluent are placed in each of two wells in a WHO plastic tray for each specimen. Three areas are marked off on a microscope slide with a felt pen.

Two drops of serum or CSF are added to the left-hand area of the slide and two drops to the first well. After mixing, two drops of this 1:5 dilution are added to the centre area of the slide and two to the second well, making this 1:25. Two drops of the 1:25 dilution from the second well are added to the right-hand area of the slide. One drop of sensitized latex is added to each area on the slide and the slide rocked for 3 min. A slight granularity always appears: this should be disregarded. Definite agglutination is considered positive.

When the screening test is positive or if cryptococci have been isolated, doubling dilutions are tested to obtain an end-point.

Antifungal drugs

Methods of susceptibility testing and assay

Medium
5-Fluorocytosine is sulphonamide-like in that it requires a medium free of antagonists. This excludes the use of peptone and meat extract. Some organisms, particularly cryptococci, are reluctant to grow on a medium whose only source of nitrogen is inorganic. Growth is much improved by using casamino acids, a nitrogen source free of nucleic acids and their components, supplemented by sufficient yeast extract to give good growth without antagonizing 5-fluorocytosine. Difco or Oxoid yeast extract is suitable for this; Marmite is not. A small amount of glycerophosphate is added as a buffer. This does not precipitate on autoclaving, and the sterilized medium remains clear. The other antifungal drugs have no special requirements but it is obviously simpler to use one medium for all. For preparation of the medium see p.62.

Drug solutions
5-Fluorocytosine is freely soluble in water. A stock solution containing 1000 µg/ml will keep indefinitely if frozen at −20°C. When this solution is thawed for use, a few crystals of 5-fluorocytosine may remain undissolved. Complete solution is assured by placing it in the 56°C water-bath for 10 min before use. Amphotericin B is available as Fungizone, a complex with sodium deoxycholate in 50-mg quantities. Dissolve contents of one bottle in 10 ml of sterile water (avoid saline at all stages, it precipitates the amphotericin). This stock solution will keep for 6 months if frozen at −20°C. For a working solution make two tenfold dilutions in 0.4% sodium deoxycholate solution, to give a concentration of 50 µg/ml. This keeps for several months if frozen at −20°C. Pimaricin must be dissolved in dimethyl formamide; Nystatin and the imidazole compounds in equal volumes of acetone and water. Intermediate dilutions of these drugs are opalescent, and should be prepared just before use; if kept for more than about 30 min there is an appreciable loss through precipitation on the glass of the bottle. Solutions of the imidazole compounds – 20 mg of solid weighed into a sterile bottle then taken up in 10 ml of water plus 10 ml of acetone – have been found to keep for 4 months in the refrigerator. There is no point in keeping them at −20°C since the acetone prevents them from freezing.

Working solutions for sensitivity testing
5-Fluorocytosine: 100 µg and 100 µg/ml

Miconazole, ketoconazole, pimaricin and nystatin (the last in units): 50 µg and
   5 µg/ml
Amphotericin B: 10 µg and 1 µg/ml.

Sensitivity testing
Melt the measured volumes of agar, then cool them to 48°C. Use 10-ml lots for
slopes, adding drug solutions as follows:

5-Fluorocytosine

(The drug may be added to measured volumes of agar before autoclaving.)

100 µg/ml				0.1	0.2	0.4	0.8 ml
10 µg/ml	0.1	0.2	0.5				ml
Concentration	0.1	0.2	0.5	1.0	2.0	4.0	8.0 µg/ml

Miconazole, ketoconazole, pimaricin and nystatin

50 µg/ml				0.1	0.2	0.4	0.8ml
5 µg/ml	0.1	0.2	0.4				ml
Concentration	0.05	0.1	0.2	0.5	1.0	2.0	4.0 µg/ml

Amphotericin B

10 µg/ml				0.1	0.2	0.4	0.8 ml
1.0 µg/ml	0.1	0.2	0.5				ml
Concentration	0.01	0.02	0.05	0.1	0.2	0.4	0.8 µg/ml

   The volume of antibiotic added has been neglected. The maximum error
introduced is 8%, too small a difference to affect the result. Divide each 10-ml lot
into two slopes, and include a pair of control slopes without drugs. For plates,
20-ml lots of medium are used and the above concentrations are of course doubled.

The inoculum
Grow the organism on the assay medium at 30°C. Results are less satisfactory if this
step is omitted. Yeasts will grow well enough after 24 h, and the growth should be
suspended in sterile water, then this suspension added, drop by drop, to 10 ml of
sterile water till a faint but distinct opalescence results. One drop of this diluted
suspension is run down each slope, and incubation continued at 30°C until growth
on the control slopes is adequate. *Candida* species reach this stage in 24 h:
cryptococci need a further 24 h incubation. Filamentous organisms; aspergilli,
histoplasmas and the like, are grown on the same medium again at 30°C. The young
growth is stripped off the surface of the medium after 2–3 days before spore
formation begins and macerated in a 'bijou' macerator with 3 ml of sterile water,
using 15-s bursts until enough of the growth is disintegrated to give an opalescent
suspension. The macerator bottle is allowed to stand for 10 min, so that aerosols
within may settle. The macerating dog is then removed and the cap of a sterile bijou
bottle used to cap the bottle. The contents of the bottle are now mixed gently, large
lumps are allowed to settle, then the supernatant suspension is taken off and

suspension is taken off and treated exactly like the yeast suspension, i.e. it is diluted till just opalescent. One drop is used as inoculum and the tubes are incubated as before until the growth of the controls is adequate.

If the sensitivities of several yeast strains are required at the same time, up to four organisms can be accommodated by using duplicate drops of suspension of plates marked out in sectors as in viable counting. As the drops are concentrated on a small area, a 1/25 dilution of the suspension described above is necessary to obtain comparable results. In all cases the aim should be to produce discrete colonies, not a confluent growth.

When several strains are to be tested it is convenient to prepare the slopes or plates containing 5-fluorocytosine, miconazole or ketoconazole in the afternoon, then keep them in the refrigerator overnight. If plates are used, they can be dried next morning while the amphotericin series is being prepared. 5-Fluorocytosine or miconazole incorporated in agar slopes will keep for 3 weeks in the refrigerator at 4°C with no detectable loss of potency. Where lots of tests are done, quantity production is feasible – and easier.

Short method for yeast sensitivity tests
In practice, yeast sensitivities generally fall within the four lowest drug concentrations given above. Using a square petri dish with 25 compartments, four drug concentrations and a control can be put in the five columns across one plate using volumes of 1.2 ml. If the centre row is left empty, two organisms can be tested in duplicate. Plates containing 5-fluorocytosine or ketoconazole will keep for a week in the refrigerator.

Disc test sensitivity methods
Disc methods may be used for yeasts but are not satisfactory for moulds because of the difficulty in preparing suitable seeding suspensions. Spore suspensions may give misleading results because the MICs of germinating spores may differ from those of vegetative forms.

Methods of assay

Amphotericin B, miconazole and ketoconazole assay

Standard solutions
Amphotericin B, 2.0 μg/ml; miconazole, 4.0 μg/ml or ketoconazole, 4.0 μg/ml is diluted 1/10 in horse serum. The appropriate standard is diluted in duplicate and in parallel with duplicate dilutions of the patient's serum in one of the following ways:

(1) Four rows of eleven 75 × 12 mm test-tubes with aluminium caps (Oxoid) are sterilized in a rack, 0.5 ml of the test organism suspension is added to each; 0.5 ml of standard to the first tube of rows 1 and 2 and 0.5 ml of the test serum to rows 3 and 4. Each one is then serially diluted to tube 10, leaving tube 11 as a growth control.
(2) A sterile Microtitre plate (M24AR) with lid (M42R), used with 0.1-ml volumes of diluent and sera. The whole operation is made much easier by using a four-place multichannel automatic pipette with autoclaved tips for dispensing

and diluting. The suspension may be dispensed from a 5-cm plastic petri dish or a sterile 50-ml beaker.

In either case, the results of amphotericin B assay can be read after overnight incubation. Assays of imidazoles should be left for 24 h, then the end point taken where growth is sharply reduced by comparison with the controls, neglecting traces of growth.

5-Fluorocytosine assay

5-Fluorocytosine diffuses readily in agar while amphotericin B, miconazole and ketoconazole, the other drugs commonly encountered, diffuse little if at all. Fortunately, the expected levels of these latter drugs are low and it is possible to estimate serum levels by dilution in broth in parallel with a standard prepared in horse serum.

Media
The liquid medium for dilution tests is given on p.62.

Organisms
The test organism is *Saccharomyces cerevisiae* NCPF 3178. It is the least granular when grown on glucose peptone agar and safer to use than a potential pathogen. A distinctly cloudy suspension is made in 10 ml of sterile distilled water 0.1 ml is used in broth for the liquid assay and the rest added to 110 ml of 5-fluorocytosine assay medium melted and cooled to 45°C.
Pour the inoculated medium into a bioassay plate 250-mm square (e.g. Nunc 1015).

Standard and test sera
Prepare standards containing 10, 15, 25, 40 and 60 µg/ml of 5-fluorocytosine. Keep frozen (will last for 6 months). Dilute patient's serum as follows (any amphotericin B which is present will be neutralized within the hour):
Horse serum	0.4 ml
Patient's serum	0.5 ml
Ergosterol 1% in acetone	0.1 ml

If the patient is being treated with an imidazole drug, use 0.1 ml of 10% Tween 80 in place of ergosterol.

The test
Lay the dish on a sheet of paper (template) having a pattern of 25 positions to which test or standard numbers are assigned at random. Punch holes in the medium with a 4-mm cork borer and remove the agar. Add 15 µl volumes of test and standard solutions in duplicate in the pattern shown on the template. Incubate at 30°C overnight.
Measure the diameters of the zones of inhibition. Plot the standards on one decade semilogarithmic paper and obtain the concentration of drug in the test serum by interpolation.
For more details of technical methods see Mackenzie *et al.* (1980), Mackenzie and Philpot (1981) and Campbell *et al.* (1985).

References

Campbell, C. K., Davis, C. and Mackenzie, D. W. R. (1985) Detection and isolation of pathogenic fungi. In *Isolation and Identification of Micro-organisms of Medical and Veterinary Importance* (eds C. H. Collins and J. M. Grange), Society for Applied Bacteriology Technical Series, No. 21, Academic Press, London, pp. 329–343

MacKenzie, D. W. R., Philpot, C. M. and Proctor, A. G. J. (1980) *Basic Serodiagnosis Methods for Diseases caused by Fungi and Actinomyces,* Public Health Laboratory Service Monograph No. 12, HMSO, London

Mackenzie, D. W. R. and Philpot, C. M. (1981) *Isolation and Identification of Ringworm Fungi,* Public Health Laboratory Service Monograph, No. 15, HMSO, London

Chapter 9

Counting microorganisms

It is often necessary to report on the size of the bacterial population in a sample. Unfortunately, industries and health authorities have been allowed to attach more importance to these 'bacterial counts' than is permitted by their technical or statistical accuracy (see also p.178).

If a *total count* is required, many of the organisms counted may be dead or indistinguishable from other particulate matter. A *viable count* assumes that a visible colony will develop from each organism. Bacteria are, however, rarely separated entirely from their fellows and are often clumped together in large numbers particularly if they are actively reproducing. A single colony may therefore develop from one organism or from hundreds or even thousands of organisms. Each colony develops from one *viable unit*. Because any agitation, as in the preparation of dilutions, will break up or induce the formation of clumps, it is obviously difficult to obtain reproducible results. Bacteria are seldom distributed evenly throughout a sample and as only small samples are usually examined very large errors can be introduced.

Many of the bacteria present in a sample may not grow on the medium used at the pH or incubation temperatures employed or in the time allowed.

Accuracy is often demanded where it is not needed. If it is decided that a certain product should contain less than, say, ten viable bacteria/g, this suggests that, on average, of ten tubes each inoculated with 0.1 g, seven would show growth and three would not, and out of ten tubes each inoculated with 0.01 g only one or two would show growth. If all the 0.1-g tubes or five of the 0.01-g tubes showed growth, there would be more than ten organisms/g. It does not matter whether there are 20 or 10 000: there are too many. There is no need to employ elaborate counting techniques.

In viable count methods, it is recognized that large errors are inevitable even if numbers of replicate plates are used. Some of these errors are, as indicated above, inherent in the material, others in the technique. Errors of ±90% in counts of the order of 10 000 to 100 000/ml are not unusual even with the best possible technique. It is, therefore, necessary to combine the maximum of care in technique with a liberal interpretation of results. The figures obtained from a single test are valueless. They can be interpreted only if the product is regularly tested and the normal range is known.

Physical methods are used to estimate total populations, i.e. dead and living organisms. They include direct counting and measurements of turbidity. Biological methods are used for estimating the numbers of viable units. These include the plate count, roll-tube count, drop count, surface colony count, dip slide count, contact plate, membrane filter count, most probable number estimations, and automated methods (p.135).

Counting chamber method

The Helber counting chamber is a slide 2–3 mm thick with an area in the centre called the platform and surrounded by a ditch, which is 0.02 mm lower than the reminder of the slide (Figure 9.1). The top of the slide is ground so that when an optically plane cover-glass is placed over the centre depression, the depth is uniform. On the platform an area of 1 mm^2 is ruled so that there are 400 small squares each 0.0025 mm^2 in the area. The volume over each small square is 0.02×0.0025 mm^3, i.e. 0.000 05 ml.

Add a few drops of formalin to the well mixed suspension to be counted. Dilute the suspension so that when the counting chamber is filled there will be about five or ten organisms per small square. This requires initial trial and error. The best diluent is 0.1% peptone water containing 0.1% lauryl sulphate and (unless phase contrast or dark field is used for counting) 0.1% methylene blue. Always filter before use.

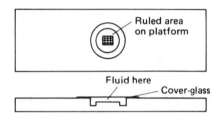

Figure 9.1 Bacterial counting chamber (depth is exaggerated)

Place a loopful of suspension on the ruled area and apply the cover-glass, which must be clean and polished. The amount of suspension must be such that the space between the platform and the cover-glass is just filled and no fluid runs into the ditch; again, this requires practice. If the cover-glass is applied properly, Newton's rings will be seen. Allow 5 min for the bacteria to settle.

Examine with a 4-mm lens with reduced light, dark field or phase contrast if available. Count the bacteria in 50–100 squares selected at random so that the total count is about 500. Divide the count by number of squares counted. Multiply by 20 000 000 and by the original dilution factor to obtain the total number of bacteria/ml. Repeat twice more and take the average of the three counts. Clumps of bacteria, streptococci, etc., can be counted as units or each cell counted as one organism.

With experience, reasonably accurate counts can be obtained but the chamber and cover-glass must be scrupulously cleaned and examined microscopically to make sure bacteria are not left adhering to either.

The Breed count

Named after its originator, this is a rough but useful technique. Breed slides have areas of 1 cm^2 marked on them. Into a square, place 0.01 ml of the fluid to be counted, for example milk, with a commercially available microsyringe or loop.

Allow to dry and stain with methylene blue. Examine with an oil immersion lens giving a known field diameter, which is usually about 0.16 mm. The area seen is thus πr^2 or 3.14×0.08^2, i.e. 0.02 mm^2, and as the total area is 100 mm^2, one field represents 1/5000 part of the whole area, or 1/5000 part of the original 0.01 ml. One organism per field equals 5000/0.01 ml or 500 000/ml. Count the number of organisms in several fields in different parts of the ruled area and use the equation

$$\text{Count/ml} = \frac{N \times 4 \times 10^4}{\pi d^2}$$

where N is the number per field and d is the diameter of the field.

Opacity tube method

International Reference Opacity Tubes, containing glass powder standards are available. These are numbered narrow glass tubes of increasing opacity and a table is provided equating the opacity of each tube with the number or organisms/ml. The unknown suspension is matched against the standards in a glass tube of the same bore. It may be necessary to dilute the unknown. These physical methods should not be used with Risk/Hazard Group 3 organisms.

Viable counts

In these techniques, the material containing the bacteria is serially diluted and some of each dilution is placed in or on suitable culture media. Each colony developing is assumed to have grown from one viable unit, which, as indicated earlier, may be one organism or a group of many.

Diluents

Some diluents, e.g. saline or distilled water, may be lethal for some organisms. Diluents must not be used direct from a refrigerator as cold-shock may prevent organisms from reproducing. Peptone water, 0.1%, is the best diluent.

Pipettes

The 10-ml and 1-ml straight-sided blow-out pipettes are commonly used and are specified for some statutory tests. Disposable pipettes are labour saving. Mouth pipetting must be expressly forbidden, regardless of the nature of the material under test. A method of controlling teats is given on p.25. Automatic pipettors or pipette pumps, described on p.26, should be used. Some of these can be pre-calibrated and used with disposable polypropylene pipette tips.

Pipettes used in making dilutions must be very clean, otherwise bacteria will adhere to their inner surfaces and may be washed out into another dilution. Siliconed pipettes may be used. Fast-running pipettes and vigorous blowing-out should be avoided: they generate aerosols. As much as 0.1 ml may remain in pipettes if they are improperly used.

Preparation of dilutions

Pipette 9-ml amounts of diluent solution into sterile test-tubes with suitable caps. These are the *dilution blanks*. Do not sterilize these after dispensing unless screw-capped bottles are used; autoclaving tubes with aluminium or polypropylene caps may alter the volume.

When diluting liquids, for example milk for bacterial counts, proceed as follows.

Mix the sample by shaking. With a straight-sided pipette dipped in half an inch only, remove 1 ml. Deliver into the first dilution blank, about half an inch above the level of the liquid. Wait 3 s, then blow out carefully to avoid aerosol formation. Discard the pipette. With a fresh pipette, dip half an inch into the liquid, suck up and down ten times to mix, but do not blow bubbles. Raise the pipette and blow out. Remove 1 ml and transfer to the next dilution blank. Discard the pipette. Continue for the required number of dilutions, and remember to discard the pipette after delivering its contents, otherwise the liquid on the outside will contribute to a cumulative error. The dilutions will be:

Tube no.	1	2	3	4	5
Dilution	1:10	1:100	1:1000	1:10 000	1:100 000
Vol. of original	0.1	0.01	0.001	0.0001	0.00001
fluid/ml	(or 10^{-1})	(10^{-2})	(10^{-3})	(10^{-4})	(10^{-5})

When counting bacteria in solid or semisolid material, weigh 10 g and place in a Stomacher or blender. Add 90 ml of diluent and homogenize.

Alternatively, cut into small pieces with a sterile scalpel, mix 10 g with 90 ml of the diluent and shake well. Allow to settle. Assume that the bacteria are now evenly distributed between the solid and liquid.

Both of these represent dilutions of 1:10, i.e. 1 ml contains or represents 0.1 g, and further dilutions are prepared as above. The dilutions will be:

Tube no.	1	2	3	4
Dilution	1:100	1:1000	1:10 000	1:100 000
Weight of original	0.01	0.001	0.0001	0.00001 g
material	(or 10^{-2})	(10^{-3})	(10^{-4})	(10^{-5} g)

Mechanical aids for preparing dilutions are available commercially.

Plate count

Melt nutrient agar or other suitable media tubed in 10-ml amounts. Cool to 45°C in a water-bath.

Set out petri dishes, two or more per dilution to be tested and label with the dilution number. Pipette 1 ml of each dilution into the centre of the appropriate dishes, using a fresh pipette for each dilution. Do not leave the dish uncovered for longer than is absolutely necessary. Add the contents of one agar tube to each dish in turn and mix as follows. Move the dish gently six times in a clockwise circle of diameter about 150 mm. Repeat counter-clockwise. Move the dish back-and-forth six times with an excursion of about 150 mm. Repeat with to-and-fro movements. Allow the medium to set, invert and incubate for 24–48 h.

Economy in pipettes
In practice, the pipette used to transfer 1 ml of a dilution to the next tube may be used to pipette 1 ml of that dilution into the petri dish.

Counting colonies

To count on a simple colony counter select plates with between 30 and 300 colonies. Place the open dish, glass side up, over the illuminated screen. Count the colonies using a 75-mm magnifier and a hand-held counter. Mark the glass above each colony with a felt tip pen. Calculate the colony or viable count/ml by multiplying the average number of colonies per countable plate by the reciprocal of the dilution. Report as 'colony forming units/ml' (cfu) or as 'viable count/ml' not as 'bacteria/g or /ml'.

If all plates contain more than 300 colonies rule sectors, e.g. of one-quarter or one-eighth of the plate, count the colonies in these and include the sector value in the calculations.

For large work loads semi- or fully automatic counters are essential. In the former the pen used to mark the glass above the colonies is connected to an electronic counter which displays the numbers counted on a small screen. In the fully automatic models a TV camera or laser beam scans the plate and the results are displayed or recorded on a screen or read-out device.

Roll-tube count

Instead of using plates, tubes or bottles containing media are inoculated with diluted material and rotated horizontally until the medium sets. After incubation, colonies are counted.

Dispense the medium in 2–4-ml amounts in 25-ml screw-capped or Astell bottles. The medium should contain 0.5–1.0% *more* agar than is usual. Melt and cool to 45°C in a water-bath. Add 0.1 ml of each dilution and rotate horizontally in cold water until the agar is set in a uniform film around the walls of the bottle. This requires some practice. An alternative method employs a slab of ice taken from the ice tray of a refrigerator. Turn it upside down on a cloth and make a groove in it by rotating horizontally on it a bottle similar to that used for the counts but containing warm water. The count tubes are then rolled in this groove. If the roll-tube method is to be used often, the Astell Roll Tube apparatus, which includes a water-bath, saves much time and labour.

Incubate roll-tube cultures inverted so that condensation water collects in the neck and does not smear colonies growing on the agar surface. To count, draw a line parallel to the long axis of the bottle and rotate the bottle, counting colonies under a low-power magnifier.

The roll-tube method is popular in the dairy and food industries. Machines for speeding up and taking the tedium out of this method are available.

A roll-tube method for counting organisms in bottles (container sanitation) is given on p.227.

Drop count method

In this method, introduced by Miles and Misra (1938) and usually referred to by their names, small drops of the material are placed on agar plates. Colonies are counted in the inoculated areas after incubation.

Prepare 50-dropper pipettes. These are pasteur pipettes passed through a Starratt Standard Wire Gauge hole No. 59 (0.95 mm) and cut off accurately above the hole. When tested they should deliver 0.02 ml, i.e. 50 drops/ml. A tolerance of ±2 drops is permissible. They are available commercially. Alternatively use hypodermic needles (19 gauge) purchased before the bevels and points are ground. Clean well with chloroform to remove grease before testing, as it affects accuracy. Attach to a short piece of glass tubing by a silicone rubber or PVC sleeve and use with a rubber teat. When tested, about 70–80% of these deliver between 48 and 52 drops/ml. Clean 50-drop pipettes of either kind in hot detergent, rinse in chloroform and sterilize in glass tubes.

Dry plates of suitable medium very well before use. Drop at least five drops from each dilution of sample from a height of not more than 2 cm (to avoid splashing) on each plate. Replace the lid but do not invert until the drops have dried. After incubation, select plates showing discrete colonies in drop areas, preferably one which gives less than 40 colonies per drop (10–20 is ideal). Count the colonies in each drop, using a hand magnifier. Divide the total count by the number of drops counted, multiply by 50 to convert to 1 ml and by the dilution used.

This method lends itself to arbitrary standards; for example, if there are less than ten colonies between five drops of sample, this represents less than 100 colonies of that material. If these drops are uncountable (more than 40 colonies), then the colony count/ml is greater than 2000.

The 'Droplette' method

This accurate and rapid method was introduced by Sharpe (1973) and co-workers (1971). Serial, replicate dilutions of the sample are made mechanically in agar medium in 0.l-ml amounts and 0.1-ml drops are placed in petri dishes automatically.

The apparatus, which offers great savings in time and labour, consists of a diluter-dispenser, which accurately and automatically meters agar, using ungraduated disposable polypropylene or pasteur pipettes in 0.1–1.7-ml amounts, a viewer, which enables the agar drops to be seen on a grid screen at about 9-cm diameter, and an electromechanical counter, which records, with a felt-tipped pen, when the colony images are touched.

Surface count method

This is not very accurate but is useful for rough estimates of bacterial numbers, e.g. in urine examinations or 'in-line' checks in food processing plant.

Place 0.1 ml or another suitable amount of the sample measured with a standard (commercial) loop or a micropipette in the centre of a well-dried plate of suitable medium and spread it with a loop or spreader all over the surface (p.85). Incubate and count colonies.

When counting colonies, difficulties arise with spreading organisms and large 'smears' of small colonies caused when a large viable unit is broken up but not dispersed during manipulation. Treat these as single units. Usually, other colonies can be seen and counted through 'smears'.

The spiral plate method

The apparatus is described on p.85. The results are said to compare favourably with those obtained by the surface spread plate method (Jarvis *et al.*, 1977) and the Droplette method.

Membrane filter counts

Liquid containing bacteria is passed through a filter that will retain the organisms. The filter is then allowed to absorb the culture medium and incubated, when the colonies which develop can be counted.

The filter-carrying apparatus is made of metal, glass or plastic and consists of a lower funnel, which carries a fritted glass platform surrounded by a silicone rubber ring. The filter disc rests on the platform and is clamped by its periphery between the rubber ring and the flange of the upper funnel. The upper and lower funnels are held together by a clamp.

The filters are thin, porous cellulose ester discs, varying in diameter and about 120 μm thick. The pores in the upper layers are 0.5–1.0 μm diameter enlarging to 3–5 μm diameter at the bottom. Bacteria are thus held back on top, but culture medium can easily rise to them by capillary action. Filters are stored interleaved between absorbent pads in metal containers. A grid to facilitate counting is ruled on the upper surface of each filter.

Membrane filter apparatus may be re-usable or sold as kits by the companies that make membranes.

Shallow metal incubating boxes with close-fitting lids, containing absorbent pads to hold the culture media, are also required.

The most convenient size of apparatus for counting colonies takes filters 47 mm in diameter, but various sizes larger or smaller are available.

Sterilization

Assemble the filter carrier with a membrane filter resting on the fritted glass platform, which in turn should be flush on top of the rubber gasket. Screw up the clamping ring, but not to its full extent. Wrap in kraft paper and autoclave for 15 min at 121°C. Alternatively, loosely assemble the filter carrier, wrap and autoclave; sterilize the filters separately interleaved between absorbent pads in their metal container also by autoclaving. This is the most convenient method when a number of consecutive samples are to be tested; in this case, sterilize also in the same way the appropriate number of spare upper funnels.

Sterilize the incubation boxes, each containing its absorbent pad by autoclaving.

Re-use of membrane filters

Membranes are expensive and can often be re-used several times provided that they have not been used for pathogens and are handled with care (entomological forceps are the best tool for this). Wash membranes well with running water. Steep in laboratory detergent solution for several hours and boil in 3% hydrochloric acid. Wash again with running water, blot and store interleaved in absorbent pads as described above.

Culture media

Ordinary or standard culture media do not give optimum results with membrane filters. It has been found necessary to vary the proportions of some of the ingredients. Membrane filter versions of standard media are noted on p.58. A 'Resuscitation' medium to revive attenuated bacteria is also desirable for some purposes.

Method
Erect the filter carrier over a filter flask connected by a non-return valve to a pump giving a suction of 25–50 mmHg. Check that the filter is in place and tighten the clamping ring. Pour a known volume, neat or diluted, of the fluid to be examined into the upper funnel and apply suction.

Pipette about 2.5–3 ml of medium on a Whatman 5 cm, No. 17, absorbent pad in an incubation container or petri dish. This should wet the pad to the edges but not overflow into the container.

When filtration is complete, carefully restore the pressure. Unscrew the clamping ring and remove the filter with sterile forceps. Apply it to the surface of the wet pad in the incubation box so that no air bubbles are trapped. Put the lid on the container and incubate. A fresh filter and upper funnel can be placed on the filter apparatus for the next sample.

For total aerobic counts, use tryptone soya membrane medium. For counting coliform bacilli, either 'presumptive' or *E. coli*, see Water examination, p.233. For counting *C. perfringens* in water see p.236.

For anaerobic counts, use the same medium incubated anaerobically, or roll the filter and submerge in a 25-mm diameter tube of melted thioglycollate agar at 45°C and allow to set. To count anaerobes causing sulphide spoilage, roll the filter and submerge in a tube of melted iron sulphite medium at 42°C and allow to set. For yeast and mould counts, use the appropriate Sabouraud-type or Czapek-type media.

Counting
Count in oblique light under low-power magnification. If it is necessary to stain colonies to see them, remove the filter from the pad and dip in a 0.01% aqueous solution of methylene blue for half a minute and then apply to a pad saturated with water. Colonies are stained deeper than the filter. The grid facilitates counting. Report as membrane colony count per standard volume (100 ml, 1 ml, etc.).

Count anaerobic colonies in thioglycollate tubes by rotating the tube under strong illumination. Count black colonies in iron sulphite medium.

Bacteria in air
Membrane filters may be used with an impinger. Methods are given on p.229.

Dip slides
This useful method, commonly used in clinical laboratories for the examination of urine, now has a more general application, especially in food and drink industries for viable coliform and yeast counts. The slides are made of plastic and are attached to the caps of screw-capped bottles. There are two kinds: one is a single- or double-sided tray containing agar culture media; the other consists of a membrane filter bonded to an absorbent pad containing dehydrated culture media. Both have ruled grids, either on the plastic or the filter, which facilitate counting.

The slides are dipped into the samples, drained, replaced in their containers and incubated. Colonies are then counted and the bacterial load estimated.

Direct epifluorescence filtration technique (DEFT)

This is a rapid, sensitive and economical method, developed in the early 1980s for estimating the bacterial content of milk and beverages. It may also be used for foods if they are first treated with proteolytic enzymes and/or surfactants.

A small volume (e.g. 2 ml) of the product is passed through a 24-mm polycarbonate membrane which is then stained by acridine orange and examined with an epifluorescence microscope. Kits are available from several companies. For detailed information see Pettifer *et al.* (1986).

Rapid automated methods

Conventional viable counts are expensive in time and labour. Holding products for 24 h or more adds to the cost of production. Rapid methods are therefore desirable, but they must be reliable, sensitive and specific. Initial and running costs must be balanced against those of conventional techniques and storage. Comparisons of the relative advantages of various methods are made by Jarvis (1985) and Jarvis and Easter (1987).

Rapid automated methods use one or other of the following: electronic particle counting; bioluminescence (as measured by bacterial ATP); changes in pH and eH by bacterial growth; changes in optical properties; detection of ^{14}C in CO_2 evolved from a substrate; microcalorimetry; changes in impedance or conductivity.

Most probable number (MPN) estimates

These are based on the assumption that bacteria are *normally* distributed in liquid media, that is, repeated samples of the same size from one source are expected to contain the same number of organisms *on average*: some samples will obviously contain a few more, some a few less. The average number is the *most probable number*. If the number of organisms is large, the differences between samples will be small; all the individual results will be nearer to the average. If the number is small the difference will be relatively larger.

If a liquid contains 100 organisms/100 ml, then 10-ml samples will contain, on average, ten organisms each. Some will contain more, perhaps one or two samples will contain as many as 20; some will contain less, but a sample containing none is most unlikely. If a number of such samples is inoculated into suitable medium, every sample would be expected to show growth.

Similarly, 1-ml samples will contain, on average, one organism each. Some may contain two or three and others will contain none. A number of tubes of culture media inoculated with 1-ml samples would therefore yield a proportion showing no growth.

Samples of 0.1 ml, however, could be expected to contain only one organism per ten samples and most tubes inoculated would be negative.

It is possible to calculate the most probable number of organisms/100 ml for any combination of results from such sample series. Tables have been prepared (Tables 9.1–9.4) for samples of 10 ml, 1 ml and 0.1 ml using five tubes or three tubes of each sample size and, for water testing, using one 50-ml, five 10-ml and five 1-ml samples.

This technique is used mainly for estimating coliform bacilli, but it can be used for almost any organisms in liquid samples if growth can be easily observed, e.g. by turbidity, acid production. Examples are yeasts and moulds in fruit juices and beverages, clostridia in food emulsions and rope-spores in flour suspensions.

'Black tube' MPN counts for anaerobes are described on p.340.

Table 9.1ᵃ MPN/100 ml, using one tube of 50 ml, and five tubes of 10 ml

50-ml tubes positive	10-ml tubes positive	MPN/100 ml
0	0	0
0	1	1
0	2	2
0	3	4
0	4	5
0	5	7
1	0	2
1	1	3
1	2	6
1	3	9
1	4	16
1	5	18+

ᵃ Tables 9.1 to 9.3 indicate the estimated number of bacteria of the coliform group present in 100 ml of water, corresponding to various combinations of positive and negative results in the amounts used for the tests. The tables are basically those originally computed by McCrady (1918) with certain amendments due to more precise calculations by Swaroop (1951). A few values have also been added to the tables from other sources, corresponding to further combinations of positive and negative results which are likely to occur in practice. Swaroop has tabulated limits within which the real density of coliform organisms is likely to fall, and his paper should be consulted by those who need to know the precision of these estimates.

These tables and the accompanying information are reproduced from *The Bacteriological Examination of Drinking Water Supplies*, Reports on Public Health and Medical Subjects, No. 71 (1982), by permission of the Controller of Her Majesty's Stationery Office, London.

Mix the sample by shaking and inverting vigorously. Pipette 10-ml amounts into each of five tubes (or three) of 10 ml of double-strength medium, 1-ml amounts into each of five (or three) tubes of 5 ml of single-strength medium and 0.1-ml amounts of (or 1 ml of a 1:10 dilution) into each of five or three tubes of 5 ml of single-strength medium. For testing water, also add 50 ml of water to 50 ml of double-strength broth.

Double-strength broth is used for the larger volumes because the medium would otherwise be too dilute.

Incubate for 24–48 h and observe growth, or acid and gas, etc. Tabulate the numbers of positive tubes in each set of five (or three) and consult the appropriate table.

Table 9.2 MPN/100 ml, using one tube of 50 ml, five tubes of 10 ml and five tubes of 1 ml

50-ml tubes positive	10-ml tubes positive	1-ml tubes positive	MPN/100 ml
0	0	0	0
0	0	1	1
0	0	2	2
0	1	0	1
0	1	1	2
0	1	2	3
0	2	0	2
0	2	1	3
0	2	2	4
0	3	0	3
0	3	1	5
0	4	0	5
1	0	0	1
1	0	1	3
1	0	2	4
1	0	3	6
1	1	0	3
1	1	1	5
1	1	2	7
1	1	3	9
1	2	0	5
1	2	1	7
1	2	2	10
1	2	3	12
1	3	0	8
1	3	1	11
1	3	2	14
1	3	3	18
1	3	4	20
1	4	0	13
1	4	1	17
1	4	2	20
1	4	3	30
1	4	4	35
1	4	5	40
1	5	0	25
1	5	1	35
1	5	2	50
1	5	3	90
1	5	4	160
1	5	5	180+

Table 9.3 MPN/100 ml, using five tubes of 10 ml, five tubes of 1 ml and five tubes of 0.1 ml

10-ml tubes positive	1-ml tubes positive	0.1-ml tubes positive	MPN/100 ml
0	0	0	0
0	0	1	2
0	0	2	4
0	1	0	2
0	1	1	4
0	1	2	6
0	2	0	4
0	2	1	6
0	3	0	6
1	0	0	2
1	0	1	4
1	0	2	6
1	0	3	8
1	1	0	4
1	1	1	6
1	1	2	8
1	2	0	6
1	2	1	8
1	2	2	10
1	3	0	8
1	3	1	10
1	4	0	11
2	0	0	5
2	0	1	7
2	0	2	9
2	0	3	12
2	1	0	7
2	1	1	9
2	1	2	12
2	2	0	9
2	2	1	12
2	2	2	14
2	3	0	12
2	3	1	14
2	4	0	15
3	0	0	8
3	0	1	11
3	0	2	13
3	1	0	11
3	1	1	14
3	1	2	17
3	1	3	20
3	2	0	14
3	2	1	17
3	2	2	20
3	3	0	17
3	3	1	20
3	4	0	20
3	4	1	25
3	5	0	25
4	0	0	13
4	0	1	17

Table 9.3 *(continued)*

10-ml tubes positive	1-ml tubes positive	0.1-ml tubes positive	MPN/100 ml
4	0	2	20
4	0	3	25
4	1	0	17
4	1	1	20
4	1	2	25
4	2	0	20
4	2	1	25
4	2	2	30
4	3	0	25
4	3	1	35
4	3	2	40
4	4	0	35
4	4	1	40
4	4	2	45
4	5	0	40
4	5	1	50
4	5	2	55
5	0	0	25
5	0	1	30
5	0	2	45
5	0	3	60
5	0	4	75
5	1	0	35
5	1	1	45
5	1	2	65
5	1	3	85
5	1	4	115
5	2	0	50
5	2	1	70
5	2	2	95
5	2	3	120
5	2	4	150
5	2	5	175
5	3	0	80
5	3	1	110
5	3	2	140
5	3	3	175
5	3	4	200
5	3	5	250
5	4	0	130
5	4	1	170
5	4	2	225
5	4	3	275
5	4	4	350
5	4	5	425
5	5	0	250
5	5	1	350
5	5	2	550
5	5	3	900
5	5	4	1600
5	5	5	1800+

Table 9.4 MPN/100 ml, using three tubes each inoculated with 10, 1.0 and 0.1 ml of sample

Tubes positive			MPN	Tubes positive			MPN	Tubes positive			MPN
10 ml	1.0 ml	0.1 ml		10 ml	1.0 ml	0.1 ml		10 ml	1.0 ml	0.1 ml	
0	0	1	3	1	2	0	11	2	3	3	53
0	0	2	6	1	2	1	15	3	0	0	23
0	0	3	9	1	2	2	20	3	0	1	39
0	1	0	3	1	2	3	24	3	0	2	64
0	1	1	6	1	3	0	16	3	0	3	95
0	1	2	9	1	3	1	20	3	1	0	43
0	1	3	12	1	3	2	24	3	1	1	75
0	2	0	6	1	3	3	29	3	1	2	120
0	2	1	9	2	0	0	9	3	1	3	160
0	2	2	12	2	0	1	14	3	2	0	93
0	2	3	16	2	0	2	20	3	2	1	150
0	3	0	9	2	0	3	26	3	2	2	210
0	3	1	13	2	1	0	15	3	2	3	290
0	3	2	16	2	1	1	20	3	3	0	240
0	3	3	19	2	1	2	27	3	3	1	460
1	0	0	4	2	1	3	34	3	3	2	1100
1	0	1	7	2	2	0	21	3	3	3	1100+
1	0	2	11	2	2	1	28				
1	0	3	15	2	2	2	35				
1	1	0	7	2	2	3	42				
1	1	1	11	2	3	0	29				
1	1	2	15	2	3	1	36				
1	1	3	19	2	3	2	44				

From Jacobs and Gerstein's *Handbook of Microbiology*, D. Van Nostrand Company, Inc., Princeton, NJ (1960). Reproduced by permission of the authors and publisher

References

DHSS (1982) *The Bacteriological Examination of Drinking Water Supplies,* Reports on Public Health and Medical Subjects No. 71, HMSO, London

Jarvis, B. (1985) A philosophical approach to rapid methods for industrial food control. In *Rapid Methods and Automation in Microbiology and Immunology* (ed. R. O. Habermeyl), Springer-Verlag, Berlin, pp. 593–602

Jarvis, B. and Easter, M. C. (1987) Rapid methods in the assessment of microbial quality. Experience and needs. In *Changing Perspectives in Applied Microbiology* (eds C. S. Gutteridge and J. R. Norris), Society for Applied Bacteriology Symposium Series No. 16, Blackwells, London, pp. 115S–126S

Jarvis, B., Lach, V. H. and Wood, J. M. (1977) Evaluation of the spiral plate maker for the enumeration of microorganisms in foods. *Journal of Applied Bacteriology,* **43,** 149–157

McCrady, M. H. (1918) Tables for rapid interpretation of fermentation test results. *Public Health Journal, Toronto,* **9,** 201–210

Miles, A. A. and Misra, S. S. (1938) The estimation of the bactericidal power of the blood. *Journal of Hygiene (Cambridge),* **38,** 732–749

Pettifer, G. L., Mansell, R., McKinnon, C. H. and Cousins, C. M. (1986) Rapid membrane filtration epifluoresent microscopy technique for direct enumeration of bacteria in raw milk. *Applied and Environmental Microbiology,* **38,** 423–429

Sharpe, A. N. (1973) Automation and instrumentation developments for the bacteriological laboratory. In *Sampling–Microbiological Monitoring of the Environment* (eds R. G. Board and D. W. Lovelock), Society for Applied Bacteriology Technical Series 7, Academic Press, London, pp. 157–132

Sharpe, A. N. and Kilsby, D. C. (1971) A rapid, inexpensive bacterial count technique using agar droplets. *Journal of Applied Bacteriology,* **34,** 435–440

Swaroop, S. (1951) Range of variations of Most Probable Numbers of organisms estimated by dilution methods. *Indian Journal of Medical Research,* **39,** 107–131

Chapter 10

Clinical material

All clinical material should be regarded as potentially infectious. Blood specimens may contain hepatitis B and/or human immunodeficiency virus. For details of special precautions see HSAC (1984); USPHS (1984); Collins (1988a,b); ACDP (1990a,b).

The laboratory investigation of clinical material for microbial pathogens is subject to a variety of external hazards. Fundamental errors and omissions may occur in any laboratory. Vigorous performance control testing programmes and the routine use of controls will reduce the incidence of error but are unlikely to eliminate it entirely.

Most samples for microbiological examination are not collected by the laboratory staff, and it must be stressed that however carefully a specimen is processed in the laboratory, the result can only be as reliable as the sample will allow. The laboratory should provide specimen containers that are suitable, i.e. leak-proof, stable, easily opened, readily identifiable, aesthetically satisfactory, suitable for easy processing and preferably capable of being incinerated. It will be apparent that some of these qualities are mutually exclusive and the best compromise to suit particular circumstances must be made. Containers are discussed on p.24. The specimen itself may be unsatisfactory because of faulty collection procedures, and the laboratory must be ready to state its requirements and give a reasoned explanation if cooperation is to be obtained from ward staff and doctors.

Specimens should be:

(1) collected without extraneous contamination;
(2) collected, if possible before starting antibiotic therapy;
(3) representative; pus rather than a swab with a minute blob on the end; faeces rather than a rectal swab. If this is not possible, then a clear note to this effect should be made on the request form, which must accompany each sample.
(4) ideally collected and sent to the laboratory in a clear, sealed plastic bag to avoid leakage during transit. The request form accompanying each specimen should not be in the same bag (bags with separate compartments for specimens and request forms are available) and clearly displayed to minimize sorting errors;
(5) delivered to the laboratory without delay. If this is not possible, appropriate specimens must be delivered in a transport medium (see p.56). For other specimens, storage at 4°C will hinder bacterial overgrowth. If urine samples cannot be delivered promptly and refrigeration is not possible, boric acid preservative may be considered as a last resort. Dip slide culture is to be preferred. Sputum and urine specimens probably account for the greater part of the laboratory work load and the problems associated with their collection are particularly difficult to overcome because the trachea, mouth and urethra have a normal flora.

It must be stressed that in any microbiological examination, only that which is sought will be found and that it is essential to examine the specimen in the light of the available clinical information (which, regrettably, is often minimal). Every specimen should be evaluated to consider bacteria, fungi, parasites and viruses. The examination of pathological material for viruses and parasites is not within the scope of this book, but the possibility of infection by these agents should be considered and material referred to the appropriate laboratory.

The laboratory has an obligation to ensure that its rules are well publicized so that the periods of acceptance of specimens are known to all. Any tests that are processed in batches should be arranged in advance with the laboratory. This may be important for microbiological assay, etc., and is essential for the monitoring of antibiotic assays (less frequently requested) for the laboratory to subculture the correct strains of organisms required.

Blood cultures

Blood must be collected with scrupulous care to avoid extravenous contamination. In patients with a recurrent fever, blood culture is an important diagnostic procedure and the highest success rate is associated with the collection of cultures just as the patient's temperature begins to increase rather than at the peak of the rigor. In patients with septicaemia, the timing is less important and a rapid answer and a susceptibility test result are essential.

Castenada blood culture bottles are biphasic, i.e. they have a slope of solid medium and a broth which may also contain Liquoid (sodium polyethanol sulphonate: Roche) which neutralizes antibacterial factors including complement. Large amounts of blood (up to 50% of the total volume of the medium) may be examined in Liquoid cultures; otherwise, it is necessary to dilute the patient's blood at least 1:20 with broth. The bottles are tipped so that the blood–broth mixture washes over the agar slope every 72 h. Colonies will appear first at the interphase and then all over the slope. The bottles should be retained for at least 4 and preferably 6 weeks. The Castenada system avoids repeated subculture and lessens the risk of laboratory contamintion. This is important because patients receiving immunosuppressive drugs are likely to be infected with organisms not usually regarded as pathogens. No organism can therefore be automatically excluded as a contaminant.

Vacuum blood collection systems, e.g. Vacutainers (Becton Dickinson) also help to reduce contamination. Some of these ensure a 10% carbon dioxide atmosphere in the container. This is important when organisms such as *Brucella* spp. are sought. If other media are used they should be incubated in an atmosphere containing at least 10% carbon dioxide.

Evacuated containers or those containing additional gas may need 'venting' with a cotton wool plugged hypodermic needle (*caution*) or a commercial device. This is important, especially for the recovery of *Candida* spp. and *Pseudomonas* spp. from neonatal samples where only a small volume of blood is available for culture.

A variety of enriched media is available for blood culture. Few workers now use Robertson's cooked meat medium. This medium has been used for many years for blood cultures but because it is turbid it must be subcultured repeatedly to see if there is any growth.

Blood cultures should be subcultured on to appropriate media when growth is visible, i.e. colonies appear on the solid phase of the Castenada bottle, or turbidity is evident. Smears stained by Gram's method or acridine orange may be useful. In any case, subculture at 24, 48 and 72 h and then at weekly intervals. Great care is needed when subcultures are made or contamination will result. This should be done in a microbiological safety cabinet. Although this protects the operator as well as the culture good technique (GMT) is still required.

Rapid systems for the detection of growth in blood cultures are now available. The 'Bactec System' (Becton Dickinson) depends on the detection of radioactive CO_2 in the headspace, released during bacterial growth from labelled glucose in the medium. This instrument saves much time and labour in subculturing 'negative' bottles. The system has recently been extended to include a non-radiometric method for detecting bacterial growth. This uses infra-red analysis of microbially-generated CO_2. The two systems have been compared by Corkhill and Rimmer (1987).

Several new commercial systems, including those of Roche, Gibco, Oxoid ('Signal'), and Dupont offer rapid and reliable detection of commonly occurring infectious agents.

Likely pathogens

Indifferent viridans streptococci, non-haemolytic streptococci, enterococci, *Brucella*, *Staphylococcus aureus*, *Neisseria meningitidis*, *Clostridium*, *Salmonella*, *Bacteroides*, 'JK' coryneforms, mycobacteria from AIDS patients. In ill or immunosuppressed patients, any organisms, especially if recovered more than once, may be pathogens. In patients with septicaemia following major surgery, a mixed growth of enteric organisms may be found including anaerobes which may require prolonged incubation and *Candida* and other fungi after open-heart surgery.

Commensals

None.

Cerebrospinal fluid (CSF)

It is important that these samples should be examined with the minimum of delay. As with other fluids, a description is important. It is essential to note the colour of the fluid; if it is turbid, then the colour of the supernatant fluid after centrifugation may be of diagnostic significance. A distinctive yellow tinge (xanthochromia) usually confirms an earlier bleed into the CSF (as in, for example, a subarachnoid haemorrhage as opposed to traumatic bleeding during the lumbar puncture) (it is necessary, of course, to compare the colour of the supernatant fluid with a water blank in a similar tube).

Before any manipulation, the specimen must be carefully examined for clots, either gross or in the form of a delicate spider's web that is usually associated with a significantly raised protein and traditionally is taken as an indication that one should look particularly for *Mycobacterium tuberculosis*. The presence of a clot will invalidate any attempt to make a reliable cell count.

If the cells and/or protein are raised ($>$ three lymphocytes/mm^3 and/or $>$ 40 mg%), a CSF sugar determination should be compared with the blood sugar level collected at the same time. This and the type of cell may indicate the type of infection.

Raised polymorphs with very low sugar – usually indicate bacterial infection or a cerebral abscess.
Raised lymphocytes with normal sugar – usually indicate viral infection.
Raised lymphocytes with lowered sugar – usually indicate tuberculosis.
Raised polymorphs or mixed cells with lowered sugar – usually indicate early tuberculosis. It is important, however, to note that there are many exceptions to these patterns.

Make three films from the centrifuged deposit, spreading as little as possible. Stain one by the Gram method, one with a cytological stain and retain the third for examination if indicated for acid-fast bacilli. A prolonged search may be necessary if tuberculosis is suspected.

Plate on blood agar and incubate aerobically and anaerobically and on a chocolate agar plate for incubation in a 5–10% carbon dioxide atmosphere. Examine after 18–24 h. If organisms are found, set up a direct sensitivity test including penicillin, ampicillin, chloramphenicol and sulphonamides. Culture for *M. tuberculosis* as indicated.

Suspensions to detect antigens from organisms commonly associated with meningitis are now available.

Likely pathogens

Haemophilus influenzae, N. meningitidis, Streptococcus pneumoniae, Listeria monocytogenes, M. tuberculosis, and in very young babies coliform bacilli, Group B streptococci and *Pseudomonas aeruginosa. Staphylococcus epidermidis* and some micrococci are often associated with infections in patients with devices (shunts) inserted to relieve excess pressure.

Opportunists

Any organism introduced during surgical manipulation involving the spinal canal may be involved in meningitis.

Commensals

None.

Dental specimens

These may be whole teeth, when there may have been an apical abscess, or scrapings of plaque or carious material. Plate on selective media for staphylococci, streptococci and lactobacilli.

Likely organisms

S. aureus, streptococci, especially *S. mutans*, *S. milleri*, lactobacilli.

Ear discharges

Swabs should be small enough to pass easily through the external meatus. Many commercially available swabs are too plump and may cause pain to patients with inflamed auditory canals.

A direct film stained by the Gram method is helpful, as florid overgrowths of faecal organisms are not uncommon and it may be relatively difficult to recover more delicate pathogens. Plate on blood agar and incubate aerobically and anaerobically overnight at 37°C. Tellurite medium may be advisable if the patient is of school age. A medium that inhibits spreading organisms is frequently of value; CLED, or MacConkey, chloral hydrate or phenethyl alcohol agar are the most useful.

Likely pathogens

S. pyogenes, *S. aureus*, *Haemophilus*, *Corynebacterium diphtheriae*, *P. aeruginosa*, coliform bacilli, anaerobic Gram-negative rods, fungi.

Commensals

Micrococci, diphtheroids, *S. epidermidis*, moulds and yeasts.

Eye discharges

It is preferable to plate material from the eye directly on culture media rather than to collect in on a swab. If this is not practicable then it is essential that the swab is placed in transport medium.

Examine a direct Gram-stained film before the patient is allowed to leave. Look particularly for intracellular Gram-negative diplococci and issue a tentative report if they are seen so that treatment can be started without delay; this is of particular importance in neonates. Infections occurring a few days after birth may be caused by *Chlamydia trachomatis*, which should be suspected if films have excess numbers of monocytes or the condition does not resolve rapidly. Direct staining with Giemsa stain or the immunofluorescence technique are indicated.

Plate on blood agar plates and incubate aerobically and anaerobically and on a chocolate agar or 10% horse blood Columbia agar plate for incubation in a 5% carbon dioxide atmosphere. It is not usually possible to do 'direct' susceptibilities as the primary growth is frequently sparse; if, however, the film shows many organisms, direct sensitivity tests including penicillin and chloramphenicol should be attempted.

Likely pathogens

S. aureus, *S. pneumoniae*, viridans streptococci, *N. gonorrhoeae*, *Haemophilus*, *Chlamydia*, *Moraxella*, rarely coliform organisms, *C. diphtheriae*, *P. aeruginosa*.

Commensals

S. epidermidis, micrococci, diphtheroids.

Faeces and rectal swabs

It is always better to examine faeces than rectal swabs. Swabs that are not even stained with faeces are useless. If swabs must be used, they should be moistened with sterile saline or water before use and care must be taken to avoid contamination with perianal flora. It may be useful to collect material from babies by dipping the swabs into a recently soiled napkin (diaper). Swabs should be sent to the laboratory in transport medium unless they can be delivered within 1 h.

Because of the random distribution of pathogens in faeces, it is customary to examine three sequential specimens.

Plate on DCA, MacConkey and bismuth sulphite agar and inoculate selenite or tetrathionate broth. Incubate for 18–24 h and subculture selenite and tetrathionate broths on to DCA. It is essential that the MacConkey agar will support the growth of *S. aureus*, which may be causative in staphylococcal enterocolitis.

Plate stools from children, young adults and diarrhoea cases on campylobacter medium plus antibiotic supplements. Incubate for 48 h at 42°C and at 37°C in 5% oxygen and 10% carbon dioxide.

If the patient is under 3 years old, plate on CLED, MacConkey agar and blood agar and look for enteropathogenic *Escherichia coli*.

Plate on TCBS and inoculate alkaline peptone water if cholera is suspected, if the patient has recently returned by air from an area where cholera is endemic or if food poisoning due to *Vibrio parahaemolyticus* is suspected.

Yersinia enterocolitica is now recognized as an enteric pathogen. Plate on Yersinia Selective Agar and incubate at 32°C for 18-24 h. See also p.289.

Plate on cycloserine cefoxitin egg yolk fructose agar (CCFA) for *Clostridium difficile*.

In cases of food poisoning due to *S. aureus*, plate faeces on blood agar and on one of the selective staphylococcal media (p.54) and inoculate salt meat broth. Incubate all cultures overnight at 37°C and subculture the salt meat broth to blood agar and selective staphylococcal medium. It should be noted that here the symptoms relate to a toxin so the organisms may not be recovered.

In food poisoning where *Clostridium perfringens* may be involved make a 1:10 suspension of faeces in nutrient broth and add 1 ml to each of two tubes of Robertson's cooked meat medium. Heat one tube at 80°C for 10 min and then cool. Incubate both tubes at 37°C overnight and plate on blood agar with and without neomycin and on egg yolk agar. Place a metronidazole disc on the streaked-out inoculum. Incubate at 37°C anaerobically overnight.

Emulsify faeces in ethanol (industrial grade) to give a 50% suspension. Mix well and stand for 1 h. Inoculate media and incubate as above. See also Chapter 28.

Likely pathogens

Salmonella, Shigella, Vibrio, enteropathogenic *E. coli, C. perfringens* (both heat resistant and non-heat resistant), *S. aureus, Bacillus cereus, B. subtilis, Campylobacter, Yersinia, Aeromonas, P. aeruginosa, C. botulinum* and *C. difficile*.

Commensals

Coliform bacilli, *Proteus, Clostridium, Bacteroides, Pseudomonas.*

Nasal swabs (see also Throat swabs)

The area sampled will influence the recovery of particular organisms, i.e. the anterior part of the nose must be sampled if the highest carriage rate of staphylococci is to be found. For *N. meningitidis*, it may be preferable to sample further in and for the recovery of *Bordetella pertussis* in cases of whooping cough a pernasal swab is essential.

Inoculate as soon as possible on charcoal agar containing 10% horse blood and 40 mg/litre cephalexin and incubate at 35°C in a humid atmosphere.

Direct films are of no value. If *M. leprae* is sought, smears from a scraping of the nasal septum may be examined after staining with ZN stain or a fluorescence stain.

Plate on blood agar and tellurite media and incubate as for throat swabs.

Likely pathogens

S. aureus, S. pyogenes, N. meningitidis, B. pertussis, Haemophilus, C. diphtheriae.

Commensals

Diphtheroids, *S. epidermidis, S. aureus, Branhamella*, aerobic spore bearers and small numbers of Gram-negative rods (*Proteus* and coliform bacilli).

Pus

As a general rule, pus rather than swabs of pus should be sent to the laboratory. Swabs are variably lethal to bacteria within a few hours because of a combination of drying and toxic components of the cotton wool released by various sterilization methods. Consequently, speed in processing is important; a delay of several hours may allow robust organisms to be recovered while the more delicate pathogens may not survive. Swabs should be moistened with broth before use on dry areas.

Examine Gram-stained films and films either direct or of concentrated material (p.347), for acid-fast rods. Because the distribution of organisms in pus is random, wash all pus samples and examine them for the 'sulphur granules' of *Actinomyces israelii*. Shake the pus with two or three changes of sterile physiological saline (it is important to avoid both contamination and aerosols in this procedure). This will produce a relatively clear supernatant fluid with the granules depositing very quickly. Plate on blood agar and incubate aerobically and anaerobically. Include a chloral hydrate plate in case *Proteus* is present. Prolonged anaerobic incubation of blood agar cultures, in a 90% hydrogen–10% carbon dioxide atmosphere possibly with neomycin or nalidixic acid may allow the recovery of strictly anaerobic *Bacteroides* spp. These organisms may be recovered frequently if care is taken and the number of sterile collections of pus consequently reduced. Culture in thioglycollate broth may be helpful. The dilution effect may overcome specific and non-specific inhibitory agents.

Culture pus from post-injection abscesses for mycobacteria (p.347).

The recovery of anaerobes will be enhanced by the addition of menadione and vitamin K to the medium. Wilkins and Chalgren's medium is also of value.

Likely pathogens

S. aureus, S. pyogenes, peptostreptococci, peptococci. *Mycobacterium, Actinomyces, Pasteurella, Yersinia, Clostridium, Neisseria, Bacteroides, B. anthracis, Listeria, Proteus, Pseudomonas, Nocardia*, fungi, other organisms in pure culture.

Commensals

None.

Serous fluids

These fluids include pleural, synovial, pericardial, hydrocele, ascitic and bursa fluids.

Record the colour, volume and viscosity of the fluid. Centrifuge at 3000 rev/min for at least 15 min preferably in a sealed bucket and note the colour of the supernatant. Transfer the supernatant fluid to another bottle and record its appearance. Make Gram-stained films of the deposit and also stain films for cytological evaluation. Plate on blood agar and incubate aerobically and anaerobically at 37°C for 18–24 h. Culture on blood agar anaerobically plus 10% carbon dioxide atmosphere and incubate for 7–10 days for *Bacteroides*. Use additional material as outlined above. Prolonged incubation may be necessary for these strictly anaerobic organisms.

Gonococcal and meningococcal arthritis must not be overlooked; inoculate one of the GC media and incubate in a 10% carbon dioxide atmosphere for 48 h at 22 and 37°C. Culture, if appropriate, for mycobacteria.

Likely pathogens

S. pyogenes, S. pneumoniae, peptostreptococci, peptococci, *S. aureus, N. gonorrhoeae, M. tuberculosis, Bacteroides*.

Commensals

None.

Sputum

Sputum is a difficult specimen. Ideally, it should represent the discharge of the bronchial tree expectorated quickly with the minimum of contamination from the pharynx and the mouth, and delivered without delay to the laboratory.

Because of the irregular distribution of bacteria in sputum it may be advisable to wash portions of purulent material to free them from contaminating mouth organisms before examining films and making cultures.

Homogenization of the specimen with Sputolysin, N-acetylcysteine or pancreatin, for example, followed by dilution enables the significant flora to be assessed. In sputum cultures not so treated initially small numbers of contaminating organisms may overgrow pathogens.

The following method is recommended.

Add about 2 ml of sputum to an equal volume of Sputolysin (Calbiochem) or Sputasol (Oxoid) in a screw-capped bottle and allow to digest for 20–30 min with occasional shaking, e.g. on a Vortex Mixer. Inoculate a blood agar plate with 0.01 ml of the homogenate, using a standard loop (e.g. a 0.01 ml plastic loop) and make a Gram-stained film. Add 0.01 ml with a similar loop to 10 ml of peptone broth. Mix this thoroughly preferably using a Vortex Mixer. Inoculate blood agar and MacConkey agar with 0.01 ml of this dilution. Incubate the plates overnight at 35°C; incubate the blood agar plates in a 10% carbon dioxide atmosphere. Chocolate agar may be used but should not be necessary for growing *H. influenzae* if a good blood agar base is used.

Five to 50 colonies of a particular organism on the dilution plate are equivalent to 10^6–10^7 of these organisms/ml of sputum and this level is significant except for viridans streptococci and *Branhamella*, one to five colonies are equivalent to 10^5–10^6 organisms/ml and this level is equivocal.

Set up direct sensitivity tests if the Gram film shows large numbers of any particular organisms.

Methods of examining sputum for mycobacteria are described in Chapter 29.

For the isolation of *Legionella* and similar organisms from lung biopsies or bronchial secretions use blood agar or one of the commercial legionella media plus supplement.

Likely pathogens

S. aureus, S. pneumoniae, H. influenzae, coliform bacilli, Klebsiella pneumoniae, Pasteurella, Mycobacterium, Candida spp., *Branhamella catarrhalis, Mycoplasma, Pasteurella/Yersinia* spp., *Y. pestis, Legionella pneumophila, Aspergillus, Histoplasma, Cryptococcus, Blastomyces* and *Pseudomonas*, particularly from patients with cystic fibrosis.

Commensals

S. epidermidis, micrococci, *Branhamella*, coliforms, *Candida* in small numbers, *S. viridans*.

Throat swabs

Collect throat swabs carefully with the patient in a clear light. It is customary to attempt to avoid contamination with mouth organisms. Evidence has been advanced that the recovery of *Streptococcus pyogenes* from the saliva may be higher than from a 'throat' or pharyngeal swab. Make a direct film, stain with dilute carbol fuchsin and examine for Vincent's organisms, yeasts and mycelium.

Plate on blood agar and incubate aerobically and anaerobically for a minimum of 18 but preferably 48 h in 7–10% CO_2 at 35°C. Plate on tellurite medium and incubate for 48 h in 7–10% CO_2 at 35°C.

Likely pathogens

S. pyogenes, Corynebacterium diphtheriae, C. ulcerans, S. aureus, Candida albicans, Neisseria meningitidis, Borrelia vincenti (H. influenzae Pittmans type b in the epiglottis is certainly a pathogen).

Commensals

Branhamella, indifferent viridans, streptococci, *S. epidermidis*, diphtheroids, *S. pneumoniae*, probably *Haemophilus influenzae*.

NB. For details of the logistics of mass swabbing, e.g. in a diphtheria outbreak, see the paper by Collins and Dulake (1983).

Tissues, biopsy specimens, post-mortem material

As a general rule, retain all material submitted from autopsy until the forensic pathologists agree to their disposal. If it is essential to grind up tissue, obtain approval.

Tissue specimens may be ground up after suitable selection procedures in a blender or, if the specimen is very small, in a Griffith's tube or with glass beads on a Vortex mixer. The Stomacher Lab-Blender is best for larger specimens as it saves time and eliminates the hazard of aerosol formation. Examine the homogenate as described under 'Pus'.

Urethral discharges

Swabs of discharge should be collected into a transport medium or plated directly at the bedside or in the clinic. Examine direct Gram-stained films. Look for intracellular Gram-negative diplococci. In the sexually active male, the commonest cause of urethral discharge is gonorrhoea. In the female, the recovery of organisms is more difficult and the interpretation of films may occasionally present problems because over-decolorized or dying Gram-positive cocci may resemble Gram-negative diplococci.

Plate on blood agar plates for aerobic and anaerobic incubation and chocolate or Columbia agar with and without antibiotics (vancomycin, colistin, nystatin and trimethoprim to prevent the spread of proteus) or on GC medium and incubate in a 5% carbon dioxide atmosphere.

Non-gonococcal urethritis is increasingly associated with chlamydial infections. These organisms (not considered in this book) may be isolated by tissue culture techniques and identified by commercial immunofluorescence kits.

Likely pathogens

N. gonorrhoeae, Mycoplasma, occasionally *S. pyogenes*, anaerobic cocci.

Commensals

S. epidermidis, micrococci, diphtheroids, small numbers of coliform organisms.

Urine

The healthy urinary tract is free from organisms over the greater part of its length; there may, however, be a few transient organisms present, especially at the lower end of the female urethra.

Collection

Catheters may contribute to urinary infections and are no longer considered necessary for the collection of satisfactory samples from females.

A 'clean catch' mid-stream urine (MSU) is the most satisfactory specimen for most purposes if it is delivered promptly to the laboratory. Failing prompt delivery, the specimen can be kept in a refrigerator at 4°C for a few hours or, exceptionally, a few crystals of boric acid may be added.

Dip slides are now widely used for assessing urinary tract infections. They do, however, preclude examination for cells and other elements.

Examination

This includes counting leucocytes, erythrocytes and casts, estimating the number of bacteria/ml and identifying them. Normally, there will be none or less than $20/mm^3$ of leucocytes or other cells and a colony count of less than 10^4 organisms/ml.

Mix the urine by rotation, never by inversion. Count the cells using a Fuchs Rosenthal or similar counting chamber and inoculate a blood agar plate and a MacConkey, CLED or EMB plate with 0.001 ml using a standard loop (Nunc disposable loops or micropipettes with disposable tips are satisfactory alternatives). Also 0.1 ml of a 1 in 100 dilution of urine may be used. Spread with a glass spreader, incubate overnight at 37°C and count the colonies.

<10 colonies	$= 10^4$ organisms/ml	— not significant
10-100 colonies	$= 10^4$ -10^5 organisms/ml	— doubtful significance
>100 colonies	$= >10^5$ organisms/ml	— significant bacteruria

If dip slides are submitted, incubate them overnight and follow the manufacturer's directions for counting.

Note that the accuracy and usefulness of bacterial counts on urines is limited by the age of the specimen. Only fresh specimens are worth examining. Other limitations are the volume of fluid recently drunk and the amount of urine in the bladder.

For methods of examining urine for tubercle bacilli, see p.348.

Automated urine examination

The Rapid Automatic Microbiological Screening system (RAMUS, Orbec), is a particle volume analyser, capable of counting leucocytes and bacteria. Counts of bacteria $>10^5$/ml are regarded as significant.

Multipoint inoculation methods are very useful in the examination of large batches of urines (Kerfoot *et al.* 1983; Cheetham and Brown, 1986; Henricksen and Moyes, 1987; Turk, 1987). The technique can incorporate tests for the presence of antibacterial substances, direct inoculation of identification media and media for those fastidious organisms that are now considered to be of importance in urinary tract infections.

Likely pathogens

E. coli, other enterobacteria, *Proteus*, staphylococci including *S. epidermidis*, *S. saprophyticus*, enterococci, *Salmonella* (rarely), leptospira, mycobacteria.

Commensals

Small numbers ($<10^4$/ml) of almost any organisms. It is unwise to be too dogmatic about small numbers of organisms isolated from a single specimen: it is better to repeat the sample.

Vaginal discharges

Puerperal infections

A high vaginal swab is usually taken. Swabs that cannot be processed very rapidly after collection should be put into transport medium. Examine direct Gram-stained films. If non-sporing square-ended Gram-positive rods, not in pairs, and which appear to be capsulated are seen they should be reported without delay, as they may be *C. perfringens*, which may be an extremely invasive organism.

Plate on blood agar and incubate aerobically and anaerobically overnight at 37°C and also plate on MacConkey or other suitable selective medium. Plate any swabs suspected of harbouring *C. perfringens* on a neomycin half-antitoxin Nagler plate for anaerobic incubation.

To isolate *Gardnerella vaginalis (Haemophilus/Corynebacterium vaginalis)* add 10% human blood to the medium. Incubate for up to 48 h at 35°C in humid conditions. Characteristic colonies have a hazy, diffuse β-haemolysis. 'Clue cells' (epithelial cells with many small rods which may be Gram-variable or Gram-negative) should correlate with the isolation of this organism.

Likely pathogens

S. pyogenes, L. monocytogenes, C. perfringens, Bacteroides, anaerobic cocci, *Candida* spp., excess numbers of enteric organisms, *G. vaginalis*.

Commensals

Lactobacilli, diphtheroids, micrococci, *S. epidermidis*, small numbers of coliform bacilli, yeasts.

Non-puerperal infections

Examine a Gram-stained smear and also a wet preparation for the presence of *Trichomonas vaginalis*. If the examination is to be immediate, then a saline suspension of vaginal discharge will be satisfactory, otherwise a swab in transport medium is essential. A wet preparation may still be examined for actively motile flagellates but culture in Trichomonas medium usually yields more reliable results, especially in cases of light infection. An alternative to the wet preparation is the examination under the fluorescence microscope of an air-dried smear stained by acridine orange.

If *N. gonorrhoeae* infection is suspected in the female, then swabs from the urethra, cervix and posterior fornix should be collected, placed into transport medium or plated immediately on selective medium. Many workers do not make films from these swabs because the difficulties of interpretation are compounded by the presence of charcoal from the buffered swabs; they rely upon cultivation using VCN (vancomycin, colistin, nystatin) or VCNT (VCN + trimethoprim) medium as a selective inhibitor of organisms other than *N. gonorrhoeae*.

Plate on blood agar and incubate, aerobically and anaerobically, Sabouraud agar (when looking for *Candida*) and chocolate agar or another suitable gonococcal medium for incubation in a 5% carbon dioxide atmosphere to encourage isolation of *N. gonorrhoeae*.

If *L. monocytogenes* infection is suspected plate on serum agar with 0.4 g/litre of nalidixic acid added.

Likely pathogens

N. gonorrhoeae, C. albicans and related yeasts in moderate numbers, *S. pyogenes, L. monocytogenes, Haemophilus, Bacteroides*, anaerobic cocci, possibly *Mycoplasma, T. vaginalis*.

Commensals

Lactobacilli, diphtheroids, enterococci, small numbers of coliform organisms.

Wounds (superficial)

Examine Gram-stained films and culture on blood agar and proceed as for pus, etc. Culture also on Lowenstein–Jensen medium and incubate at 30°C because some mycobacteria that cause superficial infections do not grow on primary isolation at 35–37°C.

Likely pathogens

S. aureus, streptococci, *M. marinum, M. ulcerans, M. chelonei*.

Contaminants

Staphylococci, pseudomonas and a wide variety of bacteria and fungi.

Wounds (deep) and burns

The collection of these specimens and their consequent rapid processing are two vital factors in the recovery of the significant organisms. Burn cases frequently become infected with staphylococci and streptococci as well as with various Gram-negative rods, especially *Pseudomonas* spp. It may be difficult to recover Gram-positive cocci for the overgrowing Gram-negative organisms.

Plate on blood agar and incubate aerobically and anaerobically. Plate also on MacConkey agar and CLED medium (to inhibit the spreading of *Proteus* spp.). Robertson's cooked meat medium may be useful.

Likely pathogens

Clostridium, S. pyogenes, S. aureus, peptococci, peptostreptococci, *Bacteroides*, Gram-negative rods.

Contaminants

Small numbers of a wide variety of organisms.

References

Cheetham, P. and Brown, S. E. (1986) Technique for the culture and direct sensitivity testing of large numbers of specimens. *Journal of Clinical Pathology,* **39,** 333–337

Collins, C. H. and Dulake, C. (1983) Diphtheria: the logistics of mass swabbing. *Journal of Infection,* **6,** 277–230

Corkhill, J. E. and Rimmer, K. (1987) Microbiological comparison of a new infra-red blood culture system (Bactec NR660) and a radiometric system (Bactec 460). *Medical Laboratory Sciences,* **44,** 150–159

Henricksen, C. and Moyes, A. (1987) A semi-automated method for the culture, identification and susceptibility testing of bacteria direct from urine specimens. *Medical Laboratory Sciences,* **44,** 50–58

Kerfoot, P., McGhie, D., Cahill, E. and Fountain, T. (1983) A mechanical batch screening method for the detection of bacteriuria. *Journal of Clinical Pathology,* **36,** 1318–1319

Turk, A. C. (1987) An improved multipoint technique for the routine microbiological examination of urine specimens. *Medical Laboratory Sciences,* **44,** 50–58

Chapter 11

Antimicrobial susceptibility and assay tests

For many years tests for the susceptibility of microorganisms to antimicrobial agents and assays of such agents in body fluids have relied on diffusion techniques of one kind or another. In recent years, however, other methods have been developed. Many of these involve mechanization or automation, and some of them are described below. This is a constantly expanding field of microbiology and it is largely in the light of the more sophisticated and objective techniques that limitations of the traditional diffusion methods have become apparent. In particular, it is now clear that it is illogical to use a biological method as a standard for comparing new methods that utilize modern machinery; to do so implies a degree of precision and accuracy which the method does not warrant.

Antibiotic susceptibility tests

The familiar diffusion techniques for testing the antibiotic susceptibility of organisms isolated from clinical specimens have suffered severely from the difficulties of standardization. It is uncertain whether tests on the same organism, done in different places, would result in consistent recommendations for treatment.

The most important efforts in the improvement of susceptibility tests have been directed towards a standardized method that is generally acceptable. As a result, in respect of diffusion tests, Sweden has adopted the method of the International Collaborative Study (Ericsson and Sherris, 1971); in the USA the Food and Drugs Administration has adopted the Kirby–Bauer technique (Bauer *et al.*, 1966); in the UK the controlled disc method of Stokes and Ridgway (1987) is generally used.

In the classic diffusion technique a source of the antimicrobial agent is applied to the surface of the medium.

The diffusion of the agent through the medium inhibits the growth of a sensitive organism growing in it or on it to a degree partially dictated by the susceptibility of the organism. It is well known, however, that a number of other factors will influence the size of the zones of inhibition, and these factors will require control.

Factors affecting zone sizes

Ingredients of culture media
These are the most important. Apart from viscosity and depth in the petri dishes many substances present in culture media may affect the zones of inhibition. These include peptone, tryptone, yeast extract and agar. They may vary in their mineral

content and this may influence the activity of the antimicrobials. It is well known that calcium, magnesium and iron affect the sizes of zones produced by tetracyclines and gentamicin. Sodium chloride reduces the activity of aminoglycosides and enhances the effect of fusidic acid. Carbohydrates may enhance the effect of nitrofurantoin or ampicillin.

Perhaps the most notable ingredients of culture media that influence susceptibility tests are those related to the drugs which act by inhibiting folate metabolism, e.g. sulphonamides and trimethoprim. The action of drugs on even the most susceptible organisms will be affected if a medium containing more than a very small amount of the end products of bacterial folic acid synthesis is used.

The most common examples of such substances are p-aminobenzoic acid and thymidine. The presence of these chemicals in culture media will often not merely decrease the size of the zone of inhibition but, depending on the amounts available, may permit the organism to grow right up to a disc, although the colonies may be smaller than usual. The effect of small amounts of thymidine in a medium, however, can usually be overcome by the addition of haemolysed horse blood. For testing the susceptibility of fastidious organisms or to perform direct susceptibility tests upon clinical specimens and at the same time to observe haemolysis, equal parts of whole blood and haemolysed blood may be added. Chocolate agar may also be used, provided that the basic medium is free from inhibitor, but all media that contain blood or blood products may produce smaller zones with antimicrobials which are particularly protein-bound.

Choice of medium
Consistent and reproducible results are obtained when media especially formulated for susceptibility testing are used. Ideally plates should be poured flat with an even depth of medium throughout and all contain the same volume of medium, e.g. 15 ml for a 90-mm petri dish.

Effect of pH
pH has an effect on the size of inhibition zones produced by some antibiotics.

The activity of aminoglycosides is enhanced in alkaline medium and reduced in acid media. The reverse effect is true for tetracyclines. Carbon dioxide in the incubator atmosphere and the presence of fermentable carbohydrates in culture media will also induce acidic conditions.

Size of inoculum
Although many antimicrobials are not markedly affected by the presence of a large number of organisms, heavy inocula reduce inhibition zones to some extent. The ideal inoculum is one which gives an even, dense growth without being confluent. The actual density of a confluent growth is impossible to assess.

Broth cultures of organisms and suitable suspensions made from solid media may be diluted accurately to give optimal inocula for susceptibility testing but in practice satisfactory results can usually be achieved by taking a loopful of a well grown liquid culture, or a suitably made suspension of organisms from a solid medium, and spreading it with a dry sterile swab (Felmingham and Stokes, 1972).

Direct susceptibility tests
Susceptibility tests may be set up on primary cultures instead of indirectly on pure suspensions prepared by selection of colonies from such cultures. These direct tests enable a report to be given 24 h earlier.

Colonies which all look alike on the usual primary culture media may show distinctly different populations on the direct sensitivity plate. This is particularly true of staphylococci. When, for instance, a single colony is picked from an aerobic blood agar plate it is a matter of chance which population is being tested and the colony chosen may not necessarily be representative of the majority. Arguments are often put forward that a large proportion of primary susceptibility tests have to be repeated because the inoculum is unsuitable, because there is an antimicrobial agent present in the original specimen, or because there is a mixed bacterial population. In practice, however, when primary susceptibility tests are carried out regularly it is surprising how often a useful answer can be given despite these drawbacks. Those who perform them regularly, moreover, argue with some justification that even when such eventualities render a test unreadable nothing is lost as far as the patient is concerned, and the only wastage is that of a small amount of laboratory time and culture medium.

Certain sources of error peculiar to primary susceptibility tests need to be noted when these tests are carried out. Firstly, there is the possibility of a penicillinase-producing organism present in a mixed culture 'protecting' an otherwise susceptible organism against the action of a penicillinase-susceptible antibiotic. Penicillinase-producing staphylococci are usually quite easily recognized because the edges of the zones of inhibition which they exhibit show no tendency to produce smaller colonies nearer to the disc and are heaped up and sharp. Any significant organism that is present in company with a penicillinase producer should be subcultured and retested in pure culture.

Bacteria which satellitise other organisms in mixed cultures (e.g. *Haemophilus* spp.) may show false susceptibility to any antimicrobial substance that prevents the growth of the colonies to which they show satellitism. When this happens the susceptibility tests should be repeated with a pure culture.

Nowadays, fewer laboratories use direct susceptibility tests, but it would seem sensible to include them in investigations on body fluids which are normally expected to be sterile, or are likely to be infected with a single species.

Swarming of proteus

Difficulties in reading sensitivity tests may arise from the presence of swarming *Proteus* species. When there is a clear edge to the zone given by the parent colonies, however, measurements may safely be taken from this and swarming nearer the disc disregarded.

The problem may be largely overcome by adding *p*-nitrophenyl glycerol (PNPG) to the medium at a concentration of 86 mg/litre. This prevents swarming without affecting growth or motility. The substance is heat stable and may be added to the medium before sterilization. The most convenient form is the Mast PNPG Selectatab.

Methicillin resistance

The diffusion method must be modified for detecting methicillin-resistant organisms. It is now known that the mechanism of resistance and its expression in the *in vitro* tests are complex; the addition of extra sodium chloride to the medium and incubation at a lower temperature are not sufficient to detect all methicillin-resistant staphylococci. In addition, the current methods are known not to be suitable for both coagulase positive and coagulase negative staphylococci and that no single medium is entirely reliable for demonstrating methicillin resistance.

Recent studies indicate that the best medium for detecting methicillin-resistant strains of *Staphylococcus aureus* is Muller–Hinton containing 5% sodium chloride; nearly all resistant strains are detected after incubation at 35°C for 18 h. Other staphylococci and apparently susceptible strains should be incubated for 40 h.

Muller–Hinton medium is not always suitable for coagulase negative staphylococci, however, and Columbia agar should be used instead. All strains not showing obvious resistance after 18 h should be reincubated for up to 40 h (see Milne *et al.*, 1987).

Only methicillin discs or strips should be used: cloxacillin and flucloxacillin may give false susceptibility results.

Methicillin-resistant coagulase-positive staphylococci are always resistant to cephalosporins.

The performance of diffusion techniques

Strength of antimicrobial agents

Until recently there has been little or no general agreement about the strengths of antimicrobial discs for use in *in vitro* susceptibility tests. Tables are now available which give recommended disc strengths, which vary according to the method of interpretation being used (Table 11.1). Some organisms, e.g. strains of *Streptococcus pneumoniae* and *Neisseria gonorrhoeae*, which are not fully susceptible to penicillin may give zones of inhibition that are difficult to interpret with standard content discs. It may be necessary to retest them with higher content discs.

Storage of discs

Appropriate storage conditions are very important if reproducible results are to be achieved. Discs should always be kept cool and dry, but should be raised to room temperature before use, and should be applied firmly to the medium to ensure proper contact and thus even diffusion. Fine pointed forceps and dissecting needles are convenient tools.

Incubation times

Ideally the incubation time should be the minimum required for the growth of the organism and consistent with the daily routine of the laboratory. Prolonged incubation of a slowly-growing organism may well give spurious results.

Controls

For the best interpretation of results and recognition of any source of error in disc diffusion susceptibility methods the correct use of controls is essential. For the Stokes method, which is described in more detail below, control organisms are present on every plate.

Almost the entire range of antibiotic susceptibilities performed in the diagnostic laboratory may be adequately controlled by using one of three organisms: *Staphylococcus aureus* (NCTC 6571: the 'Oxford Staphylococcus'), *Pseudomonas aeruginosa* (NCTC 10662), and *Escherichia coli* (NCTC 10418). These organisms may be kept on agar slopes at room temperature and subcultured about once a month. For routine daily use the organisms are most conveniently kept at 4°C on

Table 11.1 Suitable disc contents

	Organisms from sites other than urine	Organisms from urine
Amikacin	10	10
Ampicillin	10	25
Carbenicillin	100	100
Cefoxitin	30	30
Cefuroxime	10	30
Cephaloridine	10	30
Chloramphenicol	10	30
Clindamycin	2	—
Colistin	50	50
Erythromycin	10	—
Fusidate	10	—
Gentamicin	10	10
Kanamycin	10	30
Lincomycin	2	—
Methicillin	5	—
Metronidazole	2.5	—
Novobiocin	5	—
Penicillin	1[a]	10[a]
Polymyxin	300[a]	300[a]
Streptomycin	10	25
Sulphonamide	100	100
Tetracycline	10	30
Tobramycin	10	10
Trimethoprim	1.25	1.25

[a] units

Reproduced from p. 470 of *Antibiotics and Chemotherapy*, 5th edn, L. P. Garrod, H. P. Lambert and F. O'Grady, (1981). Churchill-Livingstone, London, by permission of the authors and publisher.

sterile throat (or 3-in) swabs. A jar full of such swabs can be impregnated at one time, and they keep well for at least a week. The control organisms are applied to the plates directly from the swabs. The control organism should have a growth rate similar to that of the test organism. It must be susceptible to normal doses of the antimicrobial(s) being tested. This will also be affected by the site of the infection and the concentration of the antimicrobial usually attainable at that site. More resistant organisms for instance, will be likely to respond to treatment with antimicrobials excreted by the kidneys in urinary tract infections, because much higher concentrations of such substances occur in the urine. Hence, in these circumstances, a higher content disc may be used and controlled with a more resistant organism such as *E. coli*. This does not hold good, however, when common urinary tract pathogens such as enterococci and coliform organisms are isolated from other parts of the body. Although these organisms may respond well to normal doses of antimicrobial excreted in the urine, they are usually only moderately sensitive and thus when isolated from other sites the fully susceptible control must be used to make clear in the report the higher doses are required. The Oxford Staphylococcus is used for all antimicrobials except the polymyxins. The standard *E. coli* should be used for these antimicrobials and with all organisms from the urine. The standard *P. aeruginosa* is used in tests against pure cultures of known pseudomonads that are tested against only a limited range of antimicrobials.

Technical methods

International collaborative study method

This relies on the linear relationship that for most antimicrobials exists between the diameters of the zone of inhibition and the minimum inhibitory concentration (MIC).

Determine the MICs of the antimicrobials for at least 100 organisms with widely differing susceptibilities.

At the same time and under the same strictly controlled conditions measure the diameters of the inhibition zones produced by high content discs of the same antimicrobials. Plot regression lines for each antimicrobial, with MICs on the ordinates and zone diameters on the abscissae.

Measure the diameter of the zone around the test organism and read the MIC from the graph.

Place the organism, according to the MIC reading, in one of the four classes of susceptibility calculated locally according to the original paper of Ericsson and Sherris (1971).

Kirby–Bauer method

Use Muller–Hinton agar, 5–6 mm deep. Apply carefully standardized inocula with cotton wool (throat) swabs to give a confluent growth. Apply high content discs and incubate.

Measure the zone diameters and consult the tables that give the breakpoints for each antimicrobial. These may be constructed locally. Report the organism as 'sensitive', 'intermediate' or 'resistant' (Bauer et al., 1966).

The Stokes method

This method does not require the stringent standardization demanded by others described above. Test and control organisms are compared against the same discs, on the same medium and under the same physical conditions.

For the individual disc technique inoculate the middle third of a suitable culture plate with the specimen or a suitable suspension of organisms in broth using a cotton wool (throat) swab. Inoculate the areas on either side with the control organism, leaving a small gap between the inoculated areas. Place the disc in this gap (Figure 11.1). On 90-mm round plates four discs are convenient, while on larger square dishes six discs can be accommodated. After incubation zone sizes are compared by measurement with calipers, a transparent rule, or one of the more sophisticated zone-readers.

Modifications of the Stokes method

Most of the antagonism to the use of the Stokes method has resulted from the introduction of multiple discs such as the Multodisc (Oxoid) and the Mastring (Mast). Laboratory workers quite rightly claimed that the method was unsuitable for these annular disc arrangements particularly with the original Multodisc which had a 'solid centre'. This method is perfectly suited however for rings of discs with a 'hollow centre', and particularly when the method is adapted for use with one of the rotary plating devices (p.85).

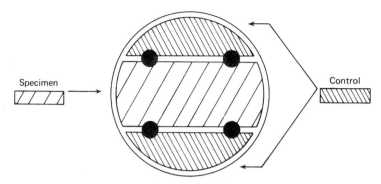

Figure 11.1 Controlled single disc susceptibility method. (From *Clinical Bacteriology*, 6th edn, 1987, (E. J. Stokes and G. L. Ridgway), Edward Arnold, London, reproduced by permission of the authors and publisher)

In the rotary plating technique a suitable inoculum of the test organism or specimen is placed in the centre of the plate and as it is rotated by the machine it is spread with a sterile swab to cover an area indicated by the template inscribed on the carrier of the machine, or by the right-angled metal guide which can be attached. In this way the gap for the discs is left between the central inoculum of the test organism and the control organism which is streaked in the same way on the periphery of the plate. The result is read as before.

One of the disadvantages of multiple disc sets is that the user is unable to change the combination of antimicrobial agents on any given set to meet the extra demands that are not unusual in the diagnostic laboratory. This means that a variety of disc sets must be kept, and from time to time unnecessary antibiotics are tested in order to obtain the result needed from an antibiotic that happens to be in the same set. Apart from any other consideration this is clearly not very cost-effective. For this reason greater flexibility is obtainable when the rotary plating technique is carried out using single discs of which many varieties are available. The manual application of the discs is an irksome chore, however, when large numbers of tests are necessary. The operation may be mechanized. Useful equipment for this purpose includes the Manual Disc Dispenser Mark 2 and the fully mechanized Discamat machine, both manufactured by Oxoid. Both use cartridges of individual discs which can be changed easily to give a variety of combinations.

When a manual dispenser is used, the central knob is depressed and the discs are released to lie above holes in the base which are just too small to allow their complete passage. When placed over a suitable plate of medium small metal rods descend upon the discs and press them into the surface of the medium, thus ensuring firm contact with the medium so that they will not fall into the lid when the plate is inverted. The Discamat works on very much the same principle, except that the machine is worked electrically and the discs are placed in position pneumatically.

The Stokes method can also be employed with paper strips instead of discs when a number of strains is to be tested against the same antibiotic. The test organisms are streaked across the plate and the strip is placed at right angles to them. The control organism is streaked across the plate in the same way. The antimicrobial may also be pipetted as a suitable solution melted in agar filled into cups or ditches cut into the medium.

Interpretation of results
Zones of inhibition produced around antimicrobial discs or strips with the test organisms must be compared with those produced by the appropriate control organisms. When zone sizes are clearly not the same to the naked eye it is necessary to measure them with calipers or a transparent rule, or with a commercial zone reader. Bearing in mind the relative densities of inocula, susceptibilities may be reported with reasonable limits, as follows.

Measure the distance from the edge of the disc to the edge of the zone of both test and control organisms.

If that of the test organism is larger than, equal to, or less than 3 mm smaller than that of the control, report as *susceptible*.

If that of the test organism is more than 3 mm but smaller than that of the control by more than 3 mm, report as *moderately susceptible*.

If that of the test organism is less than 2 mm report as *resistant*.

In this context a report that says than an organism is susceptible to a particular antimicrobial agent implies that the infection should respond to treatment with normal doses under normal circumstances. A report that says that an organism is moderately susceptible or relatively resistant implies that it might respond to treatment with high dosage. An organism which is reported resistant is unlikely to show any clinical response at all.

Why does the laboratory sometimes 'get it wrong?'

Despite all the very best efforts of the investigating laboratory using the most accurate methods possible, there are still occasions when an organism which is seen to be susceptible to a give antimocrobial agent in the laboratory causes an infection which fails to respond to treatment in spite of the patient receiving the appropriate dosage by the correct route. This is particularly difficult for a clinician to understand. There are several reasons why it may happen.

The most obvious of these, at least to the laboratory worker, is that for many reasons the antimicrobial may not have been absorbed properly and is thus not achieving a therapeutic level in the patient's tissues. It may also be that the agent is not able to reach the site of infection, particularly in the cases of osteomyelitis or lung abscesses. In extensive and contaminated wounds such as those caused by road traffic accidents or war injuries, detritus may provide a shield behind which bacteria may survive. It may also be that the laboratory has chosen an irrelevant organism for the susceptibility test. This may well happen when a specimen (e.g. sputum) has been badly taken, or has been delayed in warm conditions before reaching the laboratory. When this happens commensal organisms may outnumber the pathogen, and may be tested as the predominant flora. On occasions the original organism may be eliminated, only to be replaced by another before the clinician is able to detect any clinical change. Despite all efforts by the laboratory there are examples of the most careful culture techniques failing to demonstrate the presence of a pathogenic organism at all.

Occasionally the opposite situation occurs. An organism which has been assessed as being resistant is apparently eliminated by an antimicrobial agent to which it should, in theory, not respond. It is possible that in this case the patient was not infected at all but merely 'colonized'. It is also true to say that organisms do not always behave in the human body in exactly the same way as they do in the laboratory. A classic example of this is shown by typhoid bacilli which are present in blood cultures from patients who have been given tetracycline, but fail to grow, although therapeutically tetracycline may not be used successfully in the treatment of the disease. It is, of course, also true that many infections resolve spontaneously without chemotherapy.

Further information

Further detailed information about antimicrobial susceptibility testing may be found in the books by Garrod *et al.* (1981), Stokes and Waterworth (1972), Lorian (1986) and Stokes and Ridgway (1987).

Assay of antimicrobial agents in body fluids

It is necessary to ascertain the concentration of antimicrobials in the body fluids to ensure that the levels are adequate to deal with the infecting organism but not so high that the patient will suffer ill effects, e.g. with aminoglycosides that there is no damage to the eighth nerve in patients with renal problems.

Traditionally antimicrobial agent assays have been performed by either tube dilution or diffusion methods. As with susceptibility tests, however, many other methods of performing assays have been demonstrated since publication of the last edition of this book, and the most useful of these, in our judgement, will be considered below.

Collection of specimens

If proper assessment of the results is to be made it is essential that the laboratory knows how much time has elapsed between the administration of the last dose of the agent and the collection of the specimen. Blood samples for assay are usually taken just before a dose of the antimicrobial is due to be given. The laboratory must be told if the patient is receiving any other antimicrobial. Blood from a patient with renal insufficiency may contain antibiotic many days after the last dose was given.

Choice of organisms

A list of organisms, with their NCTC numbers, is given in Table 11.2.

Stock solutions and standard dilutions of antimicrobial agents

These may be prepared in water, phosphate buffer at a suitable pH or in dimethyl sulphoxide, although one or two antimicrobials require special solvents. Standard solutions for use in serum assays should be made in pooled human serum containing no antimicrobials, but when this is not available horse serum may be used. The pH of standard solutions is important as it affects the activity of many antimicrobials, especially tetracycline and the aminoglycosides. Standard solutions and dilutions of the test specimen should be prepared in the same buffer solution. The optimum pH for various antimicrobials is given in Table 11.2.

Table 11.2 Appropriate organisms for assaying antibacterial drugs

Drug	Organism	NCTC number	Optimum PH
Penicillins Cephalosporins }	B. subtilis S. lutea Staph. aureus	8236 8340 6571	
Carbenicillin Ticarcillin }	Ps. aeruginosa	10490	6.8
Streptomycin	B. subtilis Staph. aureus	8236 6571	
Amikacin Gentamicin Kanamycin Tobramycin }	B. subtilis Kl. edwardsii	8236 10896	7.8
Tetracycline	B. cereus Staph. aureus	10320 6571	6.6
Chloramphenicol	S. lutea	8340	
Erythromycin Clindamycin }	B. subtilis Staph. aureus	8236 6571	7.8
Fusidate	C. xerosis	9755	6.6
Vancomycin	B. subtilis Staph. aureus	8236 6571	7.8
Trimethoprim Sulphonamides }	B. pumilis	8241	7.3
Antifungal	Sacch. cerevisiae C. albicans	10716	7.2

Reproduced from p. 497 of *Antibiotics and Chemotherapy*, 5th edn, L. P. Garrod, H.P. Lambert and F. O'Grady, (1981). Churchill-Livingstone, London, by permission of the authors and publisher.

pH of specimen

When antimicrobials are being assayed in urine the pH of the specimen must be brought to the optimum for the antimicrobials concerned. Any dilutions should be prepared in the buffer used for the standard solutions.

Diffusion methods

With most antimicrobials a linear relationship exists between the diameter of the zone of inhibition and the logarithm of the concentration of the antimicrobial producing the zone. Consequently, it is usual to plot the diameters of zones of inhibition given by standard solutions of antimicrobials on semilogarithmic graph paper and then to use this graph to read the concentration of antimicrobials in the test specimen by finding the zone of inhibition it produces with the assay organisms under exactly similar conditions.

The test fluid and the standard solutions are applied either to preseeded or to surface-flooded plates. When very high levels of antimicrobial are present the specimen should be suitably diluted, as the zones of inhibition given by the specimen should ideally lie within the range given by the standard antimicrobial solutions.

The medium also affects results. It should be chosen according to the antimicrobial to be assayed.

The physical conditions of the procedure may need varying to provide optimal conditions for the antimicrobial.

The depth of the medium will affect zone sizes in any test where the antimicrobial is applied to the surface of the medium. The thinner the medium the more susceptible the reading is likely to be. The susceptibility of any test may be increased by pre-diffusion of the antimicrobial and also by incubation at 30°C. The density of the inoculum is important as a heavy inoculum will decrease the susceptibility of the test.

Plate diffusion method

No special medium or apparatus is needed. Holes are punched in nutrient agar with sterile cork borers. For small amounts of test serum the holes should not be more than 2.5 mm deep but with holes 8–9 mm diam. smaller amounts of antibiotic may be detected than is possible with any comparable diffusion.

This method, however, is not technically as simple as it seems. Considerable practice is required, particularly in cutting the holes and filling them with solutions and specimens if consistent, reproducible results are to be obtained.

Test organism

It is now customary to give patients several antimicrobial agents together and this invalidates the results of many assay methods. The organism recommended for this method is a *Klebsiella* sp. (NCTC 10896) which is resistant to nearly all antimicrobial agents except aminoglycosides and is thus seldom affected when another agent is present in the specimen.

Seeding the medium

Melt 100 ml of the medium and cool it to 48°C. Add a suspension of the indicator organism to give a concentration of about 10^5 cells/ml. Mix well and pour into a 243×18 mm petri dish. Avoid bubbles. Dry the dishes at room temperature for 1 h.

Alternatively store poured unseeded plates in plastic bags at 4°C and seed when required by making a lawn. Flood with a suspension of organisms containing about 10^6 cells/ml. Remove excess suspension with a pasteur pipette.

Test

Use pure antimicrobial agent of known potency (the simple mass of a substance does not always indicate its activity). Prepare five standard use-dilutions in human or horse serum which is known to have no inhibitory action on the indicator organism.

Make holes in the medium with an appropriate (sterile) cork borer and lift out the plugs without disturbing the surrounding medium, otherwise the solutions may seep under it and give irregular zones.

Fill the holes completely with the standard solutions and test serum or dilutions of it. Do all tests in triplicate and (ideally) include a quality control specimen of known activity.

Randomize the order of distribution (a template for this purpose is given by Stokes and Ridgway (1987) and is reproduced in Figure 11.2).

Incubate at 37°C for 18 h.

1	2	3	4	5
6	7	8	9	10
11	12	13	14	15
16	17	18	19	20
21	22	23	24	25
26	27	28	29	30

Standards (e.g. gentamicin assay)	*Well numbers*
1 mg/litre	1, 13, 25
2 mg/litre	7, 20, 21
4 mg/litre	5, 12, 23
8 mg/litre	4, 15, 22
16 mg/litre	3, 19, 27

Tests		
	a	6, 17, 28
	b	2, 14, 26
	c	8, 11, 30
	d	9, 16, 24
	e	10, 18, 29

Figure 11.2 Random code for 30-hole assay plate (From *Clinical Bacteriology*, 6th edn, 1987, (E. J. Stokes and G. L. Ridgway), Edward Arnold, London, reproduced by permission of the authors and publishers)

Reading

Measure the zones of inhibition with calipers. Plot the average diameters of the zones of inhibition of the standard solutions against the logarithm of the concentration of antimicrobial agent on semilogarithmic graph paper.

Measure the zones of inhibition of the test serum and use the average to read the concentration of antimicrobial agent from the graph. This method may provide an answer for aminoglycosides after incubation for about 4 h.

The newer methods of antibiotic assay

There are now a number of ways of performing antibiotic assays by mechanized or automated methods. These eliminate many of the shortcomings of biological assay methods, and in most cases they greatly reduce the time taken to perform the tests. The most important are described below.

High pressure liquid chromatography (HPLC)

In HPLC, the mobile phase, a liquid solvent carrying the mixture to be separated, is pumped at high pressure through an immobile stationary phase held in a relatively small-bore column where separation occurs. The individual compounds are then detected, often by absorption. They are characterized by the time they take to elute from the system (relative retention time) and are quantified by the area of height or peaks on a chart recorder, compared with known standards. Standard sera are included.

HPLC is almost universally applicable in drug assay work but it requires expensive equipment. It is precise and rapid and although of little use in clinical laboratories it can quantify the three major components of gentamicin, which is important in pharmokinetic studies.

Enzyme multiplied immunoassay technique (EMIT)

EMIT is a non-isotopic homogeneous immunoassay procedure which is largely used for aminoglycosides.

The antimicrobial is linked covalently to bacterial glucose-6-phosphate dehydrogenase in such a way that enzyme activity is retained. When gentamicin binds to this complex, enzymic activity is inhibited. In the test, free gentamicin in the sample competes with the conjugate for a limited amount of antibody: the more gentamicin, the greater the enzymic activity. Sample, conjugate, antibody and substrate are mixed and the enzymic activity is measured kinetically in a spectrophotometer. As the enzyme acts on the substrate NAD is reduced to NADH, the UV absorbance of which at 340 nm is greater than that of NAD, so the rate of increase of absorbance indicates the rate of the enzyme reaction. Human glucose-6-phosphate dehydrogenase does not interfere as it requires NAP, not NAD.

The method can be automated and is rapid. For full details see Anderson (1987).

Abbott TDX polarization fluoroimmunoassay system

This combines competitive protein binding with fluorescence polarization to give direct measurement without the need for prior separation.

The label used is the fluorochrome fluorescein. In the fluorimeter the fluorochrome is illuminated with polarized light of a particular wavelength. The change in the angle of polarized fluorescent light emitted by the fluorescein is measured in an optical system that employs a liquid crystal polarizer. This change reflects the binding of the tracer to the antibody. Tracer not bound to antibody emits non-polarized fluorescence.

Substrate-labelled fluoroimmunoassay (SLFA: Ames TDA)

This resembles EMIT in being an enzyme immunoassay but differs in that the labelled drug reagent is an enzyme substrate, not an enzyme, and the amount of degraded substrate after incubation for a fixed time with the enzyme is measured with a fluorimeter.

The substance to be assayed is chemically labelled with a derivative of the fluorogen umbelliferyl-β-D-galactoside. This is not itself fluorescent but when hydrolysed by β-galactosidase the product fluoresces. When specific antibody to the substance being assayed is introduced it will prevent the hydrolysis.

In the actual test there is a fixed amount of fluorogenic drug reagent, a limiting amount of antibody, the sample, and β-galactosidase. The assay substance competes with the fluorigenic reagent for antibody-binding sites and any reagent that does not bind to antibody is hydrolysed by the enzyme. Therefore the degree of fluorescence is proportional to the amount of substance being assayed.

References

Anderson, A. (1987) Emit homogeneous enzyme immunoassay: past, present and future. In *Immunological Techniques in Microbiology* (ed. J. M. Grange, A. Fox and N. L. Morgan) Society for Applied Bacteriology Technical Series No. 24, Blackwells, London, pp. 211–216

Bauer, A. W., Kirby, W. M. M., Sherris, J. C. and Turck, M. (1966) Antibiotic susceptibility testing by a standardized single disc method. *American Journal of Clinical Pathology*, **45**, 493–496

Ericsson, H. M. and Sherris, J. C. (1971) Antibiotic sensitivity testing. Report of an International Collaborative Study. *Acta Pathologia Microbiologia Scandinavica*, Supplement No. 217, Section B

Felmingham, D. and Stokes, E. J. (1972) Sterile swabs as essential equipment for rapid, reliable antibiotic sensitivity testing. *Medical Laboratory Technology*, **29**, 189–200

Garrod, L. P., Lambert, H. P. and O'Grady, F. (1981) *Antibiotics and Chemotherapy*, 5th edn, Churchill-Livingstone, London, pp. 490 and 497

Lorian, V. (ed.) (1986) *Antibiotics in Laboratory Medicine,* 2nd edn, Williams and Wilkins, Baltimore

Milne, L. M., Curtis, G. D. W., Crow, M. and Krakk, W. A. G. (1987) Comparison of culture media for detecting methicillin resistance in *Staphylococcus aureus* and coagulase negative staphylococci. *Journal of Clinical Pathology*, **40**, 1178–1181

Stokes, E. J. and Ridgway, G. L. (1987) *Clinical Microbiology,* 6th edn, Edward Arnold, London, pp. 213 and 236

Stokes, E. J. and Waterworth, P. M. (1972) *Antibiotic Sensitivity Tests by Diffusion Methods,* Broadsheet No. 55 (Revised), Association of Clinical Pathologists, London

Chapter 12

Bacterial food poisoning: food-borne disease

Before commencing investigations into outbreaks of suspected food poisoning, obtain the following information:

(1) number of people at risk, i.e. how many ate the suspected food;
(2) number of persons actually ill;
(3) nature of illness: (a) vomiting, (b) diarrhoea, (c) nausea, (d) headache, (e) disturbance of the central nervous system;
(4) exact time food was eaten and exact time of onset of symptoms;
(5) description of *all* food eaten in 24 h before the onset of symptoms, and as much information as possible about the food consumed during the previous 4 days.

This information will indicate whether the incident merits laboratory investigation. A very small number of people alleged to have been ill when a very large number is at risk is probably not an outbreak of food poisoning. Similarly, stories of single cases of poisoning due to foreign bodies, canned fruit, fresh vegetables, unusual articles of diet or food which has 'gone off' can usually be discredited. In nurseries or institutions, a wide variation in the times of onset usually suggests viral gastroenteritis rather than bacterial food poisoning. When symptoms appear 30 min or so after eating, chemical rather than bacterial poisoning is indicated. It may not necessarily be the last meal which causes bacterial illness.

The information will suggest one of the types of food poisoning described below and thus indicate the bacteriological examinations required (Table 12.1).

Types of food poisoning

Salmonella food poisoning

This is due to *infection* with a salmonella organism and usually the food must contain several million organisms to initiate infection. Diarrhoea and vomiting occur from 12–36 h after eating infected food. Illness may last from 2–7 days. Not all the persons who ate the food will develop symptoms; some may become symptomless excreters. Foods to suspect include poultry, meat products, eggs and egg products. *S. enteritidis* (phage type 4) is currently causing much concern. These foods become infected from animal or human intestinal sources, directly or indirectly. The salmonellas responsible for enteric fever *(S. typhi* and *S. paratyphi*

169

Table 12.1 Food poisoning: agents, symptoms and likely foods

Incubation period	Organisms	Vomiting	Diarrhoea	Cramps	Fever	Duration	Examples of foods
15 min – 4 h	B. cereus (emetic)	+++	(+)[a]	(+)[a]	–	short	Rice
4 – 6 h	S. aureus	++	+	++	–	6 – 12 h	Cold meats
							Dairy produce
8 – 22 h	C. perfringens	–	++	++	–	12 – 24 h	Reheated meat-based products
8 – 48 h	V. parahaemolyticus	–	+++	–	+	2 – 3 days	Seafoods
8 – 12 h	B. cereus (diarrhoeal)	–	++	++	–	1 – 2 days	Soups, milk products
12 – 48 h	Salmonella	(+)	++	=	+	48 h – 7 days	Poultry, meat, duck eggs
24 – 72 h	C. botulinum	–	–	Disturbed CNS	–	3 days	(Home) canned/bottled vegetables, meat, fish
2 – 11 days	Campylobacter	-	+++	++	++	3 days – 3 weeks	Milk, water, ? poultry

[a](+) may or may not be present

A, B and C) do not usually infect animals. They are transmitted from person to person by the faeces-food-oral route.

Vibrio parahaemolyticus food poisoning

This is rare in the UK but it is one of the commonest causes of food poisoning in Japan and has been incriminated in outbreaks in the USA and Australia.

It is closely associated with eating raw and processed fish products. The organism has been frequently isolated from sea foods, particularly those harvested from warm coastal waters. Symptoms of infection may appear from 2 to 96 h after ingestion, depending on the size of the infecting dose, type of food and acidity of the stomach. They vary from acute gastroenteritis with severe abdominal pain to mild diarrhoea.

Escherichia coli food poisoning

Large numbers of E. coli are present in the human and animal intestines. They are often found in raw foods where they may indicate poor quality. At least four different groups (p.263) may be involved in food poisoning. E. coli may cause travellers' diarrhoea when the source may be contaminated food or water.

Staphylococcal food poisoning

Staphylococcus aureus produces several enterotoxins, labelled A–E which will withstand heating at 100°C for 30 min. The toxin, present in food in which the organisms have grown, causes an *intoxication* type of disease with onset in 4–6 h and symptoms of diarrhoea and vomiting lasting 6–8 h. Foods likely to be infected are those which contain a high proportion of salt, meat such as ham, synthetic creams, sauces (particularly Hollandaise) which are never heated above 110°F (40°C), custards, trifles and occasionally fresh fish or canned vegetables (due to leaker spoilage). Infection occurs usually in the kitchen from infected cuts and abrasions, boils and other lesions or from nasal carriers.

Clostridium perfringens (welchii) food poisoning

Our knowledge of the role of C. perfringens in food poisoning has altered in recent years. Originally, only C. perfringens forming spores resistant to boiling for 1 h was thought to cause this type of food poisoning. These were invariably non-haemolytic. Recently, both haemolytic and non-haemolytic heat-sensitive strains have been incriminated in food poisoning outbreaks. The incubation period varies from 8–22 h. Young and long-stay patients and hospital staff tend to become colonized with C. perfringens. Counts of 10^5/g in the faeces of sufferers are not uncommon.

Because the heat resistance of C. perfringens spores varies, the organisms survive some cooking processes. Much depends upon the amount of heat that reaches the cells and the period of exposure. If the spores survive and are given suitable conditions, they will germinate and multiply. It is a common practice to cook a large joint of meat, allow it to cool, slice it when cold and re-heat it before serving. This is hazardous unless cooling is rapid and re-heating is thorough and above 80°C. Growth of heat-sensitive strains in cooked food may be the result of too low a cooking temperature or may be due to recontamination after cooking. The usual vehicles for 'welchii' food poisoning are meat and poultry. Stews, made-up meat dishes and large deep-frozen birds are suspect. *C. perfringens* is commonly found in soil; small numbers are normally present in human faeces and in some foods. The symptoms are abdominal cramps and diarrhoea, lasting for 12–24 h, and are caused by the production of an enterotoxin.

Bacillus cereus food poisoning

Cases of food poisoning associated with eating fried rice have been reported from Scandinavia, the USA and Europe. Large numbers of *B. cereus* were isolated from the rice.

Rice for fried rice dishes is often boiled in bulk and left to cool at room temperature. Refrigeration causes the grains to stick together. Before serving, egg is added to the cooked rice, which may have been standing in a warm kitchen for many hours. The mixture is then lightly cooked in a frying pan and served. During the initial cooking, vegetative organisms are destroyed; during standing, spores germinate in ideal conditions for growth. Subsequent heating does not destroy any toxins that may have formed.

B. cereus food poisoning can be confused with staphylococcal or perfringens food poisoning. There are two types, diarrhoeal and emetic.

	Diarrhoeal	Emetic
Incubation	8–12 h	15 min–4 h
Symptoms	diarrhoea, nausea	vomiting (v.common)
		diarrhoea (common)
Duration of illness	12–24 h	12–48 h
Isolation from faeces	sometimes (low numbers)	sometimes
Isolation from vomit	—	large numbers
Nature of toxin	protein	small peptide

The emetic type is more common. Occasionally other *Bacillus* spp. may be responsible for similar symptoms. See Gilbert *et al.* (1981).

Botulism

This form of intoxication, caused by the pre-formed toxin of *Clostridium botulinum*, is rare in the UK, but more prevalent in Europe and the USA where home canning of meat and vegetables is common practice. Between 12 and 36 h after eating food containing the toxin, patients develop symptoms including thirst, vomiting, double vision and pharyngeal paralysis, lasting 2–6 days and usually fatal. Vehicles are usually home-preserved meat or vegetables, contaminated with soil or animal faeces and insufficiently heated. Correct commercial canning processes destroy the organisms. Conditions must be optimum for toxin formation: complete anaerobiosis, neutral pH, absence of competing organisms.

Recently there have been several reports (from outside the UK) of botulism caused by the Type E organism that is usually associated with fish and fish products. This type is frequently present in estuarine mud.

Campylobacter food poisoning

These organisms have been growing steadily in importance since 1977 and are now the most common cause of gastroenteritis. They are not a new cause of infection but efficient methods for their isolation have demonstrated their ubiquity. Campylobacters are part of the normal flora of many mammals and birds. Poultry appear to be responsible for some sporadic outbreaks but outbreaks associated with inadequately pasteurized milk have been reported. Water supplies are also a source of infection. The organisms are heat sensitive and, in common with salmonellas, control measures such as adequate cooking and prevention of recontamination are indicated. They do not normally multiply in food, except possibly in vacuum packs.

The incubation period has been estimated to be between 2 and 11 days. Symptoms commonly include abdominal pain, diarrhoea, fever, headache, malaise, musculoskeletal pain, rigors and delirium. Vomiting occurs in less than 30% of cases. Sometimes 'flu-like' symptoms are present.

Yersinia enterocolitica food poisoning

This is an uncommon cause of food poisoning but its ability to grow at 4°C makes it unusual and a potential hazard. The symptoms are diarrhoea, fever, abdominal pain and vomiting.

Listeriosis

This food-borne disease, caused by Listeria monocytogenes, affects mainly young children, pregnant women and the elderly. The organisms are present in small numbers in many foods, but are unlikely to cause disease unless they have the opportunity to multiply. This may happen at quite low temperatures, e.g. in improperly maintained cold storage cabinets and refrigerators.

Methods for the examination of foods are given below.

Mycotoxins and aflatoxins

Certain fungi, mainly Aspergillus spp. form toxins in nuts and grains during storage. These toxins can cause serious disease in man and animals. The toxins are detected chemically and commercial kits are available. For information about these toxins see Moss et al. (1989).

Dose of infecting organisms

To cause food poisoning, a sufficiently large number of organisms must contaminate the food, which must be physically and chemically suitable for their growth. Enough time must elapse between contamination of the food and its ingestion to enable the organisms to form a large enough population to cause disease.

The important factors to consider in investigating infected foods are: source of organism; nature of food, e.g. moisture content; temperature of food; time between preparation and serving.

Examination of pathological material and food

Do a total viable count on all foods suspected of causing food poisoning. In most cases, this will be high. Relate findings to conditions of storage after serving. If possible, also count the number of causative pathogens.

Salmonella infections

Plate the stools of patients, suspected carriers and food handlers on DCA and bismuth sulphite medium and inoculate selenite medium. Isolation from food and identification of these bacteria is described in Chapter 21. Salmonellas in foods may also be detected by an ELISA technique (Clayden et al., 1987).

V. parahaemolyticus infections

Plate stools on thiosulphate citrate bile salt sucrose agar (TCBS) and inoculate alkaline peptone water (APW). Incubate overnight at 37°C. Subculture enrichment cultures on TCBS.

Homogenize 10–25 g of food in 0.1% peptone water solution: make direct cultures on TCBS and add homogenate to an equal volume of double-strength APW. Make tenfold dilutions in single-strength concentrations of these media. Incubate and plate out as for faeces. Methods for the identification of V. parahaemolyticus are given in Chapter 20.

Staphylococcal food poisoning

There is no need to examine the stools of all the patients; a sample of 10–25% is sufficient. Inoculate Robertson's cooked meat medium containing 7–10% sodium chloride (salt-meat medium) with 2–5 g of faeces or vomit and incubate overnight and plate on blood agar or one of the special staphylococcal media. Examine all food handlers for lesions on exposed parts of the body and swab these as well as the noses, hands and fingernails of all kitchen staff. Swab chopping boards and utensils which have crevices and cracks likely to harbour staphylococci. Use ordinary pathological laboratory swabs; cotton wool wound on a thin wire or stick, sterilized in a test-tube (these swabs can be obtained from most laboratory suppliers). Plate on the solid medium and break off the cotton wool swab in cooked meat medium. Methods for the identification of staphylococci are given in Chapter 24.

It is important that all strains of S. aureus isolated from food handlers, implements, foodstuffs and patients are sent to a staphylococcal reference expert. Not all strains of S. aureus can cause food poisoning and some of the other strains are usually encountered in this kind of investigation. A test kit for the detection of S. aureus enterotoxin in foods is available from Oxoid (Berry et al., 1987).

Examination for Clostridium perfringens

Examine the stools from a proportion of cases (10–25% if the outbreak is large). Heat-sensitive C. perfringens are found in almost all stools from normal people but large numbers may be demonstrated if the individual is a victim of C. perfringens food poisoning. A quantitative or, more practically, a semiquantitative examination is therefore necessary. Make a 1:10 suspension of faeces in nutrient broth and add 1 ml to each of two tubes of Robertson's cooked meat medium. Heat one tube at 80°C for 10 min and then cool. Incubate tubes at 37°C overnight and plate on blood agar with and without neomycin and on egg yolk agar. Place a metronidazole disc on the streaked-out inoculum. Incubate at 37°C anaerobically overnight.

Emulsify faeces in ethanol (industrial grade) to give a 50% suspension. Mix well and stand for 1 h. Inoculate media and incubate as above. See also Chapter 28.

Outbreaks involving both heat-sensitive and heat-resistant strains have occurred. Prepare a 10% suspension of food in 0.1% peptone water. Make 10^{-1}, 10^{-2} and 10^{-3} dilutions and inoculate all four on neomycin blood agar plate using the Miles and Misra technique (p.131). Incubate as for faeces. There is no point in heating cooked foods because spores will have germinated. Cooked meat medium containing 70–100 µg/ml of neomycin sulphate should also be inoculated because *C. perfringens* may be present only in small numbers in cooked foods.

All strains isolated from food poisoning outbreaks should be sent to a reference laboratory.

It is pointless to examine stools of food handlers and 'contacts' in outbreaks of *C. perfringens* food poisoning as the disease is not spread by human agency.

Test kits for the detection of *C. perfringens* enterotoxin are available (Oxoid) and are useful in food poisoning investigations (Berry *et al.*, 1987).

Examination for *Bacillus cereus*

Soak 20 g of material in 90 ml 0.1% peptone water for 50 min at room temperature. Add a further 90 ml of 0.1% peptone water solution. For other materials add 10 ml or 10 g to 90 ml diluent. Homogenize in a Stomacher. Prepare further dilutions and plate 0.1 ml amounts on *B. cereus* selective agar. Incubate at 37°C for 24 h then at 25–30°C for a further 24 h. This medium shows egg yolk reaction, mannitol fermentation and allows sporulation to occur. (The heat resistance of spores grown on laboratory media is lower than in foods.) Do confirmatory tests if necessary (see Chapter 28).

Laboratory investigation of botulism

This is difficult. Methods are given on p.334. It is best to consult a reference expert.

Examination for campylobacters

Faeces
Inoculate blood agar base plus 7% lysed horse blood and campylobacter selective supplement. Incubate at 37°C in 10–15% CO_2 in hydrogen and nitrogen for 48 h.

Foods
These may have been frozen, chilled or heated, all of which may cause sublethal damage to the organisms.

Milk   Filter 200 ml through a cotton wool plug. Place the plug in a sterile jar containing 100 ml of campylobacter enrichment medium without antibiotics. Incubate at 37°C for 2 h. Then add selective antibiotics and incubate at 43°C for 38 h. Subculture from the surface layer to selective campylobacter agar and incubate plates at 43°C for 36–48 h under microaerophilic conditions.

Water   Filter 100 ml through a 0.45 µm membrane and place the membrane face down on selective campylobacter agar. Incubate at 43°C for 24 h under microaerophilic conditions. Remove the filter and reincubate plates at 43°C for 36–48 h.

Place the filter in 100 ml of enrichment broth and proceed as for milk.

Meat, poultry etc.   Add 10 g to 100 ml of campylobacter selective broth plus antibiotic supplements and incubate at 37°C for 2 h. Then continue incubation at 43°C and proceed as for milk.

ELISA techniques are now available for the detection of campylobacters in food (Fricker and Park, 1987).

Examination for enterococci

Plate suspect food on one of the media described on pp.305–306.

Examination for *Yersinia enterocolitica*

Prepare a suspension of faeces or food in M/15 phosphate buffer. Incubate some at 32°C for 24 h. Refrigerate the remaining suspension at 4°C for up to 21 days and subculture at intervals on yersinia selective agar. Media used for other enteric pathogens can be reincubated at 28–30°C for 24 h and examined for *Y. enterocolitica* but some strains of this organism are inhibited by bile salts and deoxycholate.

Detailed identification of food poisoning bacteria from food

Methods are described under the appropriate headings in this book.

For further information on food poisoning see Hobbs and Gilbert (1978), DHSS (1982) and Corry *et al.* (1982).

References

Berry, P. R., Weinecke, A. A., Rodhouse, J. C. and Gilbert, R. J. (1987) Use of commercial tests for the detection of *Clostridium perfringens and Staphylococcus aureus* enterotoxins. In *Immunological Techniques in Microbiology* (eds J. M. Grange, A. Fox and N. L. Morgan), Society for Applied Bacteriology Technical Series No. 24, Blackwells, Oxford, pp. 245–250

Clayden, J. A., Alcock, S. J. and Stringer, M. F. (1987) Enzyme linked immunosorbent assays for the detection of salmonellas in food. In *Immunological Techniques in Microbiology* (eds J. M. Grange, A. Fox and N. L. Morgan), Society for Applied Bacteriology Technical Series No. 24, Blackwells, Oxford, pp. 217–229

Corry, J. E. L., Roberts, D. and Skinner, F. A. (eds) (1982) *Isolation and Identification Methods for Food Poisoning Organisms,* Society of Applied Bacteriology Technical Series No. 17, Academic Press, London

DHSS (1982) *The Investigation and Control of Food Poisoning in England and Wales,* Memorandum 188 Med, HMSO, London

Fricker, C. R. and Park, R. W. A. (1987). Competitive ELISA, co-agglutination and passive haemagglutination for the detection and serotyping of campylobacters. In *Immunological Techniques in Microbiology* (eds J. M. Grange, A. Fox and N. L. Morgan), Society for Applied Bacteriology Technical Series No. 24, Blackwells, Oxford, pp. 195–210

Gilbert, R. J., Turnbull, P. C. B., Parry, J. M. and Kramer, J. M. (1981) *Bacillus cereus* and other bacillus species: their part in food poisoning and other clinical infections. In *The Aerobic Endosporing Bacteria: Classification and Identification* (eds R. C. M. Berkeley and M. Goodfellow), Academic Press, London, pp. 297–314

Hobbs, B. C. and Gilbert, R. J. (1978) *Food Poisoning and Food Hygiene,* 4th ed, Edward Arnold, London

Moss, M. O., Jarvis, B. and Skinner, F. A. (1989) *Filamentous Fungi in Foods and Feeds. Journal of Applied Bacteriology* Symposium Supplement No 18, 67, 1S–144S.

Chapter 13

Food – general principles

Although the bacteriology of food is under consideration by several EC committees (Codex Alimentarius Commission) there is still no general agreement on methods for laboratory examination. In the UK there are statutory methods for testing milk and officially recognized methods for the examination of water but as yet no official or even recommended methods for examining foods. The techniques used have evolved as a result of cooperation between microbiologists in the food industry and those in the public health services.

In the USA the American Public Health Association publishes standard methods (Speck, 1984) and the Association of Official Analytical Chemists produces a Bacteriological Analytical Manual (AOAC, 1978). There are also the publications of the International Commission on Microbiological Specifications for Food (cited elsewhere in the text).

Four kinds of microbiological investigations are usually undertaken in the bacteriological examination of food and drink:

(1) estimation of the viable count, i.e. the numbers of colony-forming units (cfu) of bacteria, yeasts and moulds;
(2) estimation of the numbers of coliform bacilli, *E. coli* (Enterobacteriaceae). This may indicate the standard of hygiene;
(3) detection of specific organisms known to be associated with spoilage. It is becoming apparent that this may be more important than the total or viable count;
(4) detection of food poisoning organisms.

There is no single method which is completely satisfactory for the examination of all foodstuffs. The methods selected and the examinations performed will depend on local conditions, e.g. availability of staff, space and materials, numbers of samples to be examined and the time allowed to obtain a result.

It may even be desirable in some circumstances to examine a large number of samples by a slightly suboptimal method than a smaller number by an exhaustive method.

Many organisms present in foods may be sublethally damaged, e.g. by heat, cold or adverse conditions such as low pH and moisture content, high salt or sugar content. On the other hand these conditions may favour the growth of yeasts and moulds. In this case mycotoxins might be present.

Standards

Several national and international committees are studying this subject. Criteria for any legislation will have to include protection of the public and a fair deal for the manufacturer and make reasonable use of laboratory staff, time, facilities and money. Tests must be standardized and reproducible. Standards must be set by people with laboratory experience. In practice, some degree of tolerance may be desirable, i.e. allowing a small proportion of samples to show deficiencies. In establishing any criteria for standards, it is vital to take account of the point of sampling. Products sampled in the factory should have a lower bacterial count than those samples at the point of sale (see Mossel, 1982).

Suggested standards are given where appropriate in this section.

Sampling

The usefulness of laboratory tests is largely dependent upon correct sampling procedures. A statistical sampling plan must be applied to each situation so that samples submitted to the laboratory are fully representative of the 'lot'. The choice of plan, whether it be a class attribute or random number scheme, will depend upon the known facts about the material, the degree of hazard to the consumer and economic feasibility. The subject is fully discussed in a publication of the International Commission of Microbiological Specifications for Foods (ICMSF, 1986). In factory sampling, it is more important to take a small number of samples at differing times during the day than to take many samples at one time. Where resources are limited, concentrate on the finished product but be ready to work back through a process taking 'in-line' samples if levels of microorganisms begin to increase. Use screw-capped aluminium or polypropylene containers, large mono-containers or polythene bags; never use glass jars where any breakages could contaminate the product or cause injury to personnel. Instruments for removing portions of bulk material vary according to the nature of the substance. For frozen samples, use a wrapped sterilized brace and bit or a clean chopper washed in methylated spirit and flamed. Take care to avoid cuts by ice splinters. Use individually wrapped and sterilized spoons, spatulas or wooden tongue depressors for softer materials. With prepackaged material, examine several complete packs as offered for sale.

Transport and storage of samples

Samples should be handled correctly before examination. They should be carried in insulated containers cooled with an ice pack. The ice should not have melted when the samples arrive at the laboratory. They should be transferred to a refrigerator pending examination.

Note the visual appearance of samples, and the odour, if any.

Total and viable counts

These are useful monitors in processing and may reflect poor handling or storage at retail level.

Methods of performing these counts are given in Chapter 9. No single medium or incubation temperature will give the whole picture and these must be selected according to known storage conditions of the food, for example for chilled meat and frozen fish, a temperature in the psychrophilic range is used. Many spoilage psychrophiles will not grow in the mesophilic range. For these, surface plating is best.

If the food has been refrigerated, the organisms may be 'cold-shocked' and may grow only on media that are enriched. Lower incubation temperatures, e.g. 15–20°C may be necessary. In general, 30°C for 48 h is satisfactory, but the incubation time must be extended if lower incubation temperatures are used. Similarly in heated foods, non-sporing mesophiles may be impaired and they also will need enriched media. For psychrophiles 1°C for 5 days is useful.

There is often no need to set up complete plate or other counts. If a standard is set, arbitrarily or according to experience, the technique can be modified to give a 'pass' or 'fail' response.

If, for example, an upper limit of 100 000 colonies/g under given conditions of medium, time and temperature is set, then a plate count method that uses 1 ml of a 1:1000 dilution will suffice. More than 100 colonies – easily observed without laborious counting – will fail the sample. Similarly, more than 100 colonies in roll-tubes containing 0.1 ml of a 1:100 dilution or more than 40 per drop in Miles and Misra counts from a 1:50 dilution will suggest rejection. Such counts must be replicated.

Counts from the surface (p.184) are useful for carcase meat and fish where the organisms are mostly superficial. Breed smears are good rough guides (see p.128). Anaerobes may be counted by plate or 'black tube' MPN methods (p.340). Mould hyphae and yeasts may be counted directly in a Neubauer or similar counting chamber which is deeper than the Helber model. Exclude foods which depend upon a microbial population for flavour, e.g. yoghurt.

Indicator organisms

Although tests for the presence of coliform bacilli in general, and *E. coli* in particular, are very useful it may be desirable to count all the Enterobacteriaceae present because some strains of *Citrobacter* and *Klebsiella* spp. are more heat resistant than *Escherichia* spp. and their presence is a better indication of inadequate heat processing. The presence of streptococci (enterococci) is also regarded by some bacteriologists as indicating suboptimal processing and storage.

General methods

(1) Weigh 10 g of material into a sterile container or on sterile greaseproof paper. Homogenize with 90 ml of diluent in a Stomacher or other mixer. But do not compare results obtained by different methods of maceration. Do viable counts and inoculate culture media with tenfold serial dilutions of this homogenate. Table 13.1 shows the amount of food in grammes contained in dilutions of the material. If blenders are not available, good homogenates are obtained if about 20 g of coarse sterile sand are added to the mixture before shaking.

(2) When examining vegetables or materials where the microorganisms are on the surface, weigh 50 or 100 g into a sterile jar and add 100 ml of diluent. Shake for 10 s, stand for 30 min, shake again and decant. The organisms are assumed to

Table 13.1 Dilutions for plate counts

	Initial	2nd	3rd	4th
Dilution	1:10 (10^{-1})	1:100 (10^{-2})	1:1000 (10^{-3})	1:10000 (10^{-4})
10 ml contains (g)	1	0.1	0.01	0.001
1 ml contains (g)	0.1	0.01	0.001	0.0001

Tenfold serial dilutions of an initial 10 g material + 90 ml of 0.1% peptone solution in water

be washed off and disturbed in the diluent, that is, 100 ml now contains the number of bacteria present on 50 or 100 g of the foodstuff.

Plate the homogenate or washings on the various media indicated on p.56 and also add 5–10 g of original food, chopped if necessary, to 50-ml tubes of glucose tryptone broth.

Coliform bacilli, *Escherichia coli* and enterobacteria

The choice of method lies between counting the numbers present in 1 g of food or simply the presence or absence of the organisms in 1 or 0.1 g.

Coliform count
Weigh 10 g of the sample on sterile greaseproof paper and homogenize it in a Stomacher in 90 ml of peptone water diluent. Add 1 ml of this 1/10 dilution to 9 ml of diluent to make a 1/100 dilution. Add 1 ml of each dilution to a petri dish and pour on 15 ml of melted violet red bile lactose agar cooled to 50°C. Allow to set and overlay with 10 ml of the same medium. Incubate at 37°C for 18–24 h and count the dark red colonies. Calculate the number of colony-forming units/g of the food.

Presence or absence of coliform bacilli
Add 10 ml of the 1/10 food suspension to 10 ml of double strength and 1 ml to 10 ml of single strength MacConkey (or similar) broth. Incubate at 37°C overnight and plate out on MacConkey agar to confirm that coliforms are or are not present in 1 or 0.1 g.

Escherichia coli counts
Attempts to detect *E. coli* at 44°C in primary cultures often fail. In richer media, however, they usually do grow, and as very few bacteria can produce indole at this temperature this property provides a useful but rough indicator of the presence of these organisms.

Add small samples, e.g. 0.1 and 0.5 g to 15-ml tubes of tryptone or similar nutrient broth. Incubate overnight and test for indole. A positive result is presumptive evidence of *E. coli*. Plate these cultures on MacConkey or similar medium to confirm.

Presence or absence of enterobacteria
This is a modification of the method of Mossel and Harrewijn (1972). Resuscitate damaged cells by incubating 100 ml of a 10% suspension of food in tryptone soya broth at 25°C for 2 h. Add 10 ml to 10 ml of double strength EE broth, 1 ml to 10 ml of single strength EE broth and 0.1 ml to another 10 ml of single strength EE broth. Incubate at 30°C for 18–24 h. With a straight wire, inoculate freshly steamed

deep tubes (15 ml) of violet red glucose agar. Incubate at 30°C overnight. Enterobacteriaceae usually give a line of growth down the centre of the medium, surrounded by a cylinder of purple precipitate. Do oxidase tests on the growth to exclude *Aeromonas* and identify, if necessary, as described in Chapter 21.

Report Enterobacteriaceae present or absent in 1, 0.1 or 0.01 g.

Counting enterococci

Do surface counts as described on p.132 using azide blood agar, MacConkey or Slanetz and Bartley medium.

Counting clostridia

As counting these organisms usually incurs identifying them as well the techniques are described under *Clostridium*, in Chapter 28.

Yeast and mould counts

Do plate and surface counts on malt or wort agar. Incubate at 25°C for up to 5 days.

Spoilage organisms

These, and the methods for culturing them, are given under the headings of the foods concerned.

Food poisoning bacteria

'Routine' examination of all foods for salmonellas and staphylococci, or other food poisoning organisms, is hardly worthwhile. Only those foods and ingredients known to be vehicles need be tested. They are given in Chapter 12 and under the appropriate headings in the following pages.

Yeasts and moulds

Sampling
Circumstances vary. Obtain as much 'background' information as possible before deciding on the method of examination. The number of samples should be as large as is practicable.

Sample whole packages where possible and examine the casing for damage and water staining.

Visual inspection
Open carefully and look for evidence of moulds. Microscopy is rarely helpful.

Direct examination
Pre-incubation may be useful (Jarvis *et al.*, 1983). Place a filter paper soaked with glycerol in a large petri dish and sterilize it. Suspend a sample above the filter paper, e.g. on glass rods, replace the lid and incubate for up to 10 days. Examine daily for moulds.

To estimate shelf-life relative to mould growth under adverse conditions incubate unopened packages in a controlled humidity cabinet.

Culture
Prepare a 1:10 suspension in peptone water diluent containing 0.1% Tween 80. Use a Stomacher if possible. Make serial dilutions in the same diluent and plate (in duplicate) on Rose Bengal chloramphenicol agar (p.56). Incubate at 22°C for 5 days.

References

AOAC (1978) *Bacteriological Analytical Manual,* 5th edn, Association of Official Analytical Chemists, Washington DC

ICMSF (1986) *Microorganisms in Foods. 2,* International Commission on Microbiological Specifications for Foods, Blackwell, London

Jarvis, B., Seiler, D. A. L., Ould, S. J. L. and Williams, A. P. (1983) Observations on the enumeration of moulds in food and feedingstuffs. *Journal of Applied Bacteriology,* **55,** 325–326

Mossel, D. A. A. (1982) *Microbiology of Foods. The Ecological Essentials of Assurance and Assessment of Safety and Quality,* 3rd edn, University of Utrecht

Mossell, D. A. A. and Harrewijn, G. A. (1972) Les défaillances dans certain cas des milieux d'isolement des Enterobacteriaceae des aliments et des médicaments. *Alimenta,* **11,** 29–30

Speck, M. (ed.) (1984) *Compendium of Methods for the Microbiological Examination of Food,* American Public Health Association, Washington DC

Chapter 14

Meat and fish

Fresh and frozen carcase meat

Carcase sampling

The skin or surface of a carcase is usually heavily contaminated. Gram-negative, motile bacteria show a greater adherence to it than do Gram-positive species. Although deep muscle is usually sterile the meat may have a higher pH if the animal has suffered stress and it may be contaminated.

The most reliable methods are destructive, involving maceration of weighed portions.

To sample the whole carcase by a non-destructive technique use the method devised by Kitchell *et al.* (1973). Sterilize large cotton wool pads, wrapped in cotton gauze in bulk and transfer them to individual plastic bags. Moisten a pad with a 0.1% peptone water and wipe the carcase with it, using the bag as a glove. Take a dry pad and wipe the carcase again. Place both pads in the same bag, then seal and label it. Add 250 ml of diluent to the pads and knead, e.g. in a Stomacher, to extract the organisms. Prepare suitable dilutions from the extract and do total counts at 20°C. Inoculate other media as desired.

The other areas on the carcase that are most likely to be contaminated after butchering are the rump, brisket and forelegs. Murray (1969) recommended swabbing a 16-cm^2 area on each part. With good hygiene, counts of less than 150 000 on the brisket, 50 000 on the rump and 25 000 on the forelegs per 16 cm^2 at 20°C may be achieved immediately after dressing and after 3 to 4 days in chill.

'Hot boning' is now becoming more common and this may give rise to different problems.

Routine processing of boned-out meat in the laboratory

It is important to take equal quantities of both fat and lean tissues. Deep tissue near the bone should be examined if bone taint is suspected. To avoid contamination during examination, sear or paint the surface with an antiseptic dye. If an overall picture is required, take core samples from boned-out joints, macerate, dilute and plate for aerobic and anaerobic culture. Express counts as the number of colonies/g of tissue. Most counts should be of the order of, or less than, 100 000 cfu/g.

Most spoilage organisms will be growing on the surface of the meat under aerobic conditions. Scrape a convenient area, e.g. 10 cm^2 with a sterile scalpel, shake the scrapings in 90 ml of warm diluent. Prepare dilutions, plate, incubate and count.

In a meat factory, where the hygiene is good, more than 70% of samples may be expected to have count of less than 1000 cfu/cm^2.

Microbial content

In moist chill conditions, the predominant spoilage organisms are *P. (Alteromonas) putrefasciens* and other pseudomonads, *Acinetobacter, Enterobacter and Brochothrix (Microbacterium) thermosphactum*. Anaerobes do not appear to be important except where meat is held at temperatures above 25°C, when *Clostridium* spp., notably *C. perfringens*, predominate. About 10% of pork carcases contain *C. botulinum*, 66% carry *C. perfringens*. Fortunately, *C. botulinum* is a poor competitor.

Salmonellas are often found in raw meats, in meat sold as pet-food, in imported frozen boneless beef and in horseflesh. Methods for isolating them are given on p.270.

In dry chill conditions, mould spoilage is possible in chilled beef (−1°C) and frozen mutton (−5°C or less) but moulds do not grow below −10°C. 'Green spots' are usually due to *Penicillium* spp., 'white spots' to *Sporotrichum* spp., 'black spots' to *Cladosporium* spp. and 'whiskers' to *Mucor* and *Thamnidium* spp.

Comminuted meat

This is fresh meat, minced or chopped and with no added preservative. Surface organisms are therefore distributed unevenly throughout the mass and further contaminations may occur during the process.

Take several random samples.

Viable counts
Do plate counts on dilutions 10^{-4} to 10^{-8}. Counts are usually very high, but standards must be set by experience. It may be desirable to estimate the number of coliforms present and to determine what proportion of these are *E. coli*. Surface counts, MPN and membrane filter methods may all be used, but the results may not be comparable.

Microbial content
This is similar to that in fresh meat. Salmonellas and staphylococci may be present.

Fresh sausages

A fresh sausage contains comminuted meat, cereals, spices and sulphur dioxide up to a limit of 450 ppm. This checks the growth of Gram-negative species.

This product should not be confused with European or American sausages, which are smoked.

Viable counts
Remove the casing, if present, and do total viable counts using dilutions from 10^{-4} to 10^{-8}. Incubate at 30°C for 48 h and at 22°C for 3 days. Counts of several million

organisms/g may be expected. Skinless sausages may have lower counts than those with casings as the process involves blanching with hot water to remove the casings in which they are moulded.

Sausage casings, whether natural (stripped small intestine of pig or sheep) or artificial, usually do not present any bacteriological problems.

Microbial content

Spoilage organisms are those found in fresh meat and also lactobacilli, coryneforms, microbacteria, micrococci, staphylococci (including *S. aureus*) and yeasts. *B. thermosphactum* is associated with souring. Gardner's medium is useful for the isolation and enumeration of this organism. Salmonellas may be present.

We have not found the use of surfactants particularly advantageous. Sometimes they are inhibitory.

Prepacked fresh meat

Meat in oxygen-permeable wrappers has a similar flora to that of unwrapped meat. In packs with non-permeable wrappers, carbon dioxide gradually replaces oxygen; pseudomonads are suppressed and enterobacteria, *Hafnia, B. thermosphactum* and lactobacilli predominate.

Meat pies

Meat pies may be 'hot eating', e.g. steak and chicken pies, or 'cold eating', e.g. pork, veal and ham pies and sausage rolls.

'Hot eating' pies

The meat filling is pre-cooked, added to the pastry casing and the whole is cooked again.

Viable counts

Examine the meat content only. Open the pie with a sterile knife to remove the meat. Do counts on dilutions of 10^{-1} and 10^{-2} and incubate at 30°C for 48 h. A reasonable level is not more than 100 cfu/g.

Clostridia

Add 1-ml amounts of the 10^{-1} dilution to 15-ml bottles of reinforced clostridial medium melted and at 50°C. Cool, seal with petroleum jelly and incubate at 37°C for 48 h. Examine tubes for blackening. Subculture any black tubes in purple milk. Plate out on blood agar, incubated anaerobically to identify any other clostridia.

'Cold eating' pies

These contain cured meat, cereal and spices that are placed in the uncooked pastry cases. The pies are then baked and jelly made from gelatin, spices, flavouring and water is added. Provided that the jelly is heated to and maintained at a sufficiently high temperature until it reaches the pie, there should be no problem.

Bad handling, insufficient heating and re-contamination of cool jelly can cause gross contamination by both spoilage organisms and pathogens.

The pies are cooled in a pie tunnel through which cold air is blown. At this stage mould contamination is likely if the air filters are not properly cleaned. Air in pie tunnels can be monitored by exposing plates of malt agar or similar medium. Bad storage may result in outgrowth of spores of *Bacillus* and *Clostridium* spp.

Viable counts
Examine only the filling. Do counts on homogenized material using dilutions of 10^{-1} and 10^{-2}. Incubate at 30°C for 48 h. A reasonable level is less than 1000 cfu/g.

Coliform counts
Examine meat and jelly separately. Use dilutions 10^{-1} and 10^{-2}.

Poultry

Battery rearing produces birds with very tender flesh but the close confinement necessary in this method of production often results in cross-infection. Inadequate thawing of frozen birds followed by relatively little cooking has, on several occasions, resulted in outbreaks of food poisoning.

Viable counts

For the best results divide the whole carcase into portions. Otherwise, sample the skin from under the wing and around the vent. Do total and coliform counts and examine for salmonellas, staphylococci, *C. perfringens* and possibly *Campylobacter*.

To examine a large number of birds, use swabs wound on 25-cm lengths of wire. Sterilize in large test-tubes. Use one for swabbing the outside of the bird and one for swabbing the body cavity. Rinse the swabs in 100 ml of peptone water diluent and do total and coliform counts by the Miles and Misra or membrane techniques. Inoculate other media as required.

In an attempt to establish advisory standards in the poultry industry, Murray (1969) examined a 16 cm^2 area of breast skin for colony count at 22°C, coliforms at 30°C, faecal streptococci and *S. aureus* at 37°C. He found that in a well run establishment total counts of less than 250 000/cm^2, coli counts less than 1000, faecal streptococci counts of less than 5000 and *S. aureus* counts of less than 100 per 16 cm^2 could be expected. Low total counts indicate few spoilage organisms and low coli counts indicate good hygiene. Few faecal streptococci suggest good evisceration technique and good hygiene.

To test for salmonellas examine the whole carcase. Swabbing or rinsing in a plastic bag is suitable. Collect the rinsings and add them to double strength enrichment media. Proceed as for salmonella isolation (p.270).

Total counts and examinations for pathogens may be made on slush ice in the spin chillers and on fluid in the plastic bags in which the birds are packed.

Cured meat

Cured raw meat

This is pork or beef that has been treated with salt and nitrite. The haemoglobin is altered so that the characteristic pink colour is produced when the meat is cooked. This is sold as bacon, gammon, cured shoulder, salt beef or brisket.

Viable counts
Do counts on dilutions of 10^{-2} and 10^{-3}. Incubate at 30°C for 48 h. Also do counts using diluent and media containing 4% of sodium chloride to detect halophiles. Incubate these for 3 days.

Inoculate salt meat broth with 0.1-g amounts, incubate at 37°C for 18 h and plate on phenolphthalein phosphate polymyxin agar (PPPA) or milk salt agar (MSA) medium to detect *S. aureus*.

Microbial content
Besides halophilic denitrifying bacteria, lactobacilli, micrococci, staphylococci, *B. thermosphactum* and moulds may be present under chill conditions. *Pseudomonas* spp. are inhibited by salt. Above 25°C, *C. putrefasciens* may be found. This organism gives a characteristic sweet/sour odour. 'Mild cure', 'sweet cure', 'tender cure', etc., are trade terms which may indicate some degree of heat or the addition of sweetening substances. The balance of the flora may be altered by such processes.

Bone taint
This may occur in carcase leg joints or in rib areas. Organisms usually responsible are: vibrios, micrococci, streptococci, proteus, clostridia, enterobacteria, alkaligenes and arthrobacter.

Vacuum packaging

Vacuum packaging is used for a variety of products in an effort to present them to the customer in as fresh a condition as possible.

Vacuum-packed bacon

The packs used for bacon are not usually permeable to oxygen because oxidation causes fading of the colour. The initial bacterial load, the salt content and storage temperatures have a marked influence on the shelf-life.

Low salt content (5–7%) spoilage of bacon gives a characteristic scented sour odour due to the action of lactobacilli, pediococci, streptococci and leuconostoc. High salt content (12%) spoilage gives a cheesy odour which is associated with micrococci or *B. thermosphactum*. Spoilage by enterobacteria or vibrios gives a sulphurous smell.

Viable counts
Swab the outside of the pack with alcohol and open with sterile scissors. Do counts as for cured raw meat.

Microbial content
The main groups of organisms found immediately after packaging are micrococci and coagulase-negative staphylococci. During storage at 20°C, lactobacilli, Group D streptococci and pediococci become dominant. Yeasts may also be found. If large numbers of Gram-negative rods are present it is usually an indication that the initial salt content was low.

Vacuum-packed cooked meats

These meats are invariably cured products and, as with bacon, the absence of oxygen is important. These products, however, always carry a risk of infection with *C. botulinum*. Spores of this organism may survive cooking and during storage germinate and grow to produce toxin in an otherwise sterile and oxygen-free environment. Type E is known to grow at low temperatures. Fortunately, the salt and nitrite contents combined with low-temperature storage help to reduce the danger. In addition, many producers pasteurize their vacuum-packed cooked meat products in the package. This allows for spore germination after the initial cooking and killing of the vegetative forms on pasteurization. Nevertheless, *C. botulinum* has been reported in vacuum-packed frankfurters and cooked ham.

Laboratory examination
Do surface counts on blood agar and on MacConkey agar for enterococci. These are useful organisms for assessing the efficiency of pasteurization. Total counts should be less than 1000 cfu/g.

Cured, cooked meats

These meats include ham, luncheon meats, brawns, tongues, salamis and continental-type sausages. Except in delicatessen stores, these are usually sold in vacuum packs.

Viable counts
Do plate counts on dilutions of 10^{-1} to 10^{-3}, and use medium.to which 5% of horse blood has been added immediately prior to pouring. This enables the streptococci that cause 'greening' to be counted in addition to the other organisms. Counts of 1000 or less/g are reasonable.

Inoculate salt meat medium with 0.1-g amounts and incubate at 37°C for 18 h. Plate on PPPA or MSA medium to detect *S. aureus*.

Microbial content
Contaminants are usually heat-resistant organisms, including spore-bearers, group D streptococci. Coliforms, micrococci and staphylococci may be introduced after heating. In canned hams, coryneforms, lactobacilli and yeasts may be found. Staphylococci may proliferate just beneath the casing.

Brines

Injection brine

This is prepared freshly for each batch of meat.

Sampling
Collect from storage tank, store at < 7°C before counting. Do total count using nutrient agar containing 4% NaCl, incubate at 22°C for 72 h.

Advisory standard (× 10³cfu/ml)

Good	< 0.5
Fair	0.5-1.0
Poor	1.1-5.0
V. poor	> 5.0

(Gardner, 1983)

Cover brine

Sampling
Sample when fresh and between batches of meat.

Direct microscopical count (DMC)
Use a Helber haemocytometer (p.128) and examine by dark field or phase contrast microscopy, or use the DEFT method (p.134).

Total viable count
Add 4% NaCl to the medium.

E. coli count
Dilute the sample 1:10 with 4% NaCl in 0.1% peptone and use the membrane filter technique.

Advisory standard

	DMC ($\times 10^6$ cfu/ml)	Total count ($\times 10^3$cfu/ml)	E. coli ($\times 10^3$cfu/ml)
Good	<50	<50	<1
Fair	50–100	50–100	1–10
Poor	101–150	101–500	11–100
V.poor	>150	>500	>100

(Gardner, 1983)

Pseudomonas and vibrio counts may also be a useful monitor of cover brine quality.

Fresh fish

The quality of raw fish is best assessed by appearance and odour. Counts are of limited value and the flora is mainly halophilic and psychrophilic.

Counts
These are of limited value, but if necessary sample the whole surface of the fish or fillet either by swabbing a defined area (see p.183) or by washing the sample with sterile 0.1% peptone + 1% NaCl in a plastic bag. Use the washings to prepare dilutions in salt peptone and do counts on Marine agar, or use a conductance

method (p.135). Incubate at 20°C for 5 days. Coliform counts are sometimes useful. Examination for salmonellas (p.270) and *V. parahaemolyticus* (p.250) may be desirable.

Surface slime is usually heavily infected. The count is increased by careless handling and contact with dirty ice and decreased by salting in barrels, hypochlorite and ice mixtures.

Culture
Inoculate glucose tryptone media, MacConkey media, and media containing 5–10% of sodium chloride for halophilic bacteria.

Inoculate PPPA, MSA or other media for staphylococci and also plate out the salt broths on these media. *S. aureus* is not uncommon in fish.

Microbial content
Large numbers of pseudomonas, flavobacteria, coryneforms, acinetobacter, aeromonas and cytophaga, often associated with slime, and micrococci may be found. Photobacteria, luminescent in the dark, are often present. The flesh count increases rapidly after filleting but in good plants may be as low as 100 000 cfu/g.

Smoked fish

Since 1976 there has been a marked increase in the incidence of scombroid fish poisoning (non-bacterial). The symptoms resemble those of food poisoning, i.e. diarrhoea and vomiting, headache, giddiness, rash on head and neck, 10 min to a few hours after ingestion of scomboid fish, e.g. mackerel, tuna, etc. Histamine compounds and some spoilage of the fish is always involved. The condition can result from eating canned fish.

Shellfish and raw crustaceans

These are estuarine animals and are therefore liable to gross faecal pollution. Oysters, which are eaten raw, are self-cleansing 'filter-feeders' and, if placed in tanks of chlorinated sea-water, will eliminate any enterobacteria from their bodies. Cockles, whelks, winkles, prawns and shrimps are boiled for retail sale, but this cooking is usually not controlled. The presence of *E. coli* is usually accepted as evidence of less than optimum hygiene conditions of cultivation or of insufficient heat treatment.

Roll-tube method
This is applicable to both shellfish and raw crustaceans.

Place oysters or mussels in the freezing compartment of a refrigerator overnight. This makes them easy to open and no fluid will be spilled. Scrub and clean the outside of the shell and rinse in boiled water. Hold with the concave shell down and open with a sterile oyster knife. Cut the frozen flesh into small pieces (to release intestinal contents) with a sterile scalpel and place it in a sterile 200-ml Pyrex measuring cylinder. Push the flesh down with a sterile glass rod and continue until there is about 100 ml of mush. Similarly, obtain about 100 ml of the flesh of winkles, cockles or whelks. Add an equal volume of diluent, stopper and shake well, For prawns and shrimps, place 100 g in a blender with 100 ml of diluent and homogenize.

Allow the gross material to settle for 15–20 min. Melt three roll-tubes each containing 2 ml of MacConkey roll-tube agar and cool to 45–50°C. To each tube add 1 ml of the supernatant liquor and roll the rubes. When cool incubate overnight in a water-bath at 44°C inverted and completely submerged (inversion prevents water of syneresis from washing off colonies developing near the bottom of the tube).

Count the large red coliform colonies. Less than five colonies on each of the tubes inoculated with 1 ml (i.e. less than 5/g of shellfish) is regarded as satisfactory, from 5–15 colonies suspicious and more than 15 colonies as unsatisfactory.

To detect salmonellas in oysters, add 100 ml of mush to 100 ml of double-strength selenite broth (see p.270).

To detect sulphite-reducing clostridia and group D streptococci in oysters the pour plate method is recommended (Easterbrook and West, 1987).

To detect S. aureus which may be present in imported frozen prawns and shrimps (infected during handling), add 10-ml amounts of the mush to 50-ml tubes of broth containing 10% of sodium chloride. Incubate overnight at 37°C and plate on blood agar and PPPA or MSA medium (see Chapter 24). For V. parahaemolyticus add 10 g emulsified flesh to 100 ml of alkaline peptone water containing 3% salt and proceed as on p.251.

The 'percentage clean' method for oysters and mussels
Take ten shellfish, scrub and clean the outside; and wash in sterile water. Hold with the concave shell down and open with a sterile oyster knife. Take care not to spill any liquor. Cut the flesh into small pieces with a sterile scalpel and mix with the liquor already present. Add 0.2 ml of liquor from each shellfish to a tube of single-strength MacConkey broth. Add 0.2 ml from three shellfish to a tube of glucose broth and 1 ml from three others to a tube of litmus milk.

Incubate at 44°C for 24 h and examine the MacConkey tubes for acid and gas, the glucose broth microscopically for streptococci (enterococci) and the litmus milk for C. perfringens.

If coliform bacilli (44°C +) are absent from all ten shellfish, they are '100% clean'. If they are absent from eight out of ten, they are '80% clean', and so on. It is generally regarded that '80-100% clean' is satisfactory, '70% clean' is suspicious and '60% clean' or less is unsatisfactory.

Alternative methods of isolating enterococci are given in on p.305. For more information about the bacteriology of shellfish see Codex Alimentarius (1978a, b) and West and Coleman (1986).

Viruses and dinoflagellates

The possibility of contamination by viruses should be considered. Molluscs are filter feeders and depuration, which removes bacteria, may not remove viruses, nor are they always killed by heat treatment.

Frozen sea-food

Most contamination is introduced in the factory in cutting, battering, packing, etc. Enterobacteria and staphylococci also may be introduced at this stage. If the food is pre-cooked, the bacterial count is reduced but this treatment is usually not enough to kill all enterobacteria, staphylococci and anaerobes. Counts may be very high. *S. aureus*, coliforms, *V. parahaemolyticus* and salmonellas may be present. The flora is usually mixed and reflects processing rather than the raw materials.

Ready cooked deep-frozen prawns and shrimps

These are imported from the Far East and are an increasingly common article of diet in the West.

The following method, devised by Mitchell (1970) allows a simple quantitative examination to be made on this product. Chisel 20 g of fish from a frozen block into a sterile screw-capped jar. Place in a water-bath at 44°C for 10–30 min to hasten release of the juice. Prepare 1:50 and 1:500 dilutions and plate on well dried blood agar and MacConkey plates using the Miles and Misra method. Incubate overnight at 35°C. This gives the total viable and coliform counts/ml of extruded juice. Homogenize the remaining tissue preferably in a Stomacher and inoculate salt meat broth for *S. aureus*, alkaline peptone water for *V. parahaemolyticus* and selenite broth for salmonellas. Incubate overnight at 37°C and plate on suitable solid media.

The bacterial population of these crustaceans will consist of spore-forming bacteria that have survived boiling and organisms introduced after cooking.

Counts on isolated samples may not give significant results. Ideally 5–10 samples per batch should be obtained at the port of entry. Suggested standards, based on a weighing and macerating technique and using an incubation temperature of 35°C are: counts up to 100 000 cfu/g, release unconditionally; 100 000 to 1 000 000 cfu/g, release with a warning to use immediately on thawing, and over 1 000 000 cfu/g detain.

Vinegar-pickled fish

Test pH. If this is 4.5 or less no further action is necessary.

References

Codex Alimentarius Commission (1978a) *Recommended International Code of Hygiene Practice for Shrimps and Prawns*, Food and Agricultural Organization, Rome

Codex Alimentarius Commission (1978b) *Recommended International Code of Hygiene Practice for Molluscan Shellfish*, Food and Agriculture Organization, Rome

Easterbrook, J. and West, P. A. (1987) Comparison of most probable number and pour-plate procedures for the isolation and enumeration of sulphite-reducing *Clostridium* spores and group D faecal streptococci from oysters. *Journal of Applied Bacteriology*, **62**, 413–419

Gardner, G. A. (1983) Microbiological examination of curing brines. In *Sampling – Microbiological Monitoring of the Environment* (eds. R. G. Board and D. W. Lovelock), Society for Applied Bacteriology Technical Series No. 7, Academic Press, London, pp. 21-27

Kitchell, A. G., Ingram, G. C. and Hudson, W. R. (1973) Microbiological sampling in abbatoirs. In *Sampling – Microbiological Monitoring of Environments* (eds R. G. Board and D. W. Lovelock), Society for Applied Bacteriology Technical Series No. 7, Academic Press, London, pp. 43–59

Mitchell, N. J. (1970) A simplified method for quantitative microbiological examination of deep frozen seafood. *Journal of Applied Bacteriology,* **33,** 523–527

Murray, J. G. (1969) An approach to bacteriological standards. *Journal of Applied Bacteriology, 32,* 123–135

West, P. A. and Coleman, M. R. (1986) A tentative national reference procedure for the isolation and enumeration of *Escherichia coli* from bivalve molluscan shellfish by the most probable number method. *Journal of Applied Bacteriology,* **61,** 505–516

Chapter 15

Fresh, preserved and extended shelf-life foods

Fruit and vegetables

Fresh fruit and vegetables

Spoilage of fresh fruit is mostly caused by moulds. Mould and yeast counts may be indicated in fruit intended for jam making or preserving. Orange serum agar is useful for culturing citrus fruit.

Washed, peeled and chopped vegetables are common in supermarkets in the rest of Europe and are gaining acceptance in the UK. They are often 'blanched', to destroy enzymes. This helps reduce the bacterial load. Spoilage is caused by pseudomonads, lactobacilli, Lancefield Group D streptococci and leuconostocs. Salad crops, e.g. watercress, which may be grown in polluted streams, should be tested for *E. coli* and salmonellas if indicated.

Dehydrated fruit and vegetables

Counts
Examine washings (p.179). An *E. coli* count of more than 5/g is suspicious. If repeat tests give the same result the batch is unsatisfactory. Counts of lactic acid bacteria may be useful: use Rogosa agar medium and incubate at 30–32°C for 3 days.

Culture
Inoculate glucose tryptone agar, tomato juice, or Rogosa medium and incubate at 30–32°C. Incubate glucose tryptone cultures also at 55–60°C aerobically and anaerobically.

Microbial content
Counts should be low. Lactic acid bacteria, flat sour thermophiles (*B. stearother-mophilus*) and hydrogen sulphide-producing anaerobes may be present.

Blakey and Priest (1980) examined red and brown lentils, yellow and green peas, black-eyed, kidney, mung and soya beans, scotch broth mix, rice, pearled barley and chapatti flour. They found *B. cereus* at levels ranging from 1×10^2 to 6×10^4/g.

Frozen vegetables and fruit

Frozen peas offer the greatest problem as they deteriorate rapidly on thawing. Slime usually contains large numbers of leuconostocs, which imparts a yellow appearance (acid pH) but sometimes a heavy growth of coryneforms produces ammonia, which neutralizes the acid formed by the leuconostocs. If sucrose is used in the medium instead of glucose, a levan is produced by leuconostocs.

Counts on other frozen vegetables are usually low (about 100 000 cfu/g). Coliforms and enterococci are commonly found on vegetables, *E. coli* type 1 may have public health significance.

Fruits may have low yeast and lactobacilli counts. In general, they keep well at −15°C (0°F) for 2 to 3 years. Vegetables remain in good condition at this temperature for 6–12 months.

Pickles, ketchups and sauces

Pickles

Vegetables are first picked in brine. The salt is then leached out with water and they are immersed in vinegar (for sour pickles) or vinegar and sugar (for sweet pickles). Some products are pasteurized. Spoilage is due to low salt content of brine, poor quality vinegar, underprocessing and poor closures.

Counts
Use glucose tryptone agar at pH 6.8 and 4.5 and do total counts and also counts of acid-producing colonies (these have a yellow halo due to a colour change of indicator). Count lactic acid bacteria on Rogosa or other suitable medium. Estimate yeasts either by the counting chamber method (stain with 1:5000 erythrosin) or on malt agar.

Culture
Use glucose tryptone and Rogosa or Mann, Rogosa and Sharp (MRS) agar. Grow suspected film-yeasts in a liquid mycological medium, containing 5 and 10% of sodium chloride for three days at 30°C. For obligate halophiles, use a broth medium containing 15% of sodium chloride.

Microbial content
Pickles are high-acid foods. Counts are usually low, for example 1000 cfu/g. Yeasts are a frequent source of spoilage and may be either gas-producing or film-producing. In the former case, enough gas may be generated to burst the container. Bacterial spoilage may be due to acid-producing or acid-tolerant bacteria such as acetic acid bacteria, lactic acid bacteria and aerobic spore-bearers. Infected pickles are often soft and slimy.

Fermented pickles of the sauerkraut type contain large numbers of lactic acid bacteria (*Lactobacillus* and *Leuconostoc*) which are responsible for their texture and flavour.

Acid-forming bacteria are active at salt concentrations below 15%. Above this concentration, obligate halophiles are found.

Ketchups and sauces

The most common cause of spoilage is *Zygosaccharomyces (Saccharomyces) baillii*. This organism grows at pH 2, at <5°C up to 37°C, in 50–60% glucose and is heat resistant to 65°C.

Sugar and confectionery

Sugars, molasses and syrups

The importance of microorganisms is related to the use of the products, for example flat-sour bacteria are less important in bakery than in canning. Do total counts, examine for thermophiles (*B. stearothermophilus, C. nigrificans, C. thermosaccharolyticum*) and for yeasts and moulds. Osmophilic yeasts, aspergillus, penicillium, etc., may cause inversion.

Chocolate

This has been shown to be the vehicle of salmonella infection in a number of cases. (See the review by D'Aoust, 1977.)

Shave the chocolate into nutrient broth and selenite, incubate overnight and plate out on DCA medium. Do not place large lumps of chocolate into liquid medium. Automated methods, especially those employing impedimetric principles, are useful for salmonella screening in industry. Examine spoiled soft-centred chocolates for yeasts and moulds.

Cake mixes and instant desserts

Spores are likely to be present, e.g. of *C. perfringens* and *B. cereus*. Salmonellas and staphylococci may survive processing or be post-processing contaminants. Any of these organisms may multiply if the product is reconstituted and then kept under unsuitable conditions. The powdered product may also cross-contaminate other products which could provide suitable conditions for growth.

Do total viable counts; test for clostridia, *B. cereus*, coliforms and *E. coli*, and staphylococci.

Cereals and protein additives

It is important that these materials should not contribute unduly to the bacterial load of the product. Do total counts and, if the material is to be used in canned foods, do spore count. The total count at 30°C should not exceed 20 000 cfu/g. The spore counts at 30 and 55°C should not be greater than 100/g.

Flour

Contaminated flour may be responsible for the spread of infection by spoilage organisms in kitchens as well as spoilage of the bread, pastry, etc., for which it is used. Grain is naturally infected by soil, dust and rodent and bird faeces during ripening, harvesting and storing. During transport and handling, this contamination is distributed throughout the bulk. Before milling, the grain is washed, sometimes with polluted water.

Counts
Weigh 10 g of flour into a sterile jar containing coarse sand or small glass beads. Add 100 ml of 0.1% peptone water diluent and shake mechanically for 10 s, stand for 30 min, and shake again. If moisture absorption is great this dilution may need to be adjusted.

Do total bacterial counts on serial dilutions using glucose tryptone agar. Because of turbidity, it may be necessary to use the MPN method, in which case test 5×10, 5×1 and 5×0.1 ml samples of tenfold dilutions of the above homogenate in glucose tryptone broth plus bromocresol purple and record as positive any tube that shows acid production. Incubate at 32°C for 3 days.

Count anaerobes in the same way with thioglycollate broth, and coliform bacilli by the MPN method.

To count 'rope spores', heat the homogenate for 20 min at 90°C to kill vegetative bacteria, make dilutions and do MPN counts in glucose tryptone broth. Alternatively, inoculate two tubes of glucose tryptone broth each with 1 ml of the serial dilutions of heated homogenate. Incubate at 30°C for 3 days and record as positive tubes that show a pellicle. Examine the pellicle to identify B. mesentericus or B. subtilis. Record as rope spores/g or as present in so much of 1 g.

Culture
Use glucose tryptone agar for aerobic and glucose tryptone agar and iron sulphite medium for anaerobic culture. Inoculate malt extract or a similar medium.

Microbial content
Counts of 5000–500 000 are usual and E. coli is often present in 1 g or less depending on processing. Flat-sour bacteria (B. stearothermophilus) and hydrogen sulphide-producing clostridia may be found in varying numbers. 'Rope organisms' ('B mesentericus', but see p.331) are important and cause 'ropy bread'. Mould counts may be 2000 or more per g.

Pastry

Uncooked pastry, prepared for factory use or for sale to the public, may suffer spoilage due to lactobacilli. Do counts on Rogosa agar and incubate at 30°C for 4 days. Counts of up to 10 000/g are not unreasonable.

Pasta products

These are man-made, from wheat flour, semolina, farina and water. Egg (powdered or frozen), spinach, vitamins and minerals may be added. The egg in particular may contain salmonellas and while these may be destroyed in the subsequent cooking there is the possibility of cross-contamination from uncooked to cooked products.

High levels of S. aureus and preformed toxin have been reported in lasagne and high levels of those organisms in dried pasta (ICMSF, 1980).

Examination

As contamination is likely to occur during manufacture it is necessary to liberate the organisms from within the pasta.

Add 25 g of pasta to 225 ml of peptone water diluent and allow it to soften at room temperature for about 1 h. Macerate, e.g. in a stomacher, and do total and coliform counts.

Inoculate salt meat broth or Giolotti–Cantoni broth. Incubate at 37°C for 24 h and plate on Baird-Parker medium for staphylococci.

Inoculate selenite medium. Incubate at 37°C for 24 h and plate on DCA or other medium for salmonellas.

Incubate *B. cereus* selective agar and incubate at 37°C for 24 h.

If the presence of moulds is suspected it is best to obtain help from a reference expert because of the possibility of aflatoxins.

Extrusion-cooked products

These include breakfast cereals and crispbreads and may be textured with vegetable protein and bread crumbs. The liquid and solid ingredients are blended, shaped and cooked within 1–2 min.

Examine as for pasta.

Gelatin

Gelatin is often used to top-up pastry cases of cold-eating pies, in canned ham production and in ice-cream manufacture. It should be free from spores and coliforms.

If the process involves low temperature reconstitution examine for *S. aureus*, salmonellas and clostridia.

Laboratory examination

Weigh 5 g of gelatin into a bottle containing 100 ml of sterile water and allow to stand at 0–4°C for 2 h. Place the bottle in a water-bath at 50°C for 15 min and then shake well. Mix 20 ml of this solution with 80 ml of sterile water. This gives a 1:100 dilution. Use 1.0 and 0.1 for total counts by the pour-plate method. Incubate at 35°C for 48 h.

Gelatin for ice-cream manufacture

Do a semi-quantitative coliform estimation using MacConkey or similar broth. Add 10 ml of 1:100 gelatin to 10 ml of double-strength broth; add 1 ml of 1:100 gelatin to 5 ml of single-strength broth and 0.1 ml of 1:100 gelatin to 5 ml of single-strength broth. Incubate at 35°C for up to 48 h and do confirmatory tests where indicated (see p.261). Thus, the presence or absence of coliforms and *E. coli* in 0.1, 0.01 or 0.001 g of the original material can be determined. It is desirable that coliforms should be absent from 0.01 g and *E. coli* absent from 0.1 g. The total count in gelatin to be used for ice-cream manufacture should not exceed 10 000 cfu/g.

Gelatin for canned ham production

This should have a low spore count. After doing the total counts, heat the remaining 1:100 solution of gelatin at 80°C for 10 min. Plate 4 × 1 ml of this solution on standard plate count medium. Incubate two plates at 35°C and two plates at 55°C for 48 h. There should be not more than one colony per plate, i.e. 100 g of the original gelatin. The total count should not exceed 10 000 cfu/g.

Spices and onion powder

A plastic bag inverted over the hand is a satisfactory way of sampling spices. Some are toxic to bacteria and the initial dilution should be 1/100; for cloves use 1/1000, in broth. Do total viable counts and if indicated by intended usage consider inoculating *B. cereus* selective medium and one of the media for staphylococci.

The total counts on these materials vary widely. Total viable counts of 10^8/g are generally acceptable but low spore counts are important if these materials are to be used for canned foods: an acceptable level at 30 and 55°C is less than 100/g.

Coconut (desiccated)

In the 1950s and 1960s this product was often contaminated with salmonella. In spite of improved processing it is still a potential hazard in the confectionery trade. Examine samples for salmonellas by the method described on p.270.

Oily material

Mix 1.5 g of tragacanth with 3 ml of ethanol and add 10 g glucose, 1 ml of 10% sodium tauroglycocholate and 96 ml of distilled water. Autoclave at 115°C for 10 min. Add 25 g of the oily material under test and shake well. Make dilutions for examination in warm diluent.

Salad creams
These contain edible oils and spoilage may be due to the lipolytic bacteria. Test for these with tributyrin agar. Examine for coliform bacilli and thermophiles. All these should be absent. There should not be more than five yeasts or moulds/g.

Mayonnaise-based salads
Examine for lactobacilli. Yeasts may grow on the medium unless chloramphenicol, 100 mg/l, is added to the medium (Rose, 1985).

Canned, prepacked and frozen foods

The practice of prolonging the shelf-life of foods by sealing them in metal or glass containers and heat processing them has long been established but modern packaging materials have given the consumer a wider choice.

Eating habits are changing. Single portions which can be taken from the freezer and heated in a microwave oven are popular. With these, packaging is important. It should

(1) prevent contamination by microorganisms,
(2) preserve quality and nutritional value,
(3) be inert and offer no hazard in use,

(4) be economic to manufacture and distribute,
(5) be easily labelled.

Routine control of the product in the factory is the responsibility of the quality assurance and laboratory departments of the manufacturer but general laboratories may be asked to help when defects or spoilage have developed after the products have left the factory or where the product is suspected of having caused food poisoning or enteric fever. Occasionally, arbitration is needed between suppliers of ingredients and manufacturers of finished goods.

Type of containers

Cans
Tin plate and aluminium are widely used. Improved lacquering gives resistance to corrosion. Welded three-piece cans have largely superseded the soldered type. Base metal thickness has been reduced, giving lighter and cheaper cans but these have to be 'beaded' to withstand processing. Two-piece cans with 'easy-open ends' are popular for carbonated beverages. Self-heating cans are convenient for camping and picnics.

Jars and bottles
Glass closed with metal cap and resilient seal, sometimes under vacuum. They should be tamper-proof.

Trays
Aluminium or aluminium/polypropylene laminated, with lids.

Semi-rigid flexible plastic and laminate containers
These are easy to stack, light and allow a shelf-life of up to 2 years. Plastic-sided metal-ended containers are used for fluids.
    For a review of containers see Dennis (1987).

Physical defects of cans

These defects may be due to improper processing; during exhausting and autoclaving incorrect stresses may be imposed and cause 'peaking' or 'panelling', which are distortions of the ends (distinct from 'swelling') and of the body, respectively. The ends of the cans have concentric rings impressed in them to absorb the normal strain and to permit swelling in normal stresses.
    Rusting of cans, causing pinholes and consequent spoilage, is revealed by inspection. Hydrogen swell is caused by hydrogen formed when acidic foods attack the metal in places where the lacquer is defective.
    Faulty can manufacture and improper closure of seams, either side or lid, can lead to spoilage.
    Inadequate drying may result in contamination of the can contents by bacteria which enter in water droplets through minute pinholes in the seams. These holes are usually self-sealing when the can is dry. Wet cans also tend to rust.

Spoilage due to microorganisms

Leaker spoilage

When this occurs, usually only a small number of cans in each batch is affected. Minute faults may be present in some cans (see above) particularly where the end seams cross the body seams. After autoclaving and during cooling there is a negative pressure in the can and cooling water containing bacteria may be drawn in, and canned foods may thus be contaminated with pathogens. Only very few organisms, pathogens or spoilers need be drawn into a can, as they will multiply rapidly. Gas producers will manifest themselves by 'blowing' the can but organisms that produce gas in normal culture may not do so in cans. Other spoilage will be obvious when the can is opened but food contaminated with pathogens, e.g. typhoid bacilli, may appear sound and wholesome. Care must also be taken that there is not a build-up of organic matter in the line which can overcome the disinfectant and lead to bacterial multiplication in the water remaining on the cans. Personnel with septic conditions must be excluded from handling cans. Dirty or infected hands can contaminate the surface water on the cans before it is drawn into the cans through these defective seams or pinholes. Water used for cooling is usually chlorinated but some organisms are relatively resistant to the process. When the seams are dry, the chances of contamination are slight.

Underprocessing

Gross underprocessing will usually have been found by the manufacturer's tests. If this is not the case, it is common to find only one type of organism in this situation.

Some cured meats are deliberately underprocessed because they are rendered less palatable by autoclaving. Ham and mixtures of ham and other meat are usually salted and spiced and given the minimum of heat treatment. The manufacturers do not claim sterility and the label on the can invariably recommends cold or cool storage. The conditions of pH, salt content and storage temperature should be such that the bacteria in the can are prevented from multiplying. There is evidence that some species of enterococci produce antibacterial substances in canned hams, which act antagonistically on some species of clostridia, lactobacilli and members of the genus *Bacillus*. This is most likely a factor in the successsful preservation of commercial products of this nature.

Poor plant hygiene, faulty design or careless operation of equipment, e.g. bad stacking of retorts, may be a contributory cause of underprocessing. Vegetative organisms are killed, except in very rare cases of gross underprocessing, but spores are unaffected; they subsequently germinate and cause spoilage. The spores of *C. botulinum* are very heat resistant and they may germinate and produce toxin. Spores of *Bacillus stearothermophilus* are even more resistant than those of *C. botulinum* and, fortunately, *C. botulinum* is a poor competitor.

Gas-producing organisms cause the can to swell. The first stage is the 'flipper' when the end of the can flips outward if the can is struck sharply. A 'springer' is caused by more gas formation. Pressing the end of the can causes the other end to spring out in a bulge. The next stage is the 'swell' or 'blower' when both ends bulge. A 'soft swell' can be pressed back but bulges again when the pressure is released. 'Hard swells' cannot be compressed.

Spoilage may not result in gas formation and may be apparent only when the can is opened. This kind of spoilage includes the 'flat-sour' defect.

Inadequate cooling
If cooling is too slow or inadequate there may be sufficient time for the highly resistant spores of *B. stearothermophilus* to outgrow and multiply. The optimum growth temperature for this organism is between 59 and 65°C. It will not grow at 28°C or at a pH of less than 5. Together with *B. coagulans* it is the chief cause of 'flat-sour' spoilage in canned foods. Most strains of *B. coagulans* will grow at 50–55°C. They will also grow in more acid conditions.

Preprocessing spoilage
This can occur when the material to be canned is mishandled before processing, e.g. if precooked meats with a large number of surviving spores are kept for too long at a high ambient temperature this may result in the outgrowth of spores and the production of gas. When canned the organisms will be killed leaving the gas to give the appearance of a blown can.

Type of food, and organisms causing spoilage

High-acid foods
Spoilage is rare in processed foods with a pH of 3.7 or less, for example pickles and citrus fruits. Yeasts may occur when there has been serious underprocessing. These foods are usually not pressurized during heating.

Acid foods
When the pH is 3.7–4.5, as in most canned fruits, aerobic and anaerobic spore-bearers may cause spoilage but this is not common. Lactobacilli and *Leuconostoc* have been reported. Osmophilic yeasts and the mould *Byssochlamys* are sometimes found. These foods are usually not pressurized. They are too acidic for the growth of most bacteria and leaker spoilage is uncommon.

Low-acid foods
If the pH is 4.5 or above, as in canned soups, meat, vegetables and fish (usually about pH 5.0), one of the following thermophiles is usually found in spoilage due to underprocessing.

(1) *B. stearothermophilus*, causing 'flat-sour' spoilage,
(2) *C. thermosaccharolyticum*, causing 'hard swell',
(3) *C. nigrificans*, causing 'sulphur stinkers',
(4) mesophilic spore-bearers, obligate or facultative anaerobes, causing putrefaction.

Leaker spoilage may be due to a variety of organisms; aerobic and anaerobic spore-bearers, Gram-negative non-sporing rods and various cocci, including *Leuconostoc* and *Micrococcus* may be found. *S. aureus* (food poisoning type) has been isolated.

Methods of examination

Sampling
If packs are blown or swollen, examine six and take six normal ones from another batch as controls. In suspected underprocessing examine 6–12 packs from each

batch. Leaker-spoilage is likely to occur in only a very small number of packets in a batch; therefore, examine as many as possible.

Physical examination
Inspect the seams and can surfaces. A jeweller's saw is useful for cutting across seams. Note the batch or code numbers printed on the labels or stamped on the lid.

Jars
Examine cap for perforations and note code.

Trays and pouches
Inspect the seals. These are formed from a continuous weld and are less likely to leak than the double seams of cans. Furthermore there is no headspace or vacuum which could cause organisms to be sucked in if there was a small hole in the seal. Note code.

Pre-incubation
Incubate apparently sound packs at 35–37°C for 7 days. This encourages the multiplication of small numbers of organisms which might otherwise be missed in sampling the contents.

Sampling contents: cans normal in appearance
Swab the top with cotton wool and methylated spirit. Pour 1 ml of spirit on the swabbed area and flame it. Allow the spirit to burn out.

If the contents of the can are liquid, puncture the flamed surface with a 100-mm wire nail (sterilized in tins containing 10–12 nails in a hot air oven) by a sharp blow with a hammer. Remove a sample of the contents with a pasteur pipette into culture media and into a screw-capped bottle for viable counts if required.

If the contents are solid, use a punch made from brass rod 9–10 mm in diameter with one end drawn to a point. Sterilize these individually. Drive the punch well in to make a large hole. Remove a core sample with a length of glass tubing of 7–8 mm outside diameter by pushing it right to the bottom of the can. Push the core sample from the glass tube into a screw-capped bottle with a piece of glass rod of suitable thickness. These core and rod samplers can be sterilized together in copper pipette drums. In addition to taking core samples, it is desirable to sample jelly adjacent to the seam of the can. To do this, remove the end of the can, previously punctured with a sterile domestic can opener and tip the contents on to a sterile tray. Take care not to disturb the material under the seam. Note the appearance of this material and sample with a cotton wool swab.

Jars normal in appearance
Sterilize as for cans and pierce with a sterile nail. It is very important to release any vacuum. Carefully remove the cap so that the sealing gasket is undisturbed. This should be examined thoroughly for evidence of improper seating or twisting. Liners are sometimes misplaced causing inadequate closure of the jar. Look for damage to the rim of the jar. Examine contents for cans.

Flexible pouches and trays normal in appearance
Support in suitable racks. Sterilize with 50:50 alcohol/ether and allow to dry. Open with sterile scissors and sample with a sterile spoon. Examine contents as for cans.

Sampling blown, swollen or leaking packs

These contain gas under pressure and the contents may be offensive. Chill before opening. Place the container on a metal tray with the seam facing away from the operator. Swab with 4% iodine in 70% alcohol, allow to stand for a few minutes then dry with a sterile towel. Do not flame.

Invert a previously sterilized metal funnel or new plastic bag over pack. The diameter of the funnel should be slightly larger than that of the can. Pass a sterile brass rod with a point at one end down the funnel spout until it rests on the can; hold both firmly and puncture the can by tapping the rod with a hammer, then withdraw the rod slightly. The contents of the can may be ejected with some force but the funnel and tray will prevent broadcast. Before removing the funnel and brass rod, push the latter in and out of the hole in the can several times. Sometimes a piece of food is forced against the hole by internal gas pressure and when a sampler is inserted more gas and food are ejected.

Take samples with a pasteur pipette or core sampler as described above. After sampling, open the can with a domestic can opener and inspect the contents.

Direct film examination

Make Gram films of sample. The presence of Gram-positive rods may suggest underprocessing while cocci, yeasts, etc., indicates leaker spoilage.

Note: The organisms seen may be dead (killed during processing), so too much reliance must not be placed on this examination.

Autosterilization may also account for this phenomenon. In this instance the organisms die out during storage. When this has occurred the organisms appear degenerate and poorly stained (see Preprocessing spoilage p.202).

Culture

For general examination inoculate glucose tryptone agar (with bromocresol purple indicator) and incubate aerobically and anaerobically at 22–25, 35–37 and 55–60°C for 24–36 h.

For high-acid foods, i.e. pH 4.6 or less, inoculate four tubes of acid broth. Incubate two tubes at 55°C for 48 h and two tubes at 30°C for 96 h. Inoculate two tubes of malt extract broth and incubate these at 30°C for 96 h. Subculture and make Gram-stained smears as necessary.

Also inoculate the following media if indicated: iron sulphite medium (sulphur stinkers), blood agar and MacConkey (for putrefactive organisms, micrococci, leuconostocs, etc.), Crossley Milk Medium (putrefactive aerobic or anaerobic spore-bearers), reinforced Clostridial agar, malt extract or other mycological medium (for yeasts and fungi).

Make Gram films of colonies.

Microbial content

Identify as follows:

(1) Gram-positive rods
  (a) Thermophiles:
    (i) Aerobic: *B. stearothermophilus* (flat-sour). See p.331.
    (ii) Anaerobic: *C. thermosaccharolyticum* (hard swell). See p.339.
    (iii) Anaerobic: black colonies in iron sulphite medium: *C. nigrificans* (sulphur stinkers). See p.339.

(b) Mesophiles:
　(i) Aerobic: *Bacillus*. See Chapter 27.
　(ii) Anaerobic: *Clostridium*. See p.339.
(2) Gram-negative rods
　Pseudomonas alcaligenes or enterobacteria. See Chapters 19 and 21.
(3) Gram-positive cocci
　Micrococci, leuconostoc. See Chapters 24 and 25.
(4) Yeasts and moulds
　See Chapters 33 and 33.

Pathogens in canned food

Outbreaks of enteric (typhoid) fever and staphylococcal disease have caused food bacteriologists, public health authorities and canners to revise their opinions on the safety of canned foods, although these outbreaks are very few in proportion to the enormous amount of canned food consumed. Random or 'routine' sampling of canned foods for pathogens is an unrewarding procedure. Only low-acid foods, meat and dairy products and certain canned vegetables can support the growth of enteric organisms, staphylococci and *C. botulinum*.

Certain moulds and other organisms can raise the pH of some acid fruits, e.g. tomato juice and pears to a level at which *C. botulinum* will grow.

Examination for pathogens
When this is indicated, open the cans with sterile can openers and if the food is solid take samples under the seams, particularly where the end seams cross the side seam. Culture in selenite medium for salmonellas (continue as on p.271), in salt meat for staphylococci (continue as on p.300). For botulism, see p.334.

Frozen foods

Some frozen convenience foods are mentioned here, others – frozen meat, fish and ice-cream – are included under their appropriate headings in Chapters 14 and 15. For frozen vegetables see p.195.

Storage
Unless the cabinets are maintained at −18 to −20°C there will be difficulties due to the different melting points of the stored products, the presence of water films and the variety of microclimates. Two kinds of spoilage occur:

(1) low-temperature spoilage due to enzymes and, less often, to psychrophiles if the temperature is at or about the freezing point of water;
(2) unfreeze spoilage, when bacteria can grow because of gross temperature fluctuations. Off-odours and off-flavours and spoilage losses are likely to be doubled for each 2–3°C rise in temperature.

Counts
Count total bacteria growing on at 5–7°C in 5–7 days and 20–30°C in 2–3 days. Use enriched media because organisms will be cold shocked. Count coliforms lactic acid bacteria (except in fish), moulds and yeasts.

Culture
For total counts, inoculate one of the media for fastidious organisms mentioned on p.53 and on media for lactobacilli, fungi, staphylococci, salmonellas and enterococci. There is evidence that enterococci survive longer than coliforms in frozen foods.

Pre-cooked chilled foods
These are increasing in popularity and if handled according to the guidelines (DHSS, 1980) should not cause any problems.

Suitable tests are total viable counts and tests for coliforms, *C. perfringens*, salmonellas and *S. aureus*.

Frozen pies and complete meals
Frozen pies often have low counts (5000–30 000 cfu/g) but may contain enterobacteria and staphylococci. 'Complete meals' vary enormously in their counts and flora, usually reflecting factory conditions.

The practice of keeping complete meals in cold storage for several months appears to reduce the counts but this may be a false effect and reflect the failure to resuscitate cold-shocked organisms.

Storage life
This is usually controlled by the better manufacturers, who date-stamp their products. Chicken, pies and complete meals keep for 2–6 months. Storage life depends on the maintenance of a constant low temperature.

Arbitrary standards
Standards that have no legal status are used by the trade and public health authorities as a guide in the frozen food industry:

Plate count at 35°C in 48 h – not more than 100 000 cfu/g.

Coliform bacilli NOT present in 0.1 g.

S. aureus NOT present in 0.01 g.

For further information on the microbiology of frozen foods, see Roberts *et al.* (1981).

Cook-chill products
The safety and shelf-life of these foods depends on sublethal treatment and aseptic packaging. Major advances in transportation and handling of chilled and frozen products have increased the availability of these foods. Chilled foods account for over 40% of expenditure on foods in the UK. Temperatures should be controlled between −1 and −4°C throughout handling and storage. As many of these foods are capable of supporting the growth of food poisoning organisms it is important that they are not mishandled.

The packaging and the food are sterilized separately and united under 'commercially sterile' conditions. Flexible pouches and semi-rigid pots and cartons are used for low acid foods, e.g. ice-cream, custards, soups and sauces.

Cook-chill catering
This system is used by hospitals, works canteens, prisons, schools, banqueting and travel organizations. It demands a high level of technical control to ensure safety and quality.

Bacteriological examination
These products may be examined by the same method as other foods, and the same standards should apply.
  These foods should conform to the following standards:
Viable count after 48 h at 37°C – < 100 000/g
E. coli – < 10/g
S. aureus – < 100/g
Salmonella spp. not present in 25 g
Listeria monocytogenes not present in 25 g.
Clostridium perfringens not present in 100 g.

Baby foods

Direct breast feeding allows little chance of infection of the infant with enteric pathogens. Testing of human milk is described on p.222. With other feeds contamination may occur during the time lag between production and consumption.
  There are many milk preparations and weaning formulas, as well as dried, bottled and canned foods designed for small children. Dehydrated milk may contain organisms that have survived processing and these may multiply if the product is not stored correctly after it is reconstituted. Dirty equipment may contribute bacteria to an otherwise sterile material. Unsuitable water may be used for reconstitution.
  Central milk kitchens in maternity units should have very high standards of hygiene (see p.225 for sampling surfaces etc.). In-bottle terminal heating, with teat in place is good practice.
  Viable counts should be low and pathogens should be absent. See Robertson (1974) and Collins-Thompson (1980).
  Nasogastric feeds should conform the same standards.
  Heat-treated jars and cans should be examined as described for extended shelf-life foods on p.202.

Soft drinks

Total counts should be low and coliform bacilli absent, as in drinking water. Membrane filters can be used to test water intended for soft drink production. Millipore publish a very useful booklet on the microbiological examination of soft drinks. Yeast and mould spoilages are not uncommon and they may raise the pH and allow other less acid tolerant organisms to grow.
  In non-carbonated fruit drinks, yeasts are not inhibited by the amounts of preservatives that are permitted by law. Spoilage is generally controlled by acidity (except *Z. baillii*) (see Ketchups, p.196). In both these cases and carbonated drinks the microbial count diminishes with time.
  Lactic acid and acetic acid bacteria may grow at pH 4.0 or less in some fruit juices.

Fruit juices

Heat resistant fungi can cause problems in concentrated juices. Screen by heating at 77°C for 30 min. Cool and pour plates with 2% agar. Incubate for up to 30 days.

Dip slides, including those with medium for yeasts and fungi are useful.

Automated counting methods (e.g. ATP assay, see p.135) are useful. Fruit juices are incubated at 25°C for 24–48 h (72 h for tomato juice), the reagent is added and incubated for 45 min when the instrument gives the result.

Bottled waters

See p.237.

Milk-based drinks

See p.217

Vending machines

Water and flavoured drinks in vending machines may be of poor quality, with high counts (>1000 cfu/ml) and coliforms. This may be due to inadequate cleaning (see Hunter and Burge, 1986).

Useful references

Food microbiology is a very large subject and in addition to the references cited in this chapter and those preceding it the following are recommended: Hersom and Hulland (1980), Jowitt (1980), Roberts and Skinner (1983), Speck (1984) and ICMSF (1980).

References

Blakey, L. J. and Priest, F. G. (1980) The occurrence of *Bacillus cereus* in some dried foods including pulses and cereals. *Journal of Applied Bacteriology,* **48,** 297–302

Collins-Thompson, D. L., Weiss, K. F., Riedel, G. W. and Charbonneau, S. (1980) Microbiological guidelines and sampling plans for dried infant cereals and dried infant formulae. *Journal of Food Protection,* **43,** 613-616

D'Aoust, J. Y. (1977) Salmonella and the chocolate industry: a review. *Journal of Food Protection,* **40,** 718–726

Dennis, C. (1987) *Symposium on the Microbiological and Environmental Health Problems in Relation to the Food and Catering Industries,* Campden Food Preservation Association, Chipping Campden, pp. 131–148

DHSS (1980) *Guidelines on Pre-cooked Chilled Foods,* Department of Health and Social Security, HMSO, London

Hersom, A. C. and Hulland, E. D. (1980) *Canned Foods. Thermal Processing and Microbiology,* Churchill-Livingstone, London

Hunter, P. R. and Burge, S. H. (1986) Bacteriological quality of drinks from vending machines. *Journal of Hygiene (Cambridge),* **97,** 497–500

ICMSF (1980) Cereals and Cereal Products. International Commission on Microbiological Standards for Foods. In *Microbial Ecology of Foods* Vol. 2, Academic Press, New York

ICMSF (1980) *Microorganisms in Foods. 4.* International Commission on Microbiological Specifications for Foods, Blackwell, London

Jowitt, R. (ed.) (1980) *Hygienic Design and Operation of Food Plan,* Ellis Horwood, Chichester

Roberts, T. A. and Skinner, F. A. (eds) (1983) *Food Microbiology.* Society for Applied Bacteriology Symposium Series No. 11, Academic Press, London

Roberts, T. A., Hobbs, G., Christian, J. H. B. and Skovgaard, N. (eds) (1981) *Psychrotrophic Micro-organisms in Spoilage and Pathogenicity*, Academic Press, London

Robertson, M. H. (1974) The provision of bacteriologically safe infant feeds in hospitals. *Journal of Hygiene (Cambridge),* **73,** 297–303

Rose, S. A. (1985) A note on yeast growth in media used for the culture of lactobacilli. *Journal of Applied Bacteriology,* **59,** 153–156

Speck, M. (ed) (1984) *Compendium of Methods for the Microbiological Examination of Foods,* American Public Health Association, Washington

Chapter 16

Milk and dairy products

Milk

Statutory tests in the UK

The Milk (Special Designations) Regulations 1989 bring the tests and standards in the UK into line with those of the EC. The methylene blue test is no longer used and the bacteriological standards for raw and pasteurized milk are based on total viable counts (TVC) and the coliform test.

Raw milk
Samples taken at producers premises must have a TVC (72 h at 30°C) must be less than 20 000/ml and a coliform count (24 h at 30°C) less than 100/ml.

Pasteurized milk
This is milk that has been heated at 62.9–65.6°C for 30 min or at 71°C for 15 s and then immediately cooled to below 10°C. Samples are taken at premises where the milk is pasteurized and transported to the laboratory at 0–4°C. The TVC (as above) must be less than 30 000/ml and a 'pre-incubated plate count' of less than 100 000. The coliform count (as above) must be less than 1/ml.

Ultra-heat treated (UHT) milk
This is heated at 135°C for 1 s. Samples are taken at the processing plant for a 'colony count test' which should yield less 10 colonies/0.1 ml.

Sterilized milk
This is heated at 100°C for such a time that it will pass the turbidity test. It must also pass the same colony count test as UHT milk.

For legal work the tests must be done exactly according to the Statutory Instrument No. 2838, 1989 and British Standard 4285 BS (1984) but for routine quality control these are rather restrictive and the methods described here are adequate.

Total Viable Count

The method described on p.129 may be used but the plates should be incubated at 30°C for 72 h.

Coliform count

Make serial tenfold dilutions and do plate counts with violet red bile lactose agar. Incubate at 30°C for 24 h. Count only red colonies that are 0.5 mm in diam. or larger. Confirm if necessary by subculture in lactose bile brilliant green broth. The three-tube MPN method (p.135) is also useful here.

Preincubated plate count

This is a 'keeping quality' test for the initial numbers of psychrotrophic organisms that are responsible for spoilage in stored pasteurized milk. The milk is incubated at 6 ± 0.5°C for 5 days. A viable plate count is then done and incubated at 21°C for 25 h. Most bacteriologists do not consider this to be a reproducible test as it is difficult to ensure incubation at exactly the prescribed temperature.

Phosphatase test

The Aschaffenberg–Mullen test is used.

Test at once or refrigerate the milk overnight. Make up the reagent as follows. Dissolve 3.5 g of anhydrous sodium carbonate (AnalaR) and 1.5 g of sodium bicarbonate (AnalaR) in 1 litre of water. Store in a refrigerator. To 100 ml of this solution add 0.15 g of disodium p-nitrophenyl phosphate. Keep in the dark and in a refrigerator and use within 7 days. There must be no yellow colour.

To 5 ml of the reagent in a 152×16 mm test-tube add 1 ml of milk. Stopper with a rubber stopper and mix by inversion. Incubate for 2 h at 37°C in a water-bath. Include controls of boiled milk and also boiled milk containing about 2% of raw milk. Determine the actual amount by titration, i.e. by adding varying proportions of raw milk to pasteurized milk. Do a phosphatase test on these and make the control in the proportion which gives a reading of slightly less than 42 μg. Preserve this mixture with 0.5 ml% of saturated mercuric chloride. Store in a refrigerator. Label the bottle 'Poison'.

Compare the test sample with the boiled sample using the Lovibond Comparator and disc designed for this purpose. This is an all-purposes model supported on a stand on which the test-tubes are placed in a sloping position and viewed by reflected light. Unheated milks and insufficiently treated milks give a yellow colour due to p-nitrophenol released from the substrate by the action of phosphatase. In properly pasteurized milk, phosphatase has been destroyed and the reading on the disc will be 10 μg or less of p-nitrophenol/ml of milk.

Clean the glassware used in this test in chromic acid solution.

Colony count test for UHT and sterilized milk

Incubate the sample at 30–37°C for 24 h. Open the carton as described above and remove about 10 ml into each of two sterile test-tubes or small screw-capped bottles. Refrigerate one of these.

With a flamed standard platinum–iridium loop (BS 19) of 4-mm internal diameter, remove one loopful (about 0.01 ml) from the other bottle into 5 ml of melted yeast extract milk agar at 45–50°C in the screw-capped bottle. Mix by rotation and allow the agar to set with the bottle on its side. Incubate at 30–37°C for 48 h. Count the number of colonies. The test is satisfactory if there are less than ten; if there are more than ten colonies, repeat the test with the refrigerated sample.

Sterilized milk: the turbidity test

Weigh 4.0 g of ammonium sulphate (AnalaR) into a small flask or bottle. Add 20 ml of milk and shake for 1 min. Stand for 5 min and filter through a 12.5-cm Whatman No. 12 folded filter-paper into a test-tube. When 5 ml filtrate have collected, place the tube in a boiling water-bath for 5 min and then cool. A properly sterilized milk gives no turbidity.

Sterilization alters the protein constituents and all coagulable protein is precipitated by the ammonium sulphate. If heating is insufficient, some protein remains unaltered and is not precipitated by the ammonium sulphate. It coagulates, giving turbidity when the filtrate is boiled.

US standard methods for milk examination (APHA 1960, Speck, 1984)

Three routine tests are prescribed:

(1) plate count at 32°C;
(2) direct microscopical count, for raw milks if a high count is expected; and
(3) the coliform test.

Plate count
Use Standard Plate Count agar to prepare pour-plates. Incubate at 32±1°C for 48±3 h.

Direct microscopical count
This is recommended only for raw milks with a fairly high count. The technique is similar to the Breed count (see p.128).

Coliform test
This test is used after pasteurization in order to detect re-contamination. There are two methods.

(1) *Solid media method.* Use a pour-plate technique with either violet red bile or deoxycholate lactose agar. Mix 1.0 and 0.1 ml of milk with melted and cooled agar, allow it to set and overlay it with a further 3–4 ml of sterile medium. Cover the surface of the inoculated agar completely to prevent surface growth. Incubate at 32°C for 24±2 h and count all dark red colonies that are 5 mm or more in diameter as coliforms.
(2) *Liquid media method.* Add 10 ml to each of five tubes of double-strength brilliant green bile broth and 1 ml to each of five tubes of single-strength brilliant green bile broth. Incubate at 32°C for 48±3 h. Streak a loopful from each tube showing fermentation on to eosin methylene blue or Endo agar. Inoculate typical colonies grown on this medium to nutrient agar for Gram stain and a lactose broth to demonstrate gas production. Use the Most Probable Number (MPN) tables on p.137 to report.

Examination for psychrophiles
Do counts on Standard Methods agar as for mesophile counts but incubate at 7±1°C for 10 days.

Examination for thermoduric organisms
Pasteurize 5 ml of milk at 62.8°C for 30 min. Do pour-plates with Standard Methods agar containing 10% of sterile milk. Incubate at 23±2°C for 48 h. Express the result as Laboratory Pasteurization Count (LPC)/ml.

Microbial content

Bacteria enter milk during milking and handling. Even with the most hygienic production, some bacteria gain access. The cooling that is normal practice after milking retards the multiplication of these bacteria. Pasteurization, intended to kill pathogenic bacteria, does not necessarily reduce the count of other organisms. It may increase the numbers of thermophiles.

The 'normal microflora' depends on temperature: at 15–30°C *Streptococcus lactis* predominates and many streptococci and coryneform bacteria are present, but at 30–40°C they are replaced with lactobacilli and coliform bacilli. All of these organisms ferment lactose and increase the lactic acid content, which causes souring. The increased acid content prevents the multiplication of putrefactive organisms. Spoilage during cold storage is due to psychrophilic pseudomonads and *Alcaligenes*, psychrotrophic coliforms, e.g. *Klebsiella aerogenes* and *Enterobacter liquefaciens* which are anaerogenic at 37°C and in pasteurized milk, thermoduric coryneform organisms (*Microbacterium lactis*) may be significant. These coryneforms probably come from the animal skin or intestine and from utensils.

At temperatures above 45°C, thermophilic lactobacilli (*Lactobacillus thermophilus*) rapidly increase in numbers.

Gram-negative bacilli are rarely found in quarter samples collected aseptically. These enter from the animal skin and dairy equipment during milking and handling. *Alcaligenes* species are very common in milk. *P. fluorescens* gels UHT milk.

Undesirable microorganisms responsible for 'off flavours' and spoilage include psychrophilic *Pseudomonas*, *Achromobacter*, *Alcaligenes* and *Flavobacterium* spp., which degrade fats and proteins and give peculiar flavours. Coliform bacilli produce gas from lactose and cause 'gassy milk'. *S. cremoris*, *Alcaligenes viscosus* and certain *Aerobacter* species, all capsulated, cause 'ropy milk'. *Oospora lactis* and yeasts are present in stale milk. *P. aeruginosa* is responsible for 'blue milk' and *Serratia marcescens* for 'red milk' (differentiate from bloody milk). *B. cereus* can cause rapid decolorization of methylene blue.

Pathogenic organisms which may be present include *S. aureus*, *Campylobacter*, salmonellas, *Y. enterocolitica*, listerias, *S. pyogenes* and other streptococci from infected udders in mastitis. Tubercle bacilli may be found by culture of centrifuged milk deposits and gravity cream. *B. abortus* is excreted in milk and can be isolated by culture, or antibodies can be demonstrated by the ring test or whey agglutination test. *Rickettsia burnetti*, the agent of Q fever, may be found by animal inoculation of milk from suspected animals.

Mastitis

Looking for evidence of mastitis in bulk milk is obviously unrewarding in any but in farm or other small samples. Centrifuge 50 ml of milk and make films of the deposit. Dry in air, treat with xylene to remove fat, dry, fix and stain with methylene blue. Examine for pus cells. For bacteriological examination and identification of causative organisms, see p.305 and p.310).

Examination of dairy equipment

See Chapter 17 (Environmental microbiology) for methods.

Microbial content of dairy equipment

The types of organisms found will be similar to those in milk. Soil organisms, e.g. *Bacillus* spp., may be present. *B. cereus* reduces methylene blue very rapidly and may also cause food poisoning. Organisms originating from faeces, e.g. various coliforms and *Pseudomonas* spp., are found. Coliform bacilli also rapidly reduce methylene blue. Pseudomonads may produce oily droplets in the milk.

Natural cream

Most cream on sale in the UK is heat treated and there are statutory tests SI 1509. Pasteurized cream must satisfy the three-tube coliform test and the phosphatase test. UHT and sterilized cream the colony count test after preincubation at 37°C for 24 h.

There are no statutory tests for untreated cream but we suggest that it is tested for total count, coliforms and *E. coli*, and for *S. aureus*.

Coliform test for pasteurized cream
Place 90 ml of 2% sodium citrate in a bottle and warm it to 37°C in a water-bath. Warm the cream to the same temperature. Weigh the bottle, add 10 ml of the warm cream and mix well.

Add 10 ml of this 1:10 dilution to each of three bottles of 10 ml double strength brilliant green lactose bile broth (BGLBB) each containing a Durham's tube. Incubate at 30°C for 48 h.

The test is satisfactory if two of the three bottles show no gas production.

This is an unsatisfactory test as the opacity of the cream suspension makes it difficult to see gas in the Durham's tube.

If there is gas in at least two tubes subculture to brilliant green broth or similar medium to verify the presence of coliforms. Other organisms, e.g. *Bacillus* spp., may also give gas.

Special colony count for cream
Use the test for UHT milk.

Viable counts
Do plate counts or Miles and Misra counts on serial dilutions ranging from 10^{-2} to 10^{-6} in a 0.1% peptone water solution, using yeast extract milk agar.

If the cream is solid or difficult to pipette, make the initial 1:10 dilution by weight as described on p.221. The 1:10 dilution may be used, with appropriate media, for spiral plating for total viable, coliform and *S. aureus* counts.

Culture for pathogens
Plate serial dilutions on Baird-Parker medium for *S. aureus*.

For salmonellas pre-enrich a 1:10 mixture of cream (preferably 25 g) in buffered peptone water and proceed as on p.270.

For campylobacters use a 10 g sample and the method on p.175.

Microbial content
Fresh cream may contain large numbers of organisms, including coliforms, *B. cereus*, *E. coli*, *S. aureus*, other staphylococci, micrococci and streptococci. Most fresh cream has been pasteurized and therefore contamination by many of these organisms is due to faulty dairy hygiene.

Cream should not be used in catering in circumstances where these organisms can multiply rapidly (Davis, 1981).

Phosphatase test
Place 15 ml of the buffer reagent (p.211) into each of two tubes. Stopper and bring to 37°C in a water-bath. Weigh the tubes and add 2 g of test cream to one tube and 2 g of boiled cream to the other (control). Mix well and incubate (water-bath) for 2 h, remove and add 0.5 ml of 30% (w/v) zinc sulphate. Mix well and stand for 3 min. Add 0.5 ml of 15% (w/v) potassium ferrocyanide. Mix well and filter through a Whatman No. 40 paper. Read test, if yellow coloured, against blank using a Lovibond Comparator and APTW disc. Record results in μg of *p*-nitrophenol/ml of cream.

Less than 10 μg is satisfactory. If the reading is greater than 10 μg do the verification test for reactivation of phosphatase.

Place 10 g of cream into each of two tubes. To one add 40% (w/v) magnesium chloride according to fat content: double cream, 0.25 ml; whipping cream, 0.35 ml; single cream, 0.5 ml. To the other add nothing. Stopper and mix. Incubate both tubes at 37°C for 60 min with occasional shaking.

Double cream,	0.25 ml;
Whipping cream,	0.35 ml;
Single cream,	0.5 ml. To the other add nothing.

Remove 2 g of cream from each tube and repeat the phosphatase test. If the test (magnesium chloride treated) sample gives a higher reading than the control dilute it 1:4. If the colour is now equal or still more intense than the control reactivation has occurred and the test is reported as satisfactory. If the colour is less than that of the control the positive phosphatase test has been verified.

Imitation cream

Examine for staphylococci only, as for natural cream.

Processed milks

Dried milk

This often has a high count and, if stored under damp conditions, is liable to mould spoilage. Spray-dried milk may contain *S. aureus*.

Examination
Reconstitute in distilled water before testing. Examine direct films, do total counts as for fresh milk but prepare extra plates for incubation at 55°C for 48 h for thermophiles. Do coliform counts to detect contamination after processing. Examine for salmonellas, using at least 100 g samples. Examine for yeasts and moulds as described for butter examination (p.218).

The number of spore bearers is important in milk and milk powder intended for cheese making. Inoculate 10-ml volumes of bromocresol purple milk in triplicate with 10, 1.0 and 0.1-g amounts of milk. Heat at 80°C for 10 min, overlay with 3 ml of 2% agar and incubate for 7 days. Read by noting gas formation and use Jacobs and Gerstein's MPN tables to estimate the counts (p.140).

Condensed milk

This contains about 40% of sugar and is sometimes attacked by yeasts and moulds that form 'buttons' in the product. Examine as for butter (p.218).

Evaporated milk

This is liable to underprocessing and leaker spoilage. *Clostridium* spp. may cause hard swell and coagulation. Yeasts may cause swell. Open aseptically and examine as appropriate (see p.203).

Fermented milk products

Yoghurt, leben, kefir, koumiss, etc., are made by fermentation (controlled in factory-made products) of milk with various lactic acid bacteria, and streptococci and/or yeasts. The pH is usually about 3.0–3.5 and only the intended bacteria are usually present, although other lactobacilli, moulds and yeasts may cause spoilage. The presence of large numbers of yeasts and moulds can be indicative of poor hygiene. Plate on potato glucose agar (pH adjusted to 3.5) and incubate at 23±2°C for 5 days.

Yoghurt is made from milk with *Lactobacillus bulgaricus* and *Streptococcus thermophilus*. In the finished product there should be more than 10^8 of each/g and they should be present in equal numbers. Lower counts and unequal numbers predispose off-flavours and spoilage.

Make doubling dilutions of yoghurt in 0.1% peptone water and mix 5 ml of each with 5 ml of melted LS Differential Medium. Pour into plates, allow to set and incubate at 43°C for 48 h. Both organisms produce red colonies, because the medium contains triphenyl tetrazolium chloride (TTC). Those of *L. bulgaricus* are irregular or rhizoid, surrounded by a white opaque zone; those of *S. thermophilus* small, round and surrounded by a clear zone. Count the colonies and calculate the relative numbers of each organism. Post-pasteurization contamination with coliforms may occur. Check pH, which is usually very acid. If this is pH >5.6, examine for coliforms, *E. coli*, salmonellas and other enteric pathogens (see also Tamine (1981) and Robinson and Tamine (1981)).

Milk-based drinks

There are statutory tests in the UK. Statutory Instrument No. 1508 (1983) specifies these:

Pasteurized drinks	Coliform and or phosphatase tests
Sterilized drinks	Colony count
UHT drinks	Colony count
	Coliform test

Coliform test

Dilute 1:10 in quarter strength Ringer solution. Add 1 ml to each of three tubes containing 5 ml of single strength brilliant green lactose bile broth (BGLBB) containing Durham's tubes. Incubate at 37°C for 48 h and examine for gas production. The test is satisfactory if two of three tubes show no gas.

Subculture from tubes showing gas to brilliant green broth or similar medium. Test typical colonies on this medium for gas production in BGLBB as the original gas may have been produced from sugar in the product.

Phosphatase test
As for milk (p.211).

Colony count
As for UHT milk (p.211).

Buttermilk

Test for coliforms only. Use 10 ml of a 1:10 dilution in a pour-plate as for creams.

Butter

Soured cream is pasteurized and inoculated with the starter, usually *S. cremoris* or a *Leuconostoc* spp. Salted butter contains up to 2% of sodium chloride by weight.

Examination and culture
Emulsify 10 g of butter in 90 ml of warm (40–45°C) diluent and do plate counts. Culture on a sugar-free nutrient medium and tributyrin agar.

Microbial content
Correct flavours are due to the starters, which produce volatile acids from the acids in the soured raw material.

Rancidity may be due to *Pseudomonas* and *Alcaligenes* spp., which degrade butyric acids. Anaerobic butyric organisms (*Clostridium*) cause gas pockets. Some lactic acid bacteria (*Leuconostoc*) cause slimes and others give cheese-like flavours. Coliforms, lipolytic psychrophiles and casein-digesting proteolytic organisms, which give a bitter flavour, may be found. To detect these, plate on Standard Plate Count agar containing 10% sterile milk and incubate at 23±2°C for 48 h. Flood the plates with 1% hydrochloric acid or 10% acetic acid. Clear zones appear round the proteolytic colonies.

Yeasts and moulds may be used as an index of cleanliness in butter. Adjust the pH of potato glucose agar to 3.5 with 10% tartaric acid and inoculate with the sample. Incubate at 23±2°C for 5 days. Report the yeast and mould count/ml or /g. Various common moulds grow on the surface but often these grow only in water droplets or in pockets in the wrapper (see Murphy, 1981).

Cheese

Many different cheeses are now available. Some are traditionally made from raw milk and many contain pathogens such as *S. aureus*, brucellas, campylobacters and/or listerias.

Milk is inoculated with a culture of a starter, for example *S. lactis*, *S. cremoris*, *S. thermophilus* and rennet added at pH 6.2–6.4. To make hard cheeses, the curd is cut and squeezed free from whey, incubated for a short period, salted and pressed. The streptococci are replaced by lactobacilli (e.g. *L. casei*) naturally at this stage and ripening begins. Some cheeses are inoculated with *Penicillium* spp. (e.g. *P. roquefortii*), which give blue veining and characteristic flavours due to the formation under semi-anaerobic conditions of caproic and other alcohols.

Soft cheeses are not compressed, have a higher moisture content and are inoculated with fungi (usually *Penicillium* spp., which are proteolytic and flavour the cheese).

Types and flavours of cheeses are due to the use of different starters and ripeners and varying storage conditions.

Examination and culture
Use the sampling method described by Law *et al.* (1973). Take representative core samples aseptically, grate with a sterile food grater, thoroughly mix and sub-

sample. Fill the holes, left as a result of boring, with Hansen's paraffin cheese wax to prevent aeration, contamination and texture deterioration.

Make films, de-fat with xylol and stain. Bacteria may be present as colonies: examine thin slices under a lower-power microscope. Weigh 10 g and homogenize it in a blender with 90 ml of warm diluent. Do plate counts on yeast extract milk agar and coliform counts. Culture on whey and tributyrin agar and on media for staphylococci.

Microbial content

Apart from streptococci, lactobacilli and fungi that are deliberately inoculated or encouraged the following organisms may be found: contaminant moulds, *Penicillium, Scopulariopsis, Oospora, Mucor* and *Geotrichum* give colours and off-flavours. Putrefying anaerobes (*Clostridium* spp.) give undesirable flavours. *Rhodotorula* gives pink slime and *Torulopsis* yellow slime. Gassiness (unless deliberately encouraged by propionibacteria in Swiss cheeses) is usually due to *Enterobacter* spp., but these are not found if the milk is properly pasteurized.

Gram-negative rods, some of which may hydrolyse tributyrin, may also be present.

Psychrophilic spoilage is common and is due to *Alcaligenes* and *Flavobacterium* spp. Counts may be very high.

Bacteriophages which attack the starters and ripeners can lead to spoilage.

Propionibacteria

Although deliberately introduced into some cheeses these can cause spoilage in others. Accurate counts are difficult to do because the organisms are microaerophilic.

Prepare doubling dilutions of cheese homogenate in 0.1% peptone water and add 1 ml of each to tubes of yeast extract agar containing 2% sodium lactate or acetate. Mix and seal the surface with 2% agar. Incubate anaerobically or under carbon dioxide at 30°C for 7–10 days. Propionibacteria produce fissures and bubbles of gas in the medium. Choose a tube with a countable number of gas bubbles and calculate the numbers of presumptive propionibacteria (some other organisms may produce gas bubbles) (Harrigan and McCance (1966) modified).

Staphylococcus aureus

Outbreaks of food poisoning caused by this product are occasionally reported. Cheddar cheeses may have a high staphylococcus count. In the USA cottage cheeses have been incriminated and in some states there is a statutory limit of 50 *S. aureus*/g. Enterotoxin may be produced during the long setting period, although the organisms themselves tend to die out on subsequent storage (see also Chapman and Sharp, 1981).

Listeria monocytogenes

This is known to survive some cheese making processes and to cause human disease. For method of isolation see p.324.

Eggs

There are few bacteria in new-laid eggs and these are mostly in the yolk. Eggs are infected through the shell, which is normally impervious but may be damaged by rough handling. Removal of the 'bloom' by washing also permits the entry of microorganisms.

Counts
Do counts on shell eggs separately or in batches of ten. Homogenize the egg and macerate the shell before making counts. Thaw frozen eggs and reconstitute dried egg in warm peptone water diluent. Take frozen egg samples with a metal core-sampler.

Culture
Add 10–20 ml of egg to 100 ml of glucose tryptone broth. The egg must be well diluted because natural lysozyme prevents the growth of some bacteria which might be important when the eggs are used. Plate on glucose tryptone agar, blood agar, bile salt agar, inoculate cooked meat medium (anaerobic) and media for fungi if indicated.

Microbial content

Shell eggs
Externally, various salmonellas, especially *S. enteritidis*, coliforms, proteus, pseudomonas, *B. cereus*, yeasts, moulds and putrefactive anaerobes may be present. Internally, *S. gallinarum* and *S. pullorum*, coliforms, mycoplasmas, avian strains of mycobacteria and *Pasteurella anatipestifer* may be found. The degree of contamination is related to the rate of cooling and age of the egg, the porosity of the shell and humidity of the environment. *Proteus* causes coloured rots, e.g. 'black rot', in eggs.

Frozen egg
Total counts are high. Cultures may contain pseudomonas, aeromonas, achromobacter, alcaligenes, enterobacteria, serratia, micrococci and putrefactive anaerobes. Counts less than 50 000 cfu/g are rare; usually the count is several million and three million has been suggested as a reasonable maximum.

Frozen egg is invariably pasteurized nowadays, although this presents heat penetration problems and the α-amylase test is done to confirm adequate heat treatment by analytical chemists. Egg is incubated with a standard starch solution and the blue colour produced when iodine is added is measured. Adequate pasteurization, which reduces the bacterial content and destroys salmonellas, destroys most of the amylase. This test is less subject than bacteriological examination to sampling errors.

Dried egg
The total count is usually low (5000–10 000 cfu/g) in dried whole egg. Coliforms are rarely present in 1 g.

Dried albumen gives variable results. High-quality spray-dried material gives counts usually not exceeding 10 000 cfu/g but in some albumens counts may be as high as 10 million. Cross contamination may occur in tray drying of flake albumen.

Frozen and dried egg are recognized vehicles of salmonellas. Culture replicate samples of up to 20 g in not less than 100 ml of medium (because of lysozyme). See standard method for salmonellas, p.270.

Ice-cream and ice-lollies (water-ices)

The Food Standards (Ice-Cream) Regulations (1959) prescribe a standard of composition for ice-creams, dairy ice-cream (dairy cream ice or cream ice), 'parvex' (kosher) ice-cream and milk ice (including milk ice containing fruit, fruit pulp or fruit purée).

These regulations do not cover such articles as ice-lollies, sorbets or fruit ices and in deciding which tests are applicable to a product submitted for examination, it should be noted that if ice-cream is present as a core or covering, or if milk fat and milk solids other than fat are claimed to be present in the proportions of not less than 2.5 and 7%, respectively, the article qualifies as an ice-cream. A product containing no ice-cream or milk and consisting of fruit juices or pulp is treated as an ice-lolly.

Sampling

Send samples to the laboratory in a frozen condition, preferably packed in solid carbon dioxide. Remove the wrappers of small samples aseptically and transfer to a screw-capped jar. Sample loose ice-cream with the vendor's instruments (which can be examined separately if required, see Chapter 17). Keep larger samples in their original cartons or containers.

Hold ice-lollies by their sticks, remove the wrapper and place the lolly in a sterile screw-capped 500-ml jar. Break off the stick against the rim of the jar.

Reject samples received in a melting condition in their original wrappers.

Allow the samples to melt in the laboratory but do not allow their temperature to rise above 20°C.

Viable counts

The difficulty with bacterial counts is the measurement of volumes of a product that contains a variable amount of air. For the best results, weigh about 10 ml of ice-cream into a screw-capped jar, multiply the weight to the nearest 0.1 g by nine and add this volume of 0.1% peptone water. Shake well. From this initial 1:10 dilution make two further tenfold serial dilutions and do plate counts on 1-ml samples with Yeastrel agar. Incubate at 37°C for 48 h.

Alternatively, and for routine work, make the initial 1:10 dilutions by adding 10 ml of melted ice-cream to 90 ml of Ringer solution. Smaller amounts are less reliable.

Miles and Misra counts (p.131) are very convenient. Dilute the ice-cream 1:5 and drop three 0.02-ml amounts with a 50-dropper on blood agar plates. Allow the drops to dry, incubate at 37°C overnight and count the colonies with a hand lens.

Multiply the average number of colonies per drop by 250 (i.e. dilution factor of 5 and volume factor of 50) and report as count/ml. If there are no colonies on any drops, the count can be reported as less than 100/ml. It is difficult to count more than 40 colonies per drop; therefore, report such counts as greater than 10 000 cfu/ml.

The spiral plating method (p.85) may be used for total counts.

In counts, note the presence of *B. cereus*. This spore-bearer may be present in the raw materials. Its spores resist pasteurization.

Counts at 0–5°C for psychrophiles may be useful in soft ice-cream.

Coliform test (US)
To test for coliforms, dilute the sample 1:10 and use the pour-plate technique described for milks (p.211).

Ice-lollies

Measure the pH of the melted product. If it is 4.5 or less, no bacteriological examination is necessary. A few yeasts may be present. When the pH is higher, do plate or Miles and Misra counts and coliform counts by the MPN or drop count methods.

Aseptically separate the components of lollies that have ice-cream cores while still frozen and treat separately as above.

Ingredients and other products

Do plate counts at 5, 20 and 37°C. Although ice-cream is pasteurized and then stored at a low temperature, only high-quality ingredients with low counts will give a satisfactory product.

In cases of poor grading or high counts on the finished ice-cream, test the ingredients and also the mix at various stages of production. Swab-rinses of the plant are also useful (see Chapter 17 and Rothwell, 1981).

Human milk

The milk is usually tested before and after pasteurization. Cooling water in the pasteurization unit is potentially hazardous. There should be 10 ppm residual free chlorine to eliminate pseudomonads, etc. Mains water should be used.

Raw milk

Use dip slides (nutrient agar, blood agar, MacConkey and CLED). Incubate at 37°C overnight and count colonies on all media. Note presence and numbers of coliforms, *S. aureus*, streptococci, pseudomonads

$$\frac{\text{Colony count}}{\text{Medium surface area (cm)}} \times 100 = \text{cfu/ml}$$

Reincubate dip slides at room temperature for 24 h to detect psychrotrophs. Confirm identity of potential pathogens.

The levels for acceptance and pasteurization are:

Total count: not exceeding 10 cfu/g
Normal skin flora: not exceeding 10 cfu/g.

Absence of *S. aureus* and other potential pathogens. Milk not satisfying these criteria should be discarded.

Pasteurized milk

Do phosphatase test as for cows' milk (p.211) using boiled human milk controls.

For total count, spread-plate 0.1 ml (standard loopful) on two plates of blood agar. Incubate both at 37°C overnight, one aerobically, the other anaerobically. Count colonies and multiply by 100 to give count/ml of original milk. Satisfactory samples contain less than 100 cfu/g.

For more information see Williamson *et al.* (1978), West *et al.* (1979), Hewitt (1981-2) and DHSS (1982).

References

APHA (1960) *Standard Methods for the Examination of Dairy Products*, American Public Health Association, Washington DC

BSI (1984) *BS 4285: Microbiological Examinations for Dairy Purposes*, British Standards Institution, London

Chapman, H. R. and Sharpe, M. E. (1981) Microbiology of cheese. In *Dairy Microbiology*, Vol. 2, (ed. R. K. Robinson), Applied Science Publishers, London, pp. 157–243

Davis, J. G. (1981) Microbiology of cream and dairy desserts. In *Dairy Microbiology*, Vol. 2, (ed. R. K. Robinson), Applied Science Publishers, London

Davis, J. G. (1986) Dairy products. In *Quality Control in the Food Industry* Vol. 2, (ed. S. M. Herschdorfer), Academic Press, London, pp.47–272

DHSS (1982) *The Collection and Storage of Human Milk*, Reports on Health and Social Subjects No. 29, Department of Health and Social Security, HMSO, London

EEC (1985) Directive on health and animal problems affecting intra-community trade in heat treated milk, 85/397/EEC

Harrigan, W. F. and McCance, M. E. (1966) *Laboratory Methods in Microbiology*, Academic Press, London, p. 176

Hewitt, J. H. (1981-2) Possible bacteriological hazards associated with the use of raw (untreated) human milk for special care infants. In *Human Milk Banking*. Report Series: Refrigeration and Science Technology, International Institute of Refrigeration, Paris

Law, B. A., Sharpe, M. E., Mabbitt, L. A. and Cole, C. B. (1973) Microflora of Cheddar cheese. In *Sampling – Microbiological Monitoring of Environments* (ed. R. G. Board and D. W. Lovelock), Society for Applied Bacteriology Technical Series No. 7, Academic Press, London

Murphy, M. F. (1981) Microbiology of butter. In *Dairy Microbiology*, Vol. 2 (ed. R. K. Robinson), Applied Science Publishers, London, pp. 91–111

Robinson, R. K. and Tamine, A. Y. (1981) Microbiology of fermented milk. In *Dairy Microbiology*, Vol. 2 (ed. R. K. Robinson), Applied Science Publishers, London, pp. 245–278

Rothwell, J. (1981) Microbiology of ice cream and related products. In *Dairy Microbiology*, Vol. 2 (ed. R. K. Robinson), Applied Science Publishers, London, pp. 1–30

Speck, M. (ed.) (1984) *Compendium of Methods for the Microbiological Examination of Foods*, American Public Health Association, Washington

Tamine, A. Y. (1981) Microbiology of starter cultures. In *Dairy Microbiology*, Vol. 2 (ed. R. K. Robinson), Applied Science Publishers, London, pp. 245–278

West, P. A., Hewitt, J. H. and Murphy, O. M. (1979) The influence of methods of collection and storage on the bacteriology of human milk. *Journal of Applied Bacteriology*, **46**, 269–277

Williamson, S., Hewitt, J. H., Finucane, E. and Gamsu, H. R. (1978) Organization of banks for raw and pasteurized human milk for neonatal intensive care. *British Medical Journal*, **1**, 393–396

Chapter 17

Environmental and container microbiology

An estimate of the numbers of bacteria in the environment may be necessary for a variety of reasons:

(1) to test the standard of hygiene and efficiency of cleaning procedures in hospital wards, kitchens, canteens, operating theatres, food factories, dairies, shops, restaurants, offices or schools;
(2) to assess the level of bacteria in 'sterile' environments, e.g. in the pharmaceutical industry;
(3) to trace the route of contamination from dirty to clean situations;
(4) to educate staff.

There are four main methods of sampling:

(1) contact plates or slides using appropriate media,
(2) swabbing,
(3) rinsing,
(4) air sampling.

The choice of method will depend upon the situation. In general, contact methods are best for 'clean' surfaces but swabs for 'dirty' ones because it is possible to dilute the swab washings. It is fundamental to the success of any investigation that the microbiologist is fully conversant with the problem and is prepared to go and look (and smell) at the site and talk to the personnel concerned. For background reading see Howie (1979).

Control procedures

Agar contact for flat or nearly flat surfaces

Agar-filled contact ('Rodac') plates and agar-covered slides are available from several companies. Some dip slides may be used as contact slides. The manufacturer's instructions should be followed. It is possible to estimate total numbers, enterobacteria, staphylococci, yeasts and fungi using an agar contact method. Apart from being very easy to use they have the additional merit of encouraging cleaning staff. Before showing agar 'contacts' to staff the cultures should be sprayed with domestic hair spray (vinyl acetate in methylated spirit) to minimize the risk of infection. Cultures should always be returned to the laboratory for autoclaving or incineration.

Surface swab counts

It is generally accepted that the swabbing technique gives a count approximately ten times higher than that obtained by agar surface contact when sampling smooth surfaces.

Cut card or cellophane squares with 10-cm sides and cut squares in them with 5-cm sides. Sterilize these templates in envelopes. Place a template on the surface to be examined and swab the area within the 5×5 cm square with a cotton wool or alginate swab. Treat these swabs as described below. The count/25 cm^2 is given by the number of colonies/ml of rinse or solvent multiplied by 10. With Miles and Misra counts, it is given by the number of colonies/5 drops multiplied by 100.

The swab rinse method for crockery, cartons and containers

Dip a swab in sterile 0.1% peptone water and rub over the surface to be tested, for example the whole of the inside of the cup or glass or the whole surface of a plate. Use one swab for five such articles. Use one swab for a predetermined area of a cutting table, chopping board, etc. Swab both sides of knives, ladles, etc. Return this swab to the tube and swab the same surfaces again with another, dry swab.

To the tube containing both swabs add 10 ml of 0.1% peptone water. Shake and stand for 20–30 min. Do plate counts with 1.0 and 0.1 ml amounts using yeast extract agar. Divide the count/ml by 5 to obtain the count per article. Inoculate three tubes of single-strength MacConkey broth with 0.1-ml amounts.

There appear to be no current standards for the bacteriology of crockery and utensils used by the public in catering establishments but during the 1940s the USPHS required counts to be less than 100 per article, with no coliform bacilli.

Alginate swabs and drop counts

Buy or make the swabs of alginate wool and sterilize by autoclaving. Proceed as above but add 9 ml of Calgon Ringer's solution. Shake gently; the swabs will dissolve in 1 or 2 min. Do Miles and Misra counts (p.131) with 6 drops from each pair of swabs on one blood agar and one MacConkey plate. Two colonies or less per 6 drops on the blood agar plate is within the USPHS limits. This method also allows the organisms to be identified. Apart from coliform bacilli, the presence of respiratory organisms such as viridans and salivary streptococci, staphylococci and neisseria is evidence of inefficient sanitation.

Alginate swabs are considered by some workers to give less efficient recovery of sublethally damaged organisms.

Rinse method

For churns, bins and large utensils

Add 500 ml of 0.1% peptone water to the vessel. Rotate the vessel to wash the whole of the inner surface and then tip the rinse into a screw-capped jar.

Do counts with 1.0- and 0.1-ml amounts of the rinse in duplicate. Incubate one pair at 37°C and the other at 22°C for 48 h.

Take the mean of the 37°C and 22°C counts/ml and multiply by 500 to give the count per container.

For milk churns, counts of not more than 50 000 cfu per container are regarded as satisfactory, between 50 000 and 250 000 as fairly satisfactory and over 250 000 as unsatisfactory.

For milk, soft drink bottles and jars
If bottles are sampled after they have been through a washing plant where a row of bottles travels abreast through the machine, test all those in one such row. This shows if one set of jets or carriers is out of alignment. In any event, examine not less than six bottles. Cap or stopper them immediately. To each bottle add 20 ml of 0.1% peptone water. Close with sterile rubber bungs and roll the bottles on their sides so that all of the internal surface is rinsed. Leave them on their sides and roll at intervals for half an hour.

Pipette 5 ml from each bottle into each of two petri dishes. Add 15–20 ml of yeast extract agar, mix and incubate one plate from each bottle at 37°C and one at 22°C for 48 h. Pipette 5 ml from each bottle into 10 ml of double-strength MacConkey broth and incubate at 37°C for 48 h.

Take the mean of the 37°C and 22°C plate count and multiply by 4 to give the count per bottle. Find the average of the counts, omitting any figure that is 25 times greater than the others (indicating a possible fault in that particular line).

The figure obtained is the Average Colony Count per container. In the Ministry of Agriculture and Fisheries (1947) classification, milk bottles giving average colony counts of 200 cfu or less are satisfactory; counts from 200 to 600 cfu are regarded as fairly satisfactory, but over 600 as unsatisfactory.

Coliform bacilli should not be present in 5 ml of the rinse.

Membrane filter method
Examine clean bottles and jars by passing the rinse through a membrane filter. Place the membrane on a pad saturated with an appropriate medium, e.g. double-strength tryptone soya broth for total count. Incubate at 35°C for 18–20 h. Stain the membrane with methylene blue to assist in counting the colonies under a low-power lens. Examine another membrane for coliforms using MacConkey membrane broth or membrane enriched lauryl sulphate broth. Incubate at 35°C for 18–24 h. Staining to reveal colonies is unnecessary. Subculture suspect colonies into lactose peptone water and incubate at 37°C for 48 h to confirm gas production.

Roll-tube method for bottles

Use nutrient or similar agar for most bacteria, deMan, Rogosa and Sharp (MRS) medium for lactobacilli, reinforced clostridial medium (RCM) for clostridia and buffered yeast agar for yeasts. Increase the agar concentration by 0.5%. Use the roll-tube MacConkey agar for coliforms.

Melt the medium and cool it to 55°C. To 1-quart or 1-litre bottles add 100 ml of medium, and to smaller bottles proportionally less. Stopper the bottles with sterile rubber bungs and roll them under a cold water tap to form a film of agar over all the inner surface (see Roll-tube counts, p.131). Incubate vertically and count the colonies. For lactobacilli and clostridia, replace the bung with a cotton wool plug and incubate anaerobically.

Vats, hoppers and pipework

Large pieces of equipment are usually cleaned in place (CIP). Test flat areas by agar contact or swabbing. Take swab samples of dead ends of pipework and crevices. The dairy industry has devised a simple and effective method for testing filling equipment. Sample the first, 100th and 200th container. The first will contain any residual bacteria not killed by CIP treatment. If this has not been effective, sample 1 will give a higher count than samples 100 and 200. If all three samples are satisfactory then the cleaning was efficient.

Examination of sink waters, cloths, towels, etc.

To demonstrate unhygienic conditions and the necessity for frequent changes of washing water and cloths, examine by agar contact or sample and examine as follows.

Take 100 ml of washing or rinse water by immersing a water sample bottle (containing sodium thiosulphate in case hypochlorites are used in washing-up) into the sink and allow it to fill. Stopper and cool under a tap. Do plate and coliform counts. Test at the beginning and at intervals during the washing-up.

Spread a wash-cloth or drying cloth over the top of a screw-capped jar of known diameter. Pipette 10 ml of 0.1% peptone water on the cloth so that the area over the jar is rinsed into it. Do plate counts and compare with freshly laundered cloths. If a destructive technique is possible, cut portions of cloths, sponges or brushes with sterile scissors and add to diluent.

Mincers, grinders, etc.

After cleaning, rinse with 500 ml of 0.1% peptone water. Treat removable parts separately by rinsing them in a plastic bag in diluent. Do colony count on rinsings as for milk bottles.

Counts on working and food surfaces, floors and walls, etc.

These are used to assess hygienic conditions.

Chopping blocks

Organisms are sometimes deeply embedded in these blocks. Sample by taking scrapings from representative areas, e.g. 100 cm^2, with a sterile scalpel. Disperse the scrapings in warm diluent and do counts.

Air sampling

Contamination of air in hospitals and offices can lead to infection and cross-infection. In factories and canteens it may lead to contamination of product or food. Bacteria and fungi may be suspended in air singly, in small clumps, or as large aggregates. Smaller particles or droplets containing them (e.g. 5 μm) may remain suspended for long periods and are easily moved by air currents. Larger particles settle rapidly and contaminate surfaces.

Settle plates
Settle plates supplement surface sampling for assessing potential surface contamination. Several plates containing appropriate media are exposed for a given time and incubated. The colonies are counted. Settle plates are favoured by those who need to monitor the air over long periods, e.g. in hospital cross-infection work. Expose blood agar to test for the presence of 'presumptive' *S. aureus* and selective media, e.g. for enterococci. This method is less satisfactory than using slit samplers for testing for very small suspended particles.

Air samplers
Sampling the air itself is best for assessing the load of smaller particles. Equipment for air sampling has been described by various workers. Bourdillon *et al.* (1941) used a slit sampler beneath which an agar plate revolved. Andersen (1958) devised a sieve sampler which enumerates and facilitates the counting of viable airborne particles of differing sizes.

The Casella (UK) and Reynier (USA) models are both slit samplers. The Fisons model (UK) draws air through a membrane filter. Kitchell *et al.* (1973) tested all three and found the Casella to the most flexible and efficient, the Reynier smaller and more readily transported than the Casella, while the Fisons model was the easiest to use in field studies.

Recently, hand-held air samplers have appeared on the market. A surface air system (SAS) uses contact plates and is portable as is the Reuter Centrifugal Sampler (RCS). Air is subjected to centrifugal acceleration and particles are impacted on to an agar-coated strip which fits round a drum. These hand-held samplers are very convenient because they are portable, battery operated and therefore do not require on-site electricity. Results should be interpreted with caution but they are satisfactory for repeated monitoring. Nakla and Cummings (1981) have compared the performance of the RCS with the more conventional slit air sampler in hospital use. Clark *et al.* (1981) and Casewell *et al.* (1984, 1986) have also studied the RCS system. Millipore publish booklets describing their Sterifil system. Air is streamed into a special broth which is passed through a membrane filter. The filter is then transferred to a pad impregnated with suitable medium and incubated appropriately.

The Porton Impinger (May and Harper, 1957) resembles a chemist's Drechsel bottle. An air stream pulled through it at a standard rate by a vacuum pump impinges on the surface of a broth medium. The viable organisms in the medium are then enumerated, e.g. by membrane filtration.

References

Andersen, A. A. (1958) New sampler for collection, sizing and enumeration of viable airborne particles. *Journal of Bacteriology,* **76,** 471–480

Bourdillon, R. B., Lidwell, O. M. and Thomas, J. C. (1941) A slit sampler for collecting and counting airborne bacteria. *Journal of Hygiene (Cambridge),* **41,** 197–201

Casewell, M. W., Farmer, P. G. and Simmons, N. A. (1984) Bacterial air counts obtained with a centrifugal (RCS) sampler and a slit sampler. The influence of aerosols. *Journal of Hospital Infection,* **5,** 76–82

Casewell, M. W., Desai, N. and Lease, N. J. (1986) The use of the Reuter centrifugal air sampler for the estimation of bacterial air counts in different hospital locations. *Journal of Hospital Infection*, **7**, 250–260

Clark, S., Lach, V. and Lidwell, O. M. (1981) The performance of the Biotest RCS centrifugal air sampler. *Journal of Hospital Infection*, **2**, 181–186

Howie, J. W. (1979) Policies in the UK to ensure that a food factory does not distribute food-poisoning organisms. *Journal of Applied Bacteriology*, **47**, 233–236

Kitchell, A. G., Ingram, G. C. and Hudson, W. R. (1973) Microbiological sampling in abbatoirs. In *Sampling – Microbiological Monitoring of Environments* (eds R. G. Board and D. W. Lovelock), Society for Applied Bacteriology Technical Series No. 7, Academic Press, London, pp. 43–59

May, K. R. and Harper, G. J. (1957) The efficiency of various liquid impinger samplers in bacterial aerosols. *British Journal of Industrial Medicine*, 14, 287–297

Nakla, L. S. and Cummings, R. F. (1981) A comparative evaluation of a new centrifugal air sampler (RCS) with a slit sampler (SS) in a hospital environment. *Journal of Hospital Infection*, **2**, 261–266

Chapter 18

Water

Stored and river water may contain a wide variety of organisms, including pseudomonads, flavobacteria, micrococci, aerobic and anaerobic sporebearers, enterobacteria and streptomycetes. Piped water may, in addition, contain iron bacteria and river waters both iron bacteria and sulphate reducers.

From the public health point of view, the coliform test is the most important as the presence of these organisms, particularly *E. coli*, indicates if not actual pollution then a less than satisfactory supply. Tests for enterococci and *C. perfringens* are also useful.

As water is not the natural habitat of the enterobacteria and these organisms do not multiply in reasonably clean water, care must be taken in selecting media for their isolation. The organisms may be still viable but damaged and therefore media should not contain inhibitory substances.

In the UK, water is examined for coliforms by the multiple tube (MPN) method or the membrane filter technique. Minerals Modified Glutamate (MMG) is the medium of choice for the MPN method and lauryl sulphate has replaced Teepol in membrane culture because the latter is difficult to obtain.

Officially approved methods for examining water and for interpreting the results are published in the UK by DHSS (1985) and in the USA by the American Public Health Association (APHA, 1986).

Sampling

Drinking water

Samples are collected by health inspectors and water engineers. The laboratory should supply 120-ml glass bottles with glass, dust-proof stoppers, covered with kraft paper tied at the neck and sterilized in a hot air oven or autoclave. For samples of chlorinated waters, the bottles must contain sodium thiosulphate (0.1 ml of a 3% solution) to neutralize residual chlorine. These should be freshly prepared every 3 days. Larger bottles (300 ml) are required if the membrane filtration method is used.

Samples must be delivered to the laboratory within 6 h of sampling to ensure a meaningful report.

Plate counts

These are not usually done as a routine in the UK, except on new supplies of raw water. In the USA, plate counts are standard procedure.

Appropriate dilutions are prepared in phosphate buffer, mixed with Standard Plate Count agar and incubated at either 20°C for 48 h or 35°C for 24 h. Plates showing between 30 and 300 colonies are counted (see Chapter 9).

Coliform test: MPN method with Minerals Modified Glutamate broth

Select the range according to the expected purity of the water:

Mains chlorinated water	A and B
Piped water, not chlorinated	A, B and C
Deep well or borehole	A, B and C
Shallow well	B, C and D
No information	A, B, C and D

A: 50 ml of water to 50 ml of double-strength broth.
B: 10 ml of water to each of five tubes of 10 ml of double-strength broth.
C: 1 ml of water to each of five tubes of 5 ml of single-strength broth.
D: 0.1 ml of water to each of five tubes of 5 ml of single-strength broth.

Incubate at 35–37°C and note the numbers of tubes showing acid and gas at 48 h. Tap any tubes showing no gas. A bubble may then form in the Durham's tube. Consult the MPN tables (Tables 9.1–9.3) and read the most probable number of *presumptive* coliform bacilli/100 ml of water. Small amounts of gas occurring after 48 h in *presumptive* tubes are disregarded unless the presence of coliform bacilli is confirmed by plating.

From each tube showing acid and gas, inoculate a tube of brilliant green bile salt broth and a tube of peptone water. Incubate these at 44°C for 24 h in a reliable water-bath (Eijkman test) along with controls of known strains of *E. coli* (which grows at 44°C) and *K. aerogenes* (which does not). Plate also from positive tubes on MacConkey agar and nutrient agar.

Observe gas formation at 44°C and test the peptone–water culture for indole. Do oxidase test on growth from nutrient agar. *Aeromonas* spp., which may give acid and gas in lactose broth, are positive. Only *E. coli* produces gas *and* indole at 44°C.

Read the most probable numbers of *E. coli* ('faecal coli') from Tables 9.1–9.3.

Translation from the MPN tables of the 44°C positive tubes sometimes causes difficulty. Remember that the organisms cultured from any positive 37°C tube and grown at 44°C represent coliforms cultured from the volume of water placed in the 37°C tube. For example:

	50 ml	*10 ml*	*1 ml*	*MPN/100 ml*
Tubes positive at 37°C	1	2	2	10 'presumptive coli'
Tubes positive at 44°C	1	1	0	3 *E. coli*

For further investigation, pick colonies from the MacConkey plate into peptone water, glucose phosphate medium and citrate medium for indole, MR, VP and citrate utilization tests (Chapter 21).

Acid and gas in MMG medium, as in MacConkey broth, may occasionally be due to spore bearers, e.g. *C. perfringens* at both 37 and 44°C. These organisms do not grow in brilliant green broth or on the MacConkey plate.

Most raw waters in the UK showing acid and gas do in fact contain coliform bacilli but in about 5% of chlorinated waters acid and gas is due to *C. perfringens*.

Coliform test: US method (APHA, 1986)

In the US Standard Method, 15 fermentation tubes, each containing 20 ml of 0.5% lactose broth are used. They are inoculated with 5×10 ml, 5×1 ml and 5×0.1 ml of water sample. These are incubated at 35°C and examined for gas production after 24 h and 48 h. Gas within 48 h is presumptive evidence of coliform bacilli.

Confirmatory test

Tubes showing gas are subcultured on eosin methylene blue (EMB) agar, incubated at 35°C for 24 h and examined for typical colonies of *E. coli*. If atypical colonies are seen, the 'completed test' is carried out.

Completed test

Several colonies from the EMB plate are subcultured into lactose broth fermentation tubes and on a nutrient agar slope. Both are incubated at 35°C for 24 h. Gas in the broth and a Gram-negative non-sporing rod on the slope is evidence of coliform bacilli.

Faecal coli test

Tubes showing gas are subcultured into EC or similar medium and incubated at $44 + 0.2$°C for 48 h to test for gas production, which indicates faecal coli. Colonies from solid medium are subcultured to lactose broth and if gas is produced in these they are further subcultured into EC broth as above. For full technical methods and directions, see Standard Methods (APHA, 1986).

Coliform test: membrane filter method

Advantages of using membrane filter techniques for waters

(1) Speed of obtaining results.
(2) Saving of labour, media, glass and cost of materials if the filter is washed and re-used.
(3) Sample can be filtered on site, if the filter is placed on transport medium and posted to the laboratory, thus avoiding delay in transporting the sample.
(4) Organisms can very easily be exposed to pre-enrichment media for a short time at an advantageous temperature.

Disadvantages of using membrane filter techniques for waters

(1) There is no indication of gas production (some waters contain large numbers of non-gas producing lactose fermenters capable of growth in the medium).
(2) Membrane filtration is unsuitable for waters with high turbidity and low counts because the filter will become blocked before sufficient water can pass through it.
(3) Large numbers of non-coliform organisms capable of growing on the medium may interfere with coliform growth.

If large numbers of water samples are to be examined and much field work is involved the membrane method is undoubtedly the most convenient. The booklets published by Millipore and Gelman give valuable advice and information about all aspects of the application of this technique.

Pass two separate 100-ml volumes of the water sample through 47-mm membrane filters. If the supply is known or is expected to contain more than 100 coliform bacilli/100 ml, use 10 ml of water diluted with 90 ml of quarter-strength Ringer's solution.

Place sterile Whatman No. 17 absorbent pads in sterile petri dishes and pipette 2.5–3 ml of enriched lauryl sulphate broth over the surface. Place a membrane face up on each pad and incubate as follows.

Chlorinated samples
One membrane at 30°C for 6 h followed by 35°C for 18 h for the *presumptive* count and one membrane at 25°C for 6 h followed by 44±0.2°C for 18 h for the *E. coli* count.

Unchlorinated samples
One membrane at 30°C for 4 h followed by 35°C for 14 h for the *presumptive* count and one membrane at 30°C for 4 h followed by 44±0.2°C for 14 h for the *E. coli* count.

These times allow for 'resuscitation' of coliforms.

For incubation in water-baths at 44°C, waterproof submersible boxes are required. Several laboratory suppliers sell these. Methods of automatic changes in temperature in incubators and water-baths are mentioned on p.10.

Counting

Count the yellow colonies only and report as *presumptive* coliform and *E. coli* count/100 ml of water. Membrane counts may be higher than MPN counts because they include all organisms producing acid, not only those producing acid and gas. *C. perfringens* does not grow.

Confirmatory tests for *Escherichia coli*

Subculture colonies to lauryl sulphate tryptose broth for gas production and peptone water for indole test. Incubate broth at 44±0.2°C overnight.

Commercial methods using β-glucuronidase activity, which is shown only by *E. coli* and shigellas are now available. The API RAPIDEC coli uses glucopyranosiduronic acid as substrate. This is hydrolysed to a yellow compound by β-glucuronidase. Tests for indole and β-galactosidase are also used in this kit and improve the quality of the results. Other kits use 4-methylumbelli-feryl-β-D.glucuronide (MUG) which is hydrolysed to give a fluorescent end-product visible under UV light.

Membrane filter: US method

The methods are very similar to those above, but the filter is rinsed with three volumes of 20–30 ml of sterile buffered water before the sample is passed through it. Membrane agar medium (EMB) or membrane fluid (EMB) media are used. Pre-enrichment, where necessary, is effected on pads saturated with lactose broth at 35°C for 2 h.

Full details are given in the Standard Methods (APHA, 1986).

Faecal streptococci in water

These organisms are useful indicators when doubtful results are obtained in the coliform test. They are more resistant than *E. coli* to chlorine and are therefore useful when testing repaired mains. Group D organisms only are significant.

MPN method

Use one of the azide broths, e.g. azide glucose broth, Enterococcus Presumptive Broth or Slanetz and Bartley Broth.

Add 50 ml of water to 50 ml of double-strength medium.

Add 10 ml of water to each of five tubes of 10 ml of double-strength broth.

Add 1 ml of water to each of five tubes of 5 ml of single-strength broth.

Incubate at 37°C for 72 h. Subculture any tubes showing acid production to tubes of single-strength medium and incubate at 44–45°C for 18 h. Record tubes showing acid and consult the MPN tables (pp.137–139). Confirm by microscopic examination for short-chain streptococci. Subculture each presumptive positive tube to ethyl violet azide broth and incubate at 37°C for 24–48 h. Turbidity and a purple-stained button of growth at the bottom of the tube indicate enterococci.

Membrane method

Always use a new membrane when testing for faecal streptococci because these organisms are not always removed when membranes are cleaned. Pass 100 ml of water through a membrane filter and place the filter on a pad of one of the enterococcus membrane broths or a plate of membrane enterococcus agar. Incubate at 37°C for 4 h and then at 44–45°C for 44 h. All red or maroon colonies are presumptive positives. Carefully remove the filter and place face downwards on a plate of Mead's medium to imprint the colonies.

Remove the membrane and incubate the plate at 37°C for 18 h. *S. faecalis* gives maroon colonies surrounded by a clear zone.

US methods for faecal streptococci

These methods (APHA, 1986) are very similar to the above. In the membrane method, suspected colonies are subcultured for catalase tests and into broth incubated at 45°C for confirmation.

C. perfringens in water

MPN method

The litmus milk method is not now regarded as satisfactory for isolating this organism. The 'black tube' method is better. This is done in reinforced clostridial medium (RCM) containing 70 μg/ml of polymyxin to inhibit facultative anaerobes.

Add 50 ml of water to 50 ml of double-strength medium.

Add 10 ml of water to each of five 10-ml amounts of double-strength medium.

Add 1 ml of water to each of five 5-ml amounts of single-strength medium.

Fill the bottles almost to the neck with single-strength medium to exclude most of the air. Replace the caps and incubate at 37°C for 48 h. Tubes showing blackening are presumptive positives, but other clostridia also give this reaction. Confirm *C. perfringens* by subculturing into purple milk. Incubate at 37°C for 24 h and record as positive tubes that show stormy fermentation. Consult the MPN tables (pp.137–139).

Alternatively, plate out and identify as on p.336.

Membrane method

Pass 100 ml of the sample, or an amount suggested by experience, through a membrane filter and place the membrane face down on a plate of iron sulphite agar or Wilson and Blair medium. Pour 20 ml of the same medium, cooled to 50°C, on top. Incubate at 44°C anaerobically for 48 h and count the black colonies with haloes. These are probably *C. perfringens*. If too many clostridia are present, the whole medium will be blackened.

Other clostridia

MPN method

Heat the black tubes (above) at 75°C for 20 min to kill vegetative organisms (including *C. perfringens*, which does not form spores in the medium) and subculture into other tubes of RCM medium. This will give the MPN count of other clostridia, which can then be identified if necessary.

Microfungi and streptomycetes

These organisms cause odours and taints and often grow in scarcely used water pipes, particularly in warm situations, e.g. basements of large buildings where the drinking water pipes run near to the heating pipes.

Sample the water when it has stood in the pipes for several days. Regular running or occasional long running may clear the growth.

Centrifuge 50 ml of the sample and plate on rose bengal agar and malt agar medium. Add 100 μg/ml of kanamycin to the malt medium to suppress most eubacteria.

Alternatively, pass 100 ml or more of the sample through a membrane filter and apply this to a pad soaked in liquid media of similar composition.

Pathogens

Pass a large volume of the sample through membrane filters and add the filters to bottles of the appropriate enrichment media.

For salmonellas use selenite broth, incubate at 43°C (except for *S. typhi*, when 37°C may be better) and subculture every 12 h for 4 days on bismuth sulphite and DC agars.

For *V. cholerae* and other vibrios, use alkaline peptone water and subculture at 4, 8 and 12 h on TCBS medium.

For *Clostridium perfringens*, place the membrane in the bottle of a petri dish and pour over it reinforced clostridial medium, melted and cooled to 50°C, to a depth of at least 5 min.

For *Legionella*, pass 1–5 litres of water through a (preferably nylon) 0.22 μm membrane filter. Retain some of the filtrate.

Place the filter in 5–10 ml of the filtrate in a screw-capped bottle and shake well to dislodge the organisms.

(1) Inoculate legionella medium containing growth and selective supplements. Incubate at 37°C for up to 10 days in air of 2.5% CO_2 under humid conditions.
(2) Heat 10 ml of the concentrated sample at 50°C for 30 min. Plate and incubate as in (1). Suspend scrapings from tap washers in diluent and treat as above.

Colonies of *Legionella pneumophila* are grey-blue or lime green in colour, with entire edges. Confirm by FA technique (Smith, 1987).

Interpretation

It is not possible to assess the potability of any water supply by a single examination.

Presumptive coliforms and *E. coli* should be absent from 100 ml. Enterococci and *C. perfringens*, in the absence of coliforms, suggest contamination at a remote time. These organisms persist longer than coliforms in water.

Plate counts at 37°C should be less than 10 colonies/ml and at 22°C less than 100/ml.

Interpretation requires consideration of geographical and engineering factors as well as a laboratory report. Guidance is given by APHA (1986), EEC (1980) and DHSS (1985).

Bottled mineral waters

These are now sold extensively and may be 'still' or carbonated. The UK natural mineral water regulations (Statutory Instrument No. 71, 1985) cover the production, designation and testing. The same conditions apply throughout the EEC.

The tests used, with levels, are:

Total viable count, 20–22°C, 72 h	< 100/ml
Total viable count, 37°C, 24 h	< 10/ml

Coliforms	
E. coli	} absent from 250 ml
Enterococci	

P. aeruginosa	
Sporulated sulphate-reducing anaerobes	absent from 50 ml

The method for membrane filtration is described on p.133 and the MPN method for clostridia on p.340.

For pseudomonads place the membrane on absorbent pads saturated with modified King's A broth (p.71) or on Sartorius Nutrient Packs SM14075 (with cetrimide).

Non-pathogenic mycobacteria have been found in moderately large numbers in some bottled waters.

Swimming pools

Test the chlorine content with the portable Lovibond device at the pool side. This is often more useful than bacteriological examination, but if this is required sample from below the surface at both ends, using thiosulphate to destroy chlorine (see above). Do plate counts on 1-ml amounts and presumptive coliform tests. The plate count should be less than ten in 80% of samples and coliforms should be absent.

Staphylococci are more resistant than coliforms to chlorination. They tend to accumulate on the surface of the water in the 'grease film'. To find staphylococci take 'skin samples'. Open the sample bottle so that the surface water flows into it. Add 20-ml volumes to each of five tubes containing 20 ml of Robertson's meat medium plus 10% of sodium chloride. Incubate overnight and plate on one of the staphylococcal media.

Spa, jacuzzi and hydrotherapy pools

These have been increasingly associated with human infections, especially with P. aeruginosa (Friend and Newsom, 1986).

They are maintained at 37°C and their free chlorine (or bromine) levels should be closely monitored (small, portable devices are available).

Do total viable counts, culture for coliforms and pseudomonads.

With good management it should be possible to achieve these levels.

Total viable count, 30°C, 48 h	< 20/ml
Coliforms and E. coli	absent from 100 ml
P. aeruginosa	absent from 100 ml

Bathing beaches; other recreational waters

Test samples every 14 days for coliforms and faecal coli (membrane or MPN method). Test periodically for faecal streptococci by membrane method and examine 1 l volumes for salmonellas. (It is usual also to test 10 l volumes for enteroviruses.)

The following bacteriological standards apply in EC countries (guide levels in parentheses):

Total coliforms   less than 10 000/100 ml (guide level 500/100 ml)
Faecal coli   less than 2000/100 ml (guide level 100/100 ml)
Faecal streptococci   no mandatory standard (guide level 100/100 ml)
Salmonellas   absent from 1 l.

(EEC 1976)

Sewage

Effluent and sludge are monitored to ensure that treatment has reduced the load of pathogens, e.g. salmonellas, listerias and campylobacters, all of which can survive in water and, if sludge is spread on land, in soil.

The standards for effluent should be those for potable water.

Methods for recovering these pathogens are given on p.270, 292 and 323.

References

APHA (1986) *Standard Methods for the Examination of Water and Wastewater*, 16th edn, American Public Health Association, Washington DC

DHSS (1985) *The Bacteriological Examination of Drinking Water Supplies,* Reports on Public Health and Medical Subjects No. 71, HMSO, London

EEC (1976) European Community Council Directive No.76/166 of 8 December 1975 concerning the quality of bathing waters

EEC (1980) European Community Council Directive No. 80/778/EEC of 15 July 1980 relating to the quality of water intended for human consumption. *Official Journal of the European Communities* No. L229, 11

Friend, P. A. and Newsom, S. W. B. (1986) Hygiene for hydrotherapy pools. *Journal of Hospital Infection*, **8**, 213–216

Smith, M. G. (1987) Immunological techniques for Legionnaires disease. In *Immunological Techniques in Microbiology* (eds J. M. Grange, A. Fox and N. L. Morgan), Society for Applied Bacteriology Technical Series No. 24, Blackwell Scientific, London, pp. 123–127

Chapter 19

Pseudomonas, acinetobacter, alcaligenes, flavobacterium, chromobacterium and acetobacter

Aerobic Gram-negative non-sporing bacilli that grow on nutrient and usually on MacConkey agars are frequently isolated from human and animal material, from food and from environmental samples. Some of these bacilli are confirmed pathogens; others, formerly regarded as non-pathogenic, are now known to be capable of causing human disease under certain circumstances, e.g. in 'hospital infections', after chemotherapy or treatment with immunosuppressive drugs; others are commensals; many are of economic importance.

With the exception of the enterobacteria and some of the vibrios, which have received detailed characterization because of their importance in human disease, the taxonomic positions and the specific names of some of these organisms are uncertain. We therefore recognize the inadequacy of the list of genera and groups in Table 19.1. In the text and in other tables we give cultural and biochemical

Table 19.1 Key to some Gram-negative rods that grow on nutrient agar

	HL test	Oxidase	Arginine hydrolysis	Gelatin liquefaction	Growth on Mac-Conkey	Motility
Acinetobacter						
calcoaceticus	Ox	–	–	–	+	–
ss. *lwoffii*	None	–	–	–	+	–
Pseudomonas mallei	None	v	+	v	–	–
P. pseudomallei	Ox	+	+	+	+	+
Pseudomonas spp.	Ox or none	+	v	v	+	+
Bordetella parapertussis	None	–	–	–	+	–
B. bronchiseptica	None	+	–	–	+	+
Aeromonas	F	+	+	+	+	+
Alcaligenes faecalis	None	+	–	–	+	+
Chromobacterium						
lividum	Ox	+	–	+	v	+
C. violaceum	F	+	+	+	v	+
Enterobacteria	F	–	v	v	+	v
Flavobacterium	Ox	+	–	+	v	–
Moraxella	None	+	–	v	v	–
Pasteurella	F	+	–	–	–	–
Yersinia	F	–	–	–	+	v[a]
Vibrio	F	+	–	+	+	+

Ox, oxidative; F, fermentative; v, variable;
[a]Some species motile at 22°C

240

properties that may permit the assignment of some of the organisms to species and others to a genus or group only.

Pseudomonas

The organisms in this genus are Gram-negative non-sporing rods about 3 μm × 0.5 μm, which are motile by polar flagella, may produce a fluorescent pigment, are oxidase positive, utilize glucose oxidatively (or produce no acid) and do not produce gas.

They commonly occur in soil and water. Some species are recognized human and animal pathogens but some others, formerly regarded as saprophytes and commensals, have been incriminated as opportunist pathogens in hospital-acquired infections and have colonized distilled water supplies, soaps, disinfectants, intravenous infusions and other pharmaceuticals. Species colonizing swimming pools have been associated with ear infections.

Isolation and identification

Inoculate nutrient agar, MacConkey agar, nutrient agar containing 0.1% cetrimide and King's Medium A. Incubate at 20–25°C (food, etc.) or 35–37°C (animal material). Temperature and length of incubation are crucial for accurate identification.

Colonies on nutrient agar are usually large (2–4 mm), flat, spreading and pigmented. A greenish yellow or bluish yellow fluorescent pigment may diffuse into the medium. Occasionally, melanogenic strains are encountered. A brown pigment is formed around the colonies. Mucoid (encapsulated) strains are sometimes found in clinical material, particularly in sputum from patients with cystic fibrosis, and may be confused with klebsiellas.

Do oxidase test, reaction in Hugh and Leifson medium, inoculate Thornley's arginine broth (Moeller's method is too anaerobic), MacConkey agar and test for gelatin liquefaction (5 days' incubation). Inoculate nutrient broths and incubate at 4 and 42°C. Use a subculture from a smooth suspension for this test. If pigment (fluorescence) is seen on King's medium A test for reduction of nitrate, urease and acid production from ethanol, glucose and mannitol in ammonium salt sugars (Table 19.2).

The API 2ONE kit, designed for non-fermenting Gram-negative rods, is useful for identifying pseudomonads. Other kits include Minitek NF and Oxi-ferm.

Pseudomonads are motile (occasionally non-motile strains may be met), oxidative, non-reactive or produce alkali in Hugh and Leifson medium. The oxidase test

Table 19.2 Fluorescent pseudomonads

	Growth at		Nitrate reduction	Acid from[a]			Urease	Gelatin liquefaction
	4°C	42°C		Ethanol	Glucose	Mannitol		
P. aeruginosa	−	+	+	+	+	+	+	+
P. fluorescens	+	−	+	−	+	+	−	+
P. putida	+	−	−	+	+	−	−	−

[a] Ammonium salt sugars
(After King and Phillips, 1985)

is usually positive except for *P. maltophilia*, *P. cepacia*, *P. stutzer* and *P. alcaligenes*. Those pseudomonads which give a positive arginine test do so are not common. Those pseudomonads which give a positive arginine test do so more rapidly than other Gram-negative rods and this test is very useful for recognizing non-pigmented strains. *Alcaligenes* and *Vibrio* spp. do not hydrolyse arginine. Pseudomonads vary in their ability to liquefy gelatin and grow at 4 and 42°C (see Table 19.2).

Some fluorescent strains produce an opalescence (egg yolk reaction) on Willis and Hobbs' medium and also a lipase (pearly layer). To demonstrate the latter, flood the plate with saturated copper sulphate solution; the fatty acids released by lipolysis give a greenish blue precipitate of copper soaps.

Some varieties associated with food spoilage are psychrophiles and may be lipolytic. Brine-tolerant strains and phosphorescent strains are not uncommon.

Table 19.3 *Pseudomonas spp. Vibrio, Alcaligenes* **and** *Aeromonas*

	Fluorescent pigment	HL test	Arginine hydrolysis	Gelatin liquefaction	Growth at 4°C	Growth at 42°C
P. aeruginosa	+	Ox	+	+	−	+
P. fluorescens	+	Ox	+	+	+	−
P. putida	+	Ox	+	−	v	−
P. alcaligenes	−	Alk	−	−	−	v
P. pseudoalcaligenes	−	Alk	+	−	−	v
P. maltophilia	−	Alk	−	+	−	−
P. cepacia	−	Ox	−	+	−	v
P. mallei	−	None	+	v	−	−
P. pseudomallei	−	Ox	+	+	−	+
P. stutzeri	−	Ox	−	−	v	+
P. paucimobilis	−	Ox	−	−	−	−
P. mendocina	−	Ox	+	−	−	−
Vibrio	−	F	+	+	v	−
Alcaligenes	−	Alk or none	−	−	−	−
Aeromonas	−	F	+	+	−	−

All of these organisms are oxidase positive. *Acinetobacter* spp. are oxidase negative.
Ox, oxidative; F, fermentative; Alk, alkaline reaction; v, variable

Species of *Pseudomonas*

P. aeruginosa

Typical colonies on agar medium 18–24 h at 25–30°C, are large, flat, spreading and irregular, greyish green in colour. A variety of other colonial forms, including mucoid, dwarf and non-pigmented have been observed. The greenish pigment diffuses into the medium. Broth cultures are blue-green in colour. Two pigments are formed, pyocyanine and fluorescein; both are soluble in water but only pyocyanine is soluble in chloroform. Arginine is hydrolysed rapidly. Gelatin is liquefied. Growth takes place at 42°C but not at 4°C. The egg yolk reaction is negative. This organism is very resistant to antibiotics except aminoglysides, some recent β-lactams (e.g. ceftazidime), other β-lactams (e.g. carbenicillin, piperacillin) and quinolones.

It is a saprophyte, found in soil and water, causes spoilage of foods, including 'blue milk', is pathogenic for humans and animals ('blue pus'), often as a secondary infection, and is often found in clinical material.

It is an important agent in cross-infection in hospitals and strains may be typed by pyocin production and antisera. This is usually done at Reference Laboratories.

P. fluorescens

Colonies on agar and its biochemical properties are similar to those of *P. aeruginosa* but only one pigment (fluorescein) is produced. Growth occurs at 4°C but not at 42°C. The egg yolk reaction is positive. This organism is a saprophyte found in soil, water and sewage. It is a food spoilage organism. It may gel UHT milk if this is stored above 5°C. It is important in patients with burns and cystic fibrosis.

P. putida

Colonies resemble those of *P. aeruginosa* but only fluorescein is formed. It is a psychrotroph that may grow at 22°C and may grow at 37°C. There are two biotypes; one grows at 4°C, the other does not; both hydrolyse arginine. The egg yolk reaction is negative, gelatin is not liquefied and old cultures have a marked odour of trimethylamine (bad fish). It is an important fish pathogen and fish spoilage organism and has been isolated from human material.

P. maltophilia

Colonies resemble those of *P. aeruginosa* but a yellow or brown diffusible pigment may be produced. This species was reported as oxidase negative by Gilardi (1985), who used Difco oxidase discs, but Snell (1973) found oxidase positive strains when testing with tetramethyl-1-phenylenediamine dihydrochloride. The method of performing the tests seems to be important. This species fails to hydrolyse arginine and does not grow on cetrimide agar. It is the only pseudomonad that gives a positive lysine decarboxylase reaction. It has been isolated from a variety of sources and is the third most common non-lactose fermenter found in clinical material.

P. cepacia

Colonies may produce a yellow, water-soluble, non-fluorescent pigment, occasionally may be purple, or may be non-pigmented. Growth at 42°C has been reported. Widely distributed in nature, it occurs in hospital infections, is found in pharmaceuticals, cosmetics and water and is significant in cystic fibrosis.

P. stutzeri

Colonies are rough, dry and wrinkled and can be removed entire from the medium. Older colonies turn brown. It does not hydrolyse arginine and may not grow at 4 or 42°C or on cetrimide medium. It is salt tolerant but not halophilic. A ubiquitous organism, it has been found in clinical material but is not regarded as a pathogen.

P. mendocina

Colonies are flat and butyrous. A non-diffusible brownish or yellow pigment is produced. Found in soil and water and not regarded as pathogenic.

P. paucimobilis

Colonies may be mucoid or butyrous, producing a non-diffusible yellow pigment. Motility is poor and best seen in cultures incubated at room temperature.

P. alcaligenes and P. pseudoalcaligenes

P. alcaligenes does not grow at 42°C but some strains of *P. pseudoalcaligenes* do; neither grow at 4°C. They are ubiquitous and are opportunist pathogens. They do not liquefy gelatin and may or may not grow on cetrimide agar. *P. alcaligenes* does not hydrolyse arginine.

P. mallei

In exudates this organism may appear granular or beaded and may show bipolar staining. It gives 1-mm, shining, smooth, convex, greenish yellow buttery or slimy colonies which may be tenacious on nutrient agar, no haemolysis on blood agar and does not grow on MacConkey agar in primary culture. Growth may be poor on primary isolation. It grows at room temperature, does not change Hugh and Leifson medium, produces no acid in peptone carbohydrates, is nitrate positive and gives a variable urease reaction. The oxidase test gives variable results.

This is the causative organism of glanders (see Caution below).

P. pseudomallei

Cultures on blood agar and nutrient agar at 37°C give mucoid or corrugated, wrinkled, dry colonies in 1–2 days, and an orange pigment may develop. There is no growth on cetrimide agar. This organism is not easy to identify. It must be distinguished from non-pigmented strains of *P. aeruginosa*, *P. stutzeri* (Table 19.2) and from *Pseudomonas mallei* (Table 19.4).

Table 19.4 *Acinetobacter, Alcaligenes, Pseudomonas mallei, pseudomallei* and *Bordetella*

	Growth on MacConkey	Oxidase	HL test	Motility	Acid from Glucose	Nitrate reduction	Urease
A. calcoaceticus	+	−	Ox	−	+	−	v
ss. *lwoffii*	+	−	none	−	−	−	−
Alcaligenes spp.	+	+	alk. none	+	−	v	−
P. mallei	−	v	none	−	(+)[a]	+	v
P. pseudomallei	+	+	Ox	+	+	+	v
B. parapertussis	+	−	none	−	−	−	+
B. bronchiseptica	+	+	none	+	−	+	+

Ox, oxidative, [a](+) acid in ammonium salt glucose but not in peptone water glucose; v, variable

This is an important pathogen of humans (melioidosis) and farm animals in SE Asia, where it is endemic in rodents and is found in moist soil, on vegetables and on fruit. Cultures should be sent to a Reference Laboratory (see Caution below).

For further information about the pseudomonads see Gilardi (1985) and King and Phillips (1985).

Caution

Pseudomonas mallei and Pseudomonas pseudomallei are Risk/Hazard Group 3 pathogens and should be handled only in Biosafety Containment Level 3 laboratories.

Acinetobacter

Organisms have been moved in and out of this genus for several years. At present it seems to contain only one species, *A calcoaceticus*, which has four biotypes, *anitratus, lwoffii, haemolyticus* and *alcaligenes*, all of which are given specific status by some workers.

Isolation and identification

Plate material on nutrient, blood and MacConkey agar (and Bordetella or similar media if indicated). Incubate at 22°C and at 37°C for 24–48 h. Do oxidase and catalase tests. Inoculate nutrient broth for motility, Hugh and Leifson medium, glucose peptone water, nitrate broth and urea medium.

Acinetobacter are non-motile, obligate aerobes usually oxidase negative, catalase positive and are oxidative or give no reaction in Hugh and Leifson medium (see below and Table 19.4).

Species of *Acinetobacter*

A. calcoaceticus ss. anitratus

This gives large non-lactose fermenting colonies on MacConkey and DC agars and grows at room temperature. On Hugh and Leifson medium it is oxidative, producing acid but not gas from glucose, lactose and xylose. It is nitratase negative and gives a variable urease reaction.

Infections in a wide variety of sites in humans have been reported. Many strains are resistant to several antibiotics. Hospital cross-infections with resistant strains cause problems. Sometimes in direct films this organism may resemble *Neisseria*.

A. calcoaceticus ss. lwoffii

Colonies on blood agar are small, haemolytic and may be sticky, but this organism also grows on nutrient and MacConkey agars. There is no change in Hugh and Leifson medium; it fails to attack carbohydrates, is nitrate and urease negative. It can cause conjunctivitis. This and the other two biotypes are distinguished by the API 2-ONE kit.

For more information about infections with *Acinetobacter* see Vivian *et al.* (1981).

Alcaligenes

These organisms are widely distributed in soil, fresh and salt water and are economically important in food spoilage (especially of fish and meat). Many strains are psychrophilic. Some species have been isolated from human material and suspected of causing disease.

Isolation and identification

Plate on blood, skim milk and MacConkey agars. Incubate at 20–22°C (foodstuffs) or at 35–37°C (pathological material) for 24–48 h. Subculture white colonies of Gram-negative rods on agar slopes for oxidase test, in peptone water for motility, in Hugh and Leifson medium, glucose peptone water, bromocresol purple milk, arginine broth and gelatin.

Species of *Alcaligenes*

The genus *Alcaligenes* is restricted here to motile, oxidase positive, catalase positive Gram-negative rods which give an alkaline or no reaction in Hugh and Leifson medium, an alkaline reaction in bromocresol purple milk and grow on MacConkey agar. Arginine is not hydrolysed, gelatin is not usually liquefied. Most strains are resistant to penicillin (10 IU disc) (Tables 19.1, 19.2 and 19.4).

A. faecalis
Culturally resembles *B. bronchiseptica* but is urease negative. It is a widely distributed saprophyte and commensal. Some strains smell of apples and show a greenish discoloration around colonies on blood agar.

A. viscolactis
This dubious species is one of several that are responsible for ropy milk.

Achromobacter

This is no longer officially recognized, but food bacteriologists find it a convenient 'dustbin' for non-pigmented Gram-negative rods which are non-motile, oxidative in Hugh and Leifson medium, oxidase positive, produce acid but no gas from glucose, acid or no change in purple milk, do not hydrolyse arginine and usually liquefy gelatin.

Flavobacterium

The genus *Flavobacterium* contains many species that are difficult to identify. Gram-negative bacilli forming yellow colonies are frequently isolated from food and water samples and although some undoubtedly belong to other genera (e.g. *Aeromonas, Cellvibrio, Cytophaga, Erwinia, Klebsiella, Myxobacteria, Vibrio*) it suits the convenience of many food bacteriologists to call them 'flavobacteria'. Some of these organisms are proteolytic or pectinolytic and are associated with spoilage of fish, fruit and vegetables.

There is one species of medical importance.

Flavobacterium meningosepticum
This has been isolated from CSF in meningitis in neonates from other human material and from hospital intravenous and irrigation fluids.

It grows on ordinary media; colonies at 28 h are 1–2 mm, smooth, entire, grey or yellowish and butyrous. There is no haemolysis on blood agar, no growth on DCA and none on primary MacConkey cultures, but growth on this medium may occur after several subcultures on other media. A yellowish green pigment may diffuse into nutrient agar media.

F. meningosepticum is oxidative in Hugh and Leifson medium, non-motile, oxidase and catalase positive, liquefies gelatin in 5 days, does not reduce nitrates and does not grow on Simmons' citrate medium. Rapid hydrolysis of aesculin is a useful character. The API ZYM system is useful. Acid is produced from 10% glucose, mannitol and lactose in ammonium salt medium but not from sucrose or salicin. Indole production seems to be positive by Ehrlich's but negative by Kovac's methods (Snell, 1973). See Table 19.5 for differentiation from pseudomonads and *Acinetobacter*. For more information see Holmes (1987).

Table 19.5 *Flavobacterium meningosepticum, Pseudomonas* **and** *Acinetobacter*

Species	Oxidase	Nitrate reduction	Citrate	Acid from 10% solution in ammonium salt medium of		
				Glucose	Mannitol	Lactose
F. meningosepticum	+	−	−	+	+	+
Pseudomonas	+	+	+	+	+	+
Acinetobacter	−	−	+/−	−	−	−

Chromobacterium

The organisms in this genus are characterized by a violet pigment, but this may not be apparent until the colonies are several days old. They grow well on ordinary media, giving cream or yellowish colonies that turn purple at the edges. The best medium to demonstrate the pigment is a potato slice. The pigment is soluble in ethanol but not in chloroform or water and can be enhanced by adding mannitol or meat extract to the medium. Citrate but not malonate is utilized in basal synthetic medium, the catalase test is positive and the urease test negative. Ammonia is formed. Gelatin is liquefied. The oxidase test is positive but difficult to do if the pigment is well formed. Both species are motile (see Table 19.6).

Table 19.6 *Chromobacterium violaceum* **and** *Chromobacterium lividum*

Species	Growth at		HL test	Hydrolysis of		
	5°C	37°C		Arginine	Casein	Aesculin
C. violaceum	−	+	F	+	+	−
C. lividum	+	−	Ox	−	−	+

F = fermentative; Ox = oxidative

C. violaceum
This species is meosphilic, growing at 37°C but not at 5°C. It is fermentative in Hugh and Leifson medium, hydrolyses arginine and casein but not aesculin. It is a facultative anaerobe.

Although usually a saprophyte, cases of human and animal infection have been described in Europe, the USA and the Far East.

C. lividum
This organism is psychrophilic, growing at 5°C but not at 37°C. It is oxidative in Hugh and Leifson medium does not hydrolyse arginine or casein and hydrolyses aesulin. It is an obligate aerobe.

Other species, *C. fluviatile* and marine strains, have been described (see Sneath, 1979).

Acetic acid bacteria

These bacteria are widely distributed in vegetation. They are of economic importance in the fermentation and pickling industries, e.g. in cider manufacture

(Carr and Passmore, 1979), as a cause of ropy beer and sour wine, and of off-odours and spoilage of materials preserved in vinegar. Ethanol is oxidized to acetic acid by *Gluconobacter* spp. but *Acetobacter* spp. continue the oxidation to produce carbon dioxide and water.

Isolation and identification to genus

Identification to species is rarely necessary. It is usually sufficient to recognize acetic acid bacteria and to determine whether the organisms produce acetic acid and/or destroy it.

Inoculate wort agar or unhopped beer solidified with gelatin and yeast extract broth containing 10% glucose and 3% calcium carbonate at pH 4.5. Incubate at 25°C for 24–48 h. Colonies are large and slimy.

Subculture into yeast broth containing 2% ethanol and congo red indicator at pH 4.5. Acid is produced.

Test the ability to produce carbon dioxide from acetic acid in 2% acetic acid broth. Use either a Durham tube or the method described on p.325 for hetero-fermentative lactic acid bacilli. Inoculate lactose and starch peptone water 'sugars' at pH 4.5 with congo red indicator.

Plant bacterial masses (heavy inoculum, do not spread) on yeast extract agar at pH 4.5 containing 5% glucose and at least 3% finely divided calcium carbonate in suspension. Clear zones occur around masses in 3 weeks but if less chalk is used some pseudomonads can do this.

Culture on yeast extract agar containing 2% calcium lactate. *Acetobacter* grow well and precipitate calcium carbonate. *Gluconobacter* grow poorly and give no precipitate.

Carr and Passmore (1979) describe a medium to differentiate the two genera. It is yeast extract agar containing 2% ethanol and bromocresol green (1 ml of 2.2% solution/litre) made up in slopes. Both *Gluconobacter* and *Acetobacter* produce acid from ethanol and the indicator changes from blue-green to yellow. *Acetobacter* then utilize the acid and the colour changes back to green (See Table 19.7).

Table 19.7 Acetic acid bacteria

	Acid from ethanol	CO₂ from acetic acid	Carbonate from lactate
Gluconobacter	+	−	−
Acetobacter	+	+	+

For more information, see Carr and Passmore (1979).

References

Carr, J. G. and Passmore, S. M. (1979) Methods for identifying acetic acid bacteria. In *Identification Methods for Microbiologists*, 2nd edn (eds F. A. Skinner and D. W. Lovelock), Society for Applied Bacteriology Technical Series No. 14, Academic Press, London, pp. 33–45

Gilardi, G. L. (1985) Pseudomonas. In *Manual of Clinical Microbiology*, 4th edn (eds E. H. Lennette, A. Balows, W. J. Hauser and H. J. Shadomy), Association of American Microbiologists, Washington DC, pp. 445–249

Holmes, B. (1987) Identification and distribution of *Flavobacterium meningosepticum* in clinical material. *Journal of Applied Bacteriology* **43**, 29–42

King, A. and Phillips, I. (1985) Pseudomonads and related species. In *Isolation and Identification of Micro-organisms of Medical and Veterinary Importance* (eds. C. H. Collins and J. M. Grange), Society for Applied Bacteriology Technical Series 21, Academic Press, London, pp. 1–12

Sneath, P. H. A. (1979) Identification methods applied to Chromobacterium. In *Identification Methods for Microbiologists*, 2nd edn (eds F. A. Skinner and D. A. Lovelock), Society for Applied Bacteriology Technical Series No. 14, Academic Press, London, pp. 166–174

Snell, J. J. S. (1973) *The Distribution and Identification of Non-fermenting Bacteria*. Public Health Laboratory Service Monograph No. 4, HMSO, London

Vivian, A., Hinchcliffe, E. J. and Fewson, C. A. (1981) *Acinetobacter calcoaceticus:* some approaches to a problem. *Journal of Hospital Infection,* **2,** 199–204

Chapter 20

Vibrios, aeromonas and plesiomonas

Vibrios

Vibrios are Gram-negative, non-sporing rods, motile by polar flagella enclosed within a sheath. Some have lateral flagella and may swarm on solid media. They are catalase positive, utilize carbohydrates fermentatively and rarely produce gas. All but one species (*Vibrio metschnikovii*) are oxidase and nitrase positive. They are all

Table 20.1 *Vibrio, Plesiomonas, Aeromonas, Pseudomonas* **and enterobacteria**

	O/129[a]	Oxidase	Hugh and Leifson	Gas	Salt enhancement
Vibrio	S	+[b]	F	−[c]	+
Plesiomonas	S	+	F	−	−
Aeromonas	R	+	F	v	−
Pseudomonas	R	+	O	−	−
Enterobacteria	R	−	F	v	−

[a]150 μg: S, sensitive; R, resistant
[b]except *Vibrio metschnikovii*
[c]except some strains of *V. fluvialis*
v, different strains show different reactions

sensitive to 150 μg discs of the vibriostatic agent 2,4-diamino-6,7-di-isopropyl pteridine (0/129). Most strains liquefy gelatin and hydrolyse deoxyribonucleic acid. Vibrios occur naturally in fresh and salt water. Species include some human and fish pathogens. See Table 20.1 for differentiation between *Vibrios, Plesiomonas, Aeromonas, Pseudomonas* and enterobacteria.

Isolation

Vibrios grow readily on most ordinary media but enrichment and selective media are necessary for faeces and other material containing mixed flora.

Faeces
Add 2 g of faeces to 20-ml amounts of alkaline peptone water (APW). Inoculate thiosulphate citrate bile salt sucrose agar (TCBS) medium.

Incubate APW at 37°C for 5–8 h or at 20–25°C for 18 h and subculture to TCBS.
Incubate TCBS cultures at 37°C overnight.

Other pathological material
Vibrios may occur in wounds and other material particularly if there is a history of sea bathing. Examine primary blood agar plates (see below for colony appearance).

Water and foods
Add 20 ml to water to 100 ml of APW. Add 10 ml of food to 100 ml of APW and emulsify in a Stomacher. Incubate at 20–30°C for 18 h and subculture on TCBS as for faeces.

For specific isolation of *V. parahaemolyticus* use glucose salt Teepol broth (GSTB, p.70) or salt colistin broth (SCB).

Enumeration of vibrios
Use liquid samples neat and diluted 1:10. Make 1:10 homogenates of solid samples. Spread 50 or 100 μl on TCBS medium, incubate 20–24 h and count vibrio-like colonies. Report as count/g.

For the three-tube Most Probable Number method prepare dilutions of 10^{-1}, 10^{-2} and 10^{-3}. Add 1 ml of each dilution to each of three tubes of APW, GSTB or SCB and incubate at 30°C overnight. Subculture to TCSB agar and incubate at 30°C for 20–24 h. Examine for vibrio-like colonies, identify them and use the three-tube table (p.140).

Colonial morphology
Table 20.2 shows the colony appearances of vibrios and related organisms on TCBS medium.

Table 20.2 Growth of some vibrios and other bacteria on thiosulphate citrate bile salt agar (TCBS), 37°C for 18 h

Species	Growth	Sucrose fermented	Colour	Colony diameter (mm)
V. cholerae	+	+	Y	2–3
V. metschnikovii	+	+	Y	2–4
V. fluvialis	+	+	Y	2–3
V. furnissii	+	+	Y	2–3
V. natriegens	+	+	Y	2–3
V. alginolyticus	+	+	Y	2–5
V. mimicus	+	−	G	2–3
V. damsela	+	−	G	2–3
V. vulnificus	+	−	G	2–3
V. parahaemolyticus	+	−	G	2–5
V. campbellii	+	−	G	2–3
V. harveyi	+	v	G/Y	2–3
V. anguillarum	+	+	Y	0–3
V. hollisae	−	−	G	—
Aeromonas hydrophila	v	+	Y	0–3
Plesiomonas shigelloides	v	−	G	1
V. nigropuchritudo	−	−	G	—

+, growth or sucrose fermented   G, green
−, no growth or sucrose not fermented   Y, yellow
v, variation within strains

On non-selective media colonial morphology is variable. The colonies may be opaque or translucent, flat or domed, haemolytic or non-haemolytic, smooth or rough. One variant is rugose and adheres closely to the medium.

Electrolyte concentration

Increase the electrolyte concentration of conventional identification media by 1%. (This will not interfere with the identification of enterobacteria. Some strains grow poorly, however, and need an electrolyte supplement: NaCl 10%, MgCl$_2$, 6H$_2$O, 4%; KCl, 4% in distilled water. Add 0.1 ml to each 1.0 ml of medium.

Growth on CLED medium

Some vibrios, however, will grow without the addition of sodium chloride to the medium. Culture on an electrolyte deficient medium, e.g. CLED. This permits two groups, halophilic and non-halophilic vibrios to be distinguished. Inoculate CLED lightly with the culture and incubate at 30°C overnight.

Salt tolerance

This varies with species. Inoculate peptone water containing 0, 3, 6, 8 and 10% NaCl. This technique must be standardized to obtain consistent results.

Sensitivity to 0/129 (2,4-diamino-6,7-di-isopropyl pteridine)

This was originally used as a disc method to distinguish between *Vibrio* (sensitive) and *Aeromonas* (resistant) (Furniss *et al.*, 1978), but the use of two discs (150 µg and 10 µg) enables two groups of vibrios to be recognized. Make a lawn of the organisms on nutrient agar, and place one disc (Oxoid) of each concentration of 0/129 on it and incubate overnight. Do not use any special antibiotic sensitivity testing medium because the growth of vibrios and the diffusion characteristics of 0/129 differ from those on nutrient agar.

Oxidase test

Use Kovac's method (p.103) and test colonies from non-selective medium. Do not use colonies from TCBS or other media containing fermentable carbohydrate because changes in pH may interfere with the reaction.

Decarboxylase tests

Use Moeller's medium containing 1% additional NaCl, but do not read the results too early as there is an initial acid reaction before the medium becomes alkaline. The blank should give an acid reaction; failure to do so may suggest poor growth and the electrolyte supplement should be added. Thornley's arginine medium is particularly useful provided that the salt concentration is adequate.

VP test

Use a semisolid medium under controlled conditions. If incubation is prolonged and a sensitive method is used almost any vibrio may give positive results.

Swarming

Use Marine agar or a medium with constant characteristics, at 37°C for reliable results.

Luminescence

Some vibrios may show luminescence in the dark. It is most marked in young cultures and may be lost on continued incubation. A positive control must always be included. Growth is best examined after overnight incubation at 25°C on special nutrient agar (Furniss *et al.*, 1978). It is very important to allow at least 5 min for

one's eyes to become adapted to the dark before examining cultures for luminescence.

Identification kits

Not all of these are at present ideal for identifying vibrios. All appropriate tests
may not be included and there may not be enough data in the matrices. The API 20
system gives satisfactory results if the organisms are suspended in 1% NaCl at
pH 6.5 and a heavier than usual suspension is employed. Additional tests, e.g.
0/129 sensitivity and growth on CLED should be used.

There is a kit for detecting *V. cholerae* enterotoxin (Oxoid).

Other tests

Test for fermentation with Hugh and Leifson medium, acid production from
sucrose and arabinose in peptone water containing 1% NaCl.

Properties of vibrios, etc.

See Table 20.3.

Species of vibrios

V. cholerae

Is sensitive to 0/129, oxidase positive, decarboxylates lysine and ornithine but does
not hydrolyse arginine. It produces acid but no gas from glucose (fermentative in
Hugh and Leifson's medium) and sucrose, but no acid from arabinose or lactose. It
is non-halophilic in that it grows on CLED medium.

All strains possess the same heat-labile H antigen but may be separated into
serovars by their O antigens.

Serovar O:1

Is the causative organism of epidemic or Asiatic cholera. It is agglutinated by
specific O:1 cholera antiserum. It is possible, by using carefully absorbed sera, to
distinguish two subtypes of *V. cholerae* O:1. These are known as Ogawa and Inaba
but as they are not completely stable and variation may occur *in vitro* and *in vivo*
subtyping is of little epidemiological value, unlike phage typing (Lee and Furniss,
1981) which is of epidemiological value. Non-toxigenic strains of *V. cholerae* O:1
have been isolated and have shown distinctive phage patterns.

Although there are two biotypes of *V. cholerae* O:1, the 'classic' (non-
haemolytic) and the 'eltor' (haemolytic) the former are now virtually non-existent.
Biotyping is not, therefore, a useful epidemiological tool (See Table 20.4).

Serovars other than O:1

Have identical biochemical characteristics as *V. cholerae* O:1 and the same H
antigen but possess different O antigens and are not agglutinated by the O:1 serum.
They have been called 'non-agglutinating' (NAG) vibrios. This is an obvious
misnomer; they are agglutinated both by cholera H antiserum and by antisera
prepared against the particular O antigen they possess. Another term which has
been used is 'Non-cholera vibrio' (NCV) and as both terms have been used in
different ways this has resulted in considerable confusion. It is best if all these

Table 20.3 Properties of *Vibrio* species and allied genera likely to be encountered in clinical laboratories

| | | | | | Gas from | Acid from | | | | |
| | | | | | | | | | | |
Species	Arginine	Lysine	Ornithine	Glucose	L-Arabinose	Arbutin	Inositol	Salicin	Sucrose
V. cholerae	−	+	+	−	−	−	−	−	+
V. mimicus	−	+	+	−	−	−	−	−	−
V. metschnikovii	+	v	−	−	−	v	v	v	+
V. parahaemo-lyticus	−	+	+	−	v	−	−	−	−
V. alginolyticus	−	+	+	−	−	−	−	v	−
V. vulnificus	−	+	+	−	−	+	−	+	−
V. fluvialis	+	−	−	−	+	+	−	+	+
V. furnissii	+	−	−	+	+	−	−	−	+
V. hollisae	−	−	−	−	+	.	−	−	−
V. anguillarum	+	−	−	−	v	−	−	−	+
Aeromonas	+	v	−	v	v	v	−	v	+
Plesiomonas shigelloides	+	+	+	−	−	.	+	−	−

v, variable; ., not known; S, susceptible; R, resistant

vibrios are referred to as non-O:1 *V. cholerae*, or, if a strain has been serotyped, by that designation.

Some of these strains are undoubtedly potential pathogens and produce a toxin similar to, if not identical with, that of the cholera vibrio. Some outbreaks have occurred but most isolates have been from sporadic cases. These vibrios are widespread in fresh and brackish water in many parts of the world, including the UK. They do not, however, cause true epidemic cholera.

V. mimicus

Resembles non-O:1 *V. cholerae* but is VP negative and does not ferment sucrose. Colonies on TCBS media are therefore green. Antigenically it appears to be *V. cholerae* but it was given specific rank because of its low degree of DNA homology with *V. cholerae*. On the other hand it could be retained as a subspecies or biovar of *V. cholerae*. It has been isolated from the environment and is associated with sea foods. It has also been isolated from human faeces. Some strains produce a cholera-like toxin.

V. parahaemolyticus

Is a halophilic vibrio and will not grow on CLED. It does not ferment sucrose and therefore gives a (large) green colony on TCBS agar. O and K antigens may be

Table 20.4 Recognition of classic and eltor strains of *Vibrio* cholerae

	Classic	Eltor
Haemolysis	−	+
VP	−	+
Chick cell haemagglutination	−	+
Polymyxin (50 IU)	S	R
Classic phage IV	S	R
Eltor phage 5	R	S

S, sensitive: R, resistant

VP (24 h)	ONPG (24 h)	Growth at 43°	0/129		Growth in NaCl (%)					Oxidase (Kovacs)	NO₃ to NO₂	Growth on		Swarming
			10 µg	150 µg	0	3	6	8	10			TCBS	CLED	
+	+	+	S	S	+	+	−	−	−	+	+	Y	+	−
−	−	+	S	S	+	+	−	−	−	+	+	Y	+	−
+	+	+	S	S	v	+	+	v	−	−	−	G	+	−
−	−	+	R	S	−	+	+	+	−	+	+	G	−	−
+	+	+	R	S	−	+	+	+	+	+	+	Y	−	+
−	+	−	S	S	−	+	+	−	−	+	+	G	−	−
−	+	v	R	S	v	+	+	v	v	+	+	Y	v	−
−	+	−	R	S	v	+	+	v	−	+	+	Y	v	−
−	.	.	.	.	−	+	v	−	−	+	+	G	.	−
+	+	−	S	S	v	+	v	−	−	+	+	Y	v	−
v	+	v	R	R	+	v	−	−	−	+	+	−	+	−
−	+	+	v	S	+	v	−	−	−	+	+	−	+	−

used to serotype strains. It is recognized as the commonest cause of food poisoning in Japan. It is usually present in coastal waters, although only in the warmer months in the UK. The Kanagawa haemolysis test is said to correlate with pathogenicity. Controlled conditions for testing the haemolysis of the human red blood cells are essential. Only laboratories with sufficient experience should do this test. Most environmental strains are negative.

V. vulnificus
Resembles *V. parahaemolyticus* but ferments lactose. It has been isolated from blood cultures from patients (mostly in the USA) who are immunologically compromised or suffering from liver disease.

V. fluvialis
First reported as Group F vibrios, this is common in rivers, particularly in the brackish water of estuaries and may cause gastroenteritis in humans. *V. fluvialis* needs to be distinguished from *V. anguillarum* (Table 20.3).

V. anguillarum
Is phenotypically far from being a uniform species. It is uncommon in rivers. Some strains may be pathogenic for fish, although not especially for the eel, as its name might suggest. Many strains will not grow at 37°C and would be missed in laboratories which confine their incubation temperatures to about 37°C. There is no evidence that *V. anguillarum* has ever been pathogenic for humans.

V. furnissi
Was previously classified as *V. fluvialis* biovar II. It has been associated with human gastroenteritis in Japan and other parts of the Orient and has been isolated from the environment and sea-food.

V. metschnikovii
Is both oxidase and nitrate reduction negative but its other characters are typical of

vibrios. It includes the now illegitimate species *V. proteus*. Frequently isolated from water and shellfish, but there is little evidence of pathogenicity for humans.

V. damsela

A marine bacterium isolated from human wounds and from water and skin ulcers of damsel fish. It is arginine positive, produces gas from glucose, requires NaCl for growth, is not bioluminescent and ferments glucose, mannose and maltose. It is distinguished from other named vibrios by DNA–DNA hybridization.

V. hollisae

A halophilic vibrio isolated from human cases of gastroenteritis. It does not grow on TCBS. It is distinguished from other halophilic vibrios from human sources by its negative lysine, arginine and orthithine tests and the limited range of carbohydrates fermented.

It has not been recovered from the environment and a clear relation with human disease has not been established.

Plesiomonas shigelloides

There is no specific enrichment method or selective medium for *Plesiomonas*. It grows poorly on TCBS but well on deoxycholate and MacConkey agar. Colonies appear to be *Shigella*-like – hence the specific name – and some strains are agglutinated strongly by *Shigella sonnei* antiserum.

This organism has been variously classified and included within the genus *Vibrio*, because of its sensitivity to 0/129. It is now considered to be distinct enough to be placed in a separate genus containing one species.

It is now thought to be a cause of gastroenteritis, because most clinical isolates are from patients with diarrhoea; it is rarely isolated otherwise. Many of the isolates in the UK are from people returning from abroad.

Aeromonas

These are small Gram-negative rods which are motile by polar flagella. They are oxidase and catalase positive and resistant to 0/129. They attack carbohydrates fermentatively (gas may be produced from glucose) liquefy gelatin and reduce nitrates. The arginine dihydrolase test using arginine broth is positive. Indole and VP reaction vary with species. Some strains are psychrophilic and most grow at 10°C. See Tables 20.2 and 20.3 for differentiation from other groups.

The genus is divided into two groups. The Salmonicida group contains psychrophilic, non-motile aeromonads. They may be further divided into three subspecies which are associated with diseases of fish (see Frerichs and Hendrie, 1985) but are not pathogenic for humans. The Hydrophila group contains mostly motile mesophiles which are found in food.

Isolation

Aeromonas species will grow on non-selective media. Plate on tryptone agar or Aeromonas medium and incubate at 22°C for Salmonicida group. Plate on xylose deoxycholate citrate agar (XDC) or commercial aeromonas medium and incubate at 37°C for the Hydrophila group.

Identification

Test for growth at 20 and 37°C. Do oxidase and nitrate reduction tests, OF test, incubate glucose, arabinose, salicin and aesculin media. Do lysine decarboxylase, arginine dihydrolase and VP tests. Look for brown pigment on agar medium containing 1% tyrosine and incubated at 20°C (see Tables 20.5 and 20.6) if the Salmonicida group is suspected. Plate on nutrient or blood agar and test susceptibility to cephalothin (30 μg disc).

Table 20.5 Aeromonas groups

	Motility	Growth at 37°C	Brown pigment
Salmonicida group	−	−	+
Hydrophila group	+	+	v

Table 20.6 Hydrophila group

	Aesculin hydrolysis	Acid from		Gas from glucose	Lysine decarboxylase	VP reaction	Cephalothin 30 μg disc
		Arabinose	Salicin				
A.hydrophila[a]	+	+	+	+	+	+	R
A. caviae[b]	+ (most)	+	+	−	−	−	R
A. sobria[a]	− (most)	−	−	+	+	+	S

[a]Enterotoxigenic
[b]Not enterotoxigenic

Species of Aeromonas

There is still uncertainty about the status of the species but we favour the recognition of the four mentioned below.

Salmonicida group

Produces a brown pigment on media containing tyrosine but does not grow at 37°C and is non-motile. It causes furunculosis of fish, a disease of economic importance in fish farming.

Hydrophila group

A hydrophila, A. caviae and A. sobria are motile and usually grow at 37°C. They are more commonly isolated from the stools of patients with diarrhoea than from normal faeces in the UK but in countries where aeromonads are common in drinking water they are as frequently isolated from normal as abnormal stools. It appears that enteropathogenicity, if it does occur, is confined mainly to strains of A. hydrophila and A. sobria, as judged by virulence factors, e.g. cytotoxigenicity, haemolysis and enterotoxin production (Turnbull et al., 1984).

The role of the Hydrophila group in food poisoning is not clear at present. They grow readily, even at refrigeration temperatures, and are associated with spoilage.

For more information about the identification of organisms mentioned in this chapter see Lee and Donovan (1985).

References

Frerichs, G. N. and Hendrie, M. S. (1985) Bacteria associated with disease of fish. In *Isolation and Identification of Micro-organisms of Medical and Veterinary Importance* (eds C. H. Collins and J. M. Grange), Society for Applied Bacteriology Technical Series 21, Academic Press, London, pp. 355–371

Furniss, A. L., Lee, J. V. and Donovan, T. J. (1978) *The Vibrios.* Public Health Laboratory Service Monograph No. 11, HMSO, London

Lee, J. V. and Donovan, T. J. (1985) Vibrio, Aeromonas and Plesiomonas. In *Isolation and Identification of Micro-organisms of Medical and Veterinary Importance* (eds C. H. Collins and J. M. Grange), Society for Applied Bacteriology Technical Series 21, Academic Press, London, pp. 13–34

Lee, J. V. and Furniss, A. L. (1981) Discussion 1. The phage-typing of *Vibrio cholerae* serovar O1. In *Acute Enteric Infections in Children. New Prospects for Treatment and Prevention* (eds T. Holme *et al.*) Elsevier, Amsterdam, pp. 191–122

Turnbull, P. C. B. *et al.* (1984) Enterotoxin production in relation to taxonomic grouping and source of infection of *Aeromonas* species. *Journal of Clinical Microbiology,* **19,** 175–180

Escherichia, citrobacter, klebsiella, enterobacter, salmonella, shigella and proteus

These genera, collectively known as the enterobacteria, contain many species of small Gram-negative rods that ferment glucose to produce acid or acid and gas. They are oxidase negative; some are motile. Most are commensals or parasites in the human and animal intestine. Table 21.1 shows their general properties.

For medical and public health laboratory purposes, it is convenient to divide the enterobacteria into two groups according to the fermentation of lactose. This is an historical division, dating from the time when bacteriology was almost exclusively a medical science and the lactose fermenters were considered to include mostly saprophytic and commensal organisms while the non-lactose fermenters included the pathogens. Yersinias are now included among the enterobacteria but for practical reasons are considered in Chapter 22.

The lactose fermenters
Produce acid or acid and gas rapidly from this sugar. The genera *Escherichia, Klebsiella, Citrobacter* and *Enterobacter* are collectively known as coliform bacilli.

The non-lactose fermenters
Either completely fail to ferment lactose or ferment it late or irregularly. Included are *Salmonella, Shigella, Proteus, Providencia,* some *Citrobacter* strains and *Serratia.*

The taxonomy, biochemical properties, antigenic structure and identification procedures within the enterobacteria are complex. We have retained those names and descriptions that are still in common use in medicine and industry and introduced some changes recently made by taxonomists.

The lactose fermenting enterobacteria

These organisms are nutritionally non-exacting and will grow on simple culture media. Media containing bile, which inhibit most cocci and Gram-positive bacilli, are most useful. Lactose and an indicator in the medium allow coliform bacilli to be recognized easily.

Table 21.1 Biochemical differentiation within the enterobacteria

| | Indole | Motility | Acid from | | | | | Citrate | Urea | H₂S | Aesculin | Gelatin | PPA | ONPG | β-GUR |
			Lactose	Glucose	Mannitol	Inositol	Amygdalin								
Escherichia coli	+	+	+	+	+	–	–	–	–	–	–	–	–	+	+
Shigella	v	–	–	+	v	–	–	–	–	–	–	–	–	+	v
Salmonella	–	+	–	+	+	+	–	+	–	+	–	–	–	–	–
Citrobacter	v	+	v	+	+	–	v	+	–	+	–	–	–	+	–
Klebsiella	v	–	v	+	+	+	+	+	(+)	–	+	–	–	+	–
Enterobacter	–	+	v	+	+	+	+	+	–	–	v	–	–	+	–
Hafnia	–	+	–	+	+	–	–	+	–	–	+	–	–	+	–
Serratia	–	+	–	+	+	+	+	+	–	–	+	+	–	+	–
Proteus	v	+	–	+	v	v	v	v	+	v	–	v	+	–	–
Providencia	+	+	–	+	v	+	–	+	+	–	–	–	+	–	–
Morganella	+	+	–	+	–	–	–	v	+	–	–	–	+	–	–
Yersinia[a]	v	v	v	+	+	–	+	–	+	–	–	–	–	+	–

[a]See p. 291
ONPG, β-galactosidase; β-GUR, β-glucoronidase
v, variable; (+) late or slow

Isolation

Pathological material
Plate stools from children under 3 years old, intestinal contents of animals, urine deposits, pus, etc., on MacConkey, eosin methylene blue (EMB) or Endo agar and on blood agar. Cystine lactose electrolyte deficient medium (CLED) is useful in urinary bacteriology. Proteus does not spread on this medium. Incubate at 37°C overnight.

Foodstuffs
The organisms may be damaged and may not grow from direct plating. Make 10% suspensions of the food in tryptone soya broth. Incubate at 25°C for 2 h and then subculture into brilliant green bile broth and lauryl sulphate broth. MacConkey broth should not be used because it supports the growth of *Clostridium perfringens,* which produces acid and gas at 37°C. Minerals modified glutamate broth may be useful for the recovery of 'damaged' coliforms. Incubate at 37°C overnight and plate on MacConkey, Endo, EMB, violet red bile agar or CLED media.

Identification

Coliform bacilli show pink or red colonies, 2–3 mm in diameter, on MacConkey agar. Klebsiella colonies may be large and mucoid. On Levine EMB agar, colonies of *E. coli* are blue-black by transmitted light and have a metallic sheen by incident light. Colonies of klebsiellas are larger, brownish, convex and mucoid and tend to coalesce. On Endo medium, the colonies are deep red and colour the surrounding medium. They may have a golden yellow sheen.

Strains from clinical material
Examine a Gram-stained film: some cocci and Gram-positive bacilli grow on selective media such as MacConkey and CLED.

Inoculate peptone water to test for indole production and motility and test for β-glucuronidase β-GUR activity on medium containing *p*-nitrophenyl-β-D glucopyranosideronic acid (Mast ID37).

Or use 4-methylumbelliferyl-β-D-gluconide (MUG) which is hydrolysed to a compound which fluoresces under UV light.

Typical *E. coli* strains are indole positive, motile and give a yellow β-GUR and positive MUG reactions. Only *E. coli* and some shigella strains give a positive β-GUR.

Non-clinical strains
Examine a Gram-stained film. Inoculate lactose broth containing an indicator and a Durham's tube, and tryptone peptone water for indole test: incubate in a water bath at 44+0.2°C. Include these controls:

(1) Stock *E. coli*, which produces gas and is indole positive.
(2) Stock *K. aerogenes* which produces no gas and is indole negative.

The β-GUR test is also useful.
Confirm lactose-fermenting colonies on membrane filters and in primary broth cultures by subculturing to tryptone peptone water and lactose broth, testing for

indole and gas production at 37 and 44°C. Plate on MacConkey agar for colony appearance.

Rapid and kit methods
There are two API systems: RAPIDEC coli RAPIDEC ur, both of which include a β-GUR test.

Multipoint inoculation tests are cost-effective for testing large number of strains. A 'short set' from the MAST range may be used, e.g. β-GUR, inositol, aesculin, hydrogen sulphide or the full MAST 15 system.

Other tests
A modification of Donovan's (1966) system (p.68) uses the above 'short set' in tubed media (Table 21.2).

Table 21.2 'Short set' for coliforms

	β–GUR	Motility	Inositol	Aesculin	H₂S	Indole
E. coli	+	+	−	−	−	+
Klebsiella	−	−	+	+	−	v
Enterobacter	−	+	−	v	−	−
C. freundii	−	+	−	−	+	−
C. koseri	−	+	−	−	−	+

The indole, methyl red, Voges–Proskauer and citrate tests may still be used along with the Eijkman test.

Thermotolerant coliforms
This term is used to describe coliform bacilli that produce gas at 44°C but do not produce indole or have not been tested for it.

Antigens of enterobacteria
There are three kinds of antigens. The O or somatic antigens of the cell body are polysaccharides and are heat stable, resisting 100°C. The H or flagellar antigens are protein and destroyed at 60°C. The K and Vi are envelope, sheath or capsular antigens and heat labile. There are three kinds of K antigens: L, which is destroyed at 100°C and is an envelope, occasionally capsular; A, which is destroyed at 121°C and is capsular; and B, which is destroyed at 100°C and is an envelope. These Vi and K antigens mask the O, and agglutination with O sera will not occur unless the bacterial suspensions are heated to inactivate them.

Serological testing of E. coli
Screen first by slide agglutination with the (commercially available) antisera. Heat saline suspensions of positives in a boiling water bath for 1 h to destroy H and K antigens. Cool, and do tube O agglutination tests. A single tube test at 1:50 (with a control) is adequate. Read at 3 h and then leave on the bench overnight and read again.

Species of coliform bacilli

Escherichia coli
This species is motile, produces acid and gas from lactose at 44°C and at lower temperatures, is indole positive at 44 and 37°C, MR positive, VP negative, fails to

grow in citrate and is malonate and gluconate negative. It is H_2S negative and usually decarboxylates lysine.

These are the so-called 'faecal coli' that occur normally in the human and animal intestine and it is natural to assume that their presence in food indicates recent contamination with faeces. *E. coli* is, however, widespread in nature and although most strains probably had their origin in faeces, its presence, particularly in small numbers, does not necessarily mean that the food contains faecal matter. It does suggest a low standard of hygiene. It seems advisable to avoid calling them 'faecal coli' and to report the organisms as *E. coli*.

It is associated with human and animal infections and is the commonest cause of urinary tract infections in humans and is also found in suppurative lesions, neonatal septicaemias and meningitis. In animals it causes mastitis, pyometria in bitches, coli granulomata in fowls and white scours in calves.

As least four types of *E. coli* cause gastrointestinal disease in humans. According to Gorbach (1986, 1987) they may be described as: enteropathogenic (EPEC), enterotoxigenic (ETEC), enteroinvasive (EIEC) and verotoxigenic (VTEC). See Table 21.3.

Table 21.3 Common serotypes of *E. coli* causing gastrointestinal infections

		Common O serotypes
EPEC	Enteropathogenic (infantile)	18, 26, 44, 55, 86, 111, 114, 119, 124, 125, 126
ETEC	Enterotoxigenic	6, 8, 15, 25, 27, 63, 78, 115, 148, 153, 154
EIEC	Enteroinvasive	28ac, 112ac, 124, 136, 143, 144, 152, 164
VTEC	Verotoxigenic	157

(After Gorbach *et al.*, 1986)

The EPEC strains have been associated with outbreaks of infantile diarrhoea and are identified serologically (see above) but this is necessary only in outbreaks. It is not recommended that single cases of EPEC are investigated in this way.

ETEC strains are thought to cause gastroenteritis in both adults (travellers' diarrhoea) and children (especially in developing countries). They produce enterotoxins, one of which is heat labile (LT) and other heat stable (ST). Specialist facilities are needed to detect ST-producing strains but kits are available for LT-producers (Phadebact ETECLT; Oxoid VET-RPLA). Specific antisera are not available commercially at present.

EIEC strains cause diarrhoea similar to that in shigellosis. The strains associated with invasive enteric infections are less reactive than typical *E. coli*. They may be lysine negative, lactose negative, anaerogenic and resemble shigellas. The correlation between serotype and pathogenicity is imperfect. Reference laboratories use tissue culture and other specialized tests.

VTEC strains derive their name from their cytotoxic effect on Vero cells in tissue culture. They have been associated with haemolytic uraemic syndrome and haemorrhagic colitis. Serotype O:157 H7 is the most frequently reported. They do not ferment sorbitol and can be recognized on Sorbitol MacConkey agar (Oxoid) and tested with commercial O:157 antisera (Oxoid latex). They are also β-GUR negative.

Citrobacter (Escherichia) freundii
Is motile, indole variable, MR positive, VP negative and citrate positive. Hydrogen sulphide is produced but the lysine decarboxylase test is negative. Malonate

utilization is variable and gluconate is not oxidized. Non- or late-lactose fermenting strains occur (see below).

This organism occurs naturally in soil and is therefore a useful indicator of pollution. It can cause urinary tract and other infections in humans and animals.

Citrobacter koseri

Is indole positive, citrate positive but does not produce hydrogen sulphide.

It is found in urinary tract infections and occasionally in meningitis.

Klebsiella aerogenes (UK): Klebsiella pneumoniae (USA)

In the USA, and by the major identification kit manufacturers these organisms are called *K. pneumoniae*. It is non-motile, indole negative, MR negative, VP positive, grows in citrate medium, does not produce hydrogen sulphide but the lysine decarboxylase test is positive. It is malonate and gluconate positive. It is capsulated and most strains give large, mucoid, slimy colonies on media that contain carbohydrates. Normally present on grain and plants, it is also found in the human and animal intestine and can cause urinary tract infections. It is often isolated from the sputum of patients treated with antibiotics (Table 21.4).

Table 21.4 Lactose fermenting enterobacteria; *Klebsiella, Enterobacter;* non-lactose fermenters: *Hafnia, Serratia*

	Indole	Lysine	Ornithine	Arginine	DNase	Aesculin	Acid from	
							Adonitol	Inositol
K. aerogenes (pneumoniae)	–	+	–	–	–	+	+	+
K. oxytoca	+	+	–	–	–	+	+	+
K. ozaenae	–	v	–	v	–	+	+	v
K. rhino- scleromatis	–	–	–	–	–	+	+	+
E. cloacae	–	–	+	+	–	+	v	–
E. aerogenes	–	+	+	–	–	v	+	+
H. alvei	–	+	+	–	–	–	–	–
S. marcescens	–	+	+	–	+	+	v	v

v, variable

K. pneumoniae (Friedlander's pneumobacillus)

In the UK and in some other countries the term *K. pneumoniae* is restricted to strains of klebsiellas from severe respiratory infections that differ from *K. aerogenes* in being MR positive and VP negative and confined to capsule K types 1–3.

Some workers (see Cowan, 1974) have described biochemically atypical strains of capsule K types 1–2 from respiratory sources as the species *K. edwardsii*. This distinction is difficult to apply and is not recommended now for clinical laboratory work.

K. oxytoca

Differs from *K. aerogenes* in being indole positive (Table 21.4).

K. ozoenae

Is also difficult to identify, but is not uncommonly found in the respiratory tract associated with chronic destruction of the bronchi (Table 21.4).

K. rhinoscleromatis

Also difficult to identify, is rare in clinical material and may be found in granulomatous lesions in the upper respiratory tract (Table 21.4).

Klebsiellas may be classified serologically by their O and K antigens. For more information on this genus and associated organisms, see Shinebaum and Cooke (1983).

Enterobacter cloacae

Is motile, indole negative, MR negative, VP positive, grows in citrate and liquefies gelatin (but this property may be latent, delayed and lost). It does not produce hydrogen sulphide and the lysine decarboxylase test is negative. Growth in malonate is variable and the gluconate test is positive. It is found in sewage and in polluted water (Table 21.4).

E. aerogenes

Resembles *E. cloacae* but the lysine decarboxylase test is positive. Gelatin liquefaction is late. It is often confused with *K. aerogenes*. *E. aerogenes* is motile and urease negative; *K. aerogenes* is non-motile and urease positive.

E. agglomerans

Formerly *Erwinia herbicola*, it produces acid and occasionally gas from glucose, acid from mannitol and sometimes (late) from lactose. It liquefies gelatin but is indole, hydrogen sulphide, urease and lysine decarboxylase negative. Other reactions are variable.

The non-lactose fermenting enterobacteria

This group of organisms is, in general, no more exacting in its nutritional requirements than are the lactose fermenters and can be grown on the same media, although a few (*Salmonella typhi* and some strains of *Shigella*) are exacting for certain amino acids. The non-lactose fermenters rarely occur in pure culture but as minority populations among other enterobacteria. More selective media than MacConkey, EMB or Endo media are recommended to inhibit as many lactose fermenters as possible.

There are two fundamentally different approaches to the identification of intestinal pathogens; biochemical, followed by serological and serological confirmed by biochemical. As the final identification of *Salmonella* and *Shigella* must be serological, the second method gives the answer sooner than the first (24–48 h after receiving the sample), but it requires expert knowledge of colonial appearance, skilled slide agglutination technique and media that do not impair slide agglutinability of the organisms. On the other hand, subculturing into a set of identification media and awaiting the results delays identification unduly. Kit and automated systems are generally used, (e.g. API; Bactomatic).

Biochemical reactions are variable. Serological identification should not be delayed until biochemical tests are completed. The full biochemical reactions will, however, also be described.

Before describing methods for identifying the non-lactose fermenting enterobacteria, the antigenic structure of some of the organisms will be considered. *Salmonella* and *Shigella* are identified serologically but although the antigens of most of the other members of the family have been investigated in detail, serology is not used in routine investigations.

Salmonella antigens

The system used was initiated by Kauffmann and White and the tables which are used bear their names.

There are more than 60 somatic or O antigens and these occur characteristically in groups. Antigens 1–50 are distributed among Groups A–Z. Subsequent groups are labelled 51–61. This enables more than 1700 *Salmonella* 'species' or serotypes to be divided into about 40 groups with the commoner organisms in the first six groups.

For example, in Table 21.5 each of the organisms in Group B possesses the antigens 4 and 12 but, although 12 occurs elsewhere, 4 does not. Similarly, in Group C all the organisms possess antigen 6; some possess in addition antigen 7 and other antigens 8. In Group D, 9 is the common antigen. Unwanted antigens may be absorbed from the sera produced against these organisms and single-factor O sera are available that enable almost any salmonella to be placed by slide agglutination into one of the groups. Various polyvalent sera are also available commercially.

To identify the individual organisms in each group, however, it is necessary to determine the H or flagellar antigen. Most salmonellas have two kinds of H antigen and an individual cell may possess one or the other. A culture may therefore be composed of organisms all of which have the same antigens or may be a mixture of both. The alternative sets of antigens are called *phases* and a culture may therefore be in phase I or in phase II or in both phases simultaneously. The antigens in phase I are identified by lower-case letters; thus, the H antigen of *S. paratyphi A* was called *a*, of *S. paratyphi B*, *b*, of *S. paratyphi C*, *c*, and *S. typhi*, *d*, and so on. Unfortunately, after *z* was reached more antigens were found and so subsequent antigens were named z_1, z_2, z_3, etc. It is important to note that z_1 and z_2 are as different as are *a* and *b*; they are not merely subtypes of *z*.

Phase II antigens are given the arabic numerals *1–7*.

Some of the lettered antigens in phase I also occur in phase II as alternatives to other lettered antigens. Whereas *S. paratyphi B* has antigens *b*, and *1* and *2*, *S. worthington* has *z*, and *l* and *w*, and *S. meleagridis* has *e* and *h*, and *l* and *w*. The lettered antigens in phase II do, however, occur mostly in groups, e.g. as *e*, *h; e*, *n*, z_{15}; *e*, *n*, *x; l*, *y* etc., and lettered antigens in phase II, apart from these, are uncommon.

It is therefore necessary to identify both the phase I and the phase II antigens as well as the O antigens. It is usual to find the O group first. Various polyvalent and single-factor O antisera are available commercially for this (see note below). The H antigens are then found using the Spicer–Edwards system known as the Rapid Salmonella Diagnostic (RSD) sera. These are mixtures of antisera that enable the phase I antigens to be determined. In the British set, there are three pools of sera and *S. typhimurium* is tested for separately. In the American set, there are four pools and *S. typhimurium* is included. Both sets are described below. The phase II antigens are found by using a polyvalent serum containing factors *1–7* and then individual sera.

Another antigen is the Vi, found in *S. typhi* and a few other species. This is a surface antigen which masks the O antigen. If it is present the organisms may not agglutinate with O sera unless the suspension is boiled.

Table 21.5 Antigenic structure of some of the common salmonellas (Kauffmann-White classification)

Absorbed O antisera available for identification	Group	Name	Somatic (O) antigen	Flagellar (H) antigen	
				Phase I	Phase II
Factor 2	A	S. paratyphi A	1, 2, 12	a	—
Factor 4	B	S. paratyphi B	1, 4, 5, 12	b	1, 2
		S. stanley	4, 5, 12	d	1, 2
		S. schwarzengrund	4, 12, 27	d	1, 7
		S. saintpaul	1, 4, 5, 12	e, h	1, 2
		S. reading	4, 5, 12	e, h	1, 5
		S. chester	4, 5, 12	e, h	e, n, x
		S. abortus equi	4, 12	—	e, n, x
		S. abortus bovis	1, 4, 12, 27	b	e, n, x
		S. agona	1, 4, 12	gs	—
		S. typhimurium	1, 4, 5, 12	i	1, 2
		S. bredeney	1, 4, 12, 27	l, v	1, 7
		S. heidelberg	1, 4, 5, 12	r	1, 2
		S. brancaster	1, 4, 12, 27	z_{29}	—
Factor 7	C1	S. paratyphi C	6, 7, Vi	c	1, 5
		S. cholerae-suis*	6, 7	c	1, 5
		S. typhi-suis*	6, 7	c	1, 5
		S. braenderup	6, 7	e, h	e, n, z_{15}
		S. montevideo	6, 7	g, m, s	—
		S. oranienburg	6, 7	m, t	—
		S. thompson	6, 7	k	1, 5
		S. infantis	6, 7, 14	r	1, 5
		S. virchow	6, 7	r	1, 2
		S. bareilly	6, 7	y	1, 5
Factor 8	C2	S. tennessee	6, 7	z_{29}	—
		S. muenchen	6, 8	d	1, 2
		S. newport	6, 8	e, h	1, 2
		S. bovis morbificans	6, 8	r	1, 5
		S. hadar	6, 8	z_{10}	e, n, z
Factor 9	D	S. typhi	9, 12, Vi	d	—
		S. enteritidis	1, 9, 12	g, m	—
		S. dublin	1, 9, 12	g, p	—
		S. panama	1, 9, 12	lv	1, 5
Factors 3, 10	E1	S. anatum	3, 10	e, h	1, 6
		S. meleagridis	3, 10	e, h	1, w
		S. london	3, 10	l, v	1, 6
		S. give	3, 10	l, v	1, 7
Factor 19	E4	S. senftenburg	1, 3, 19	g, s, t	—
Factor 11	F	S. aberdeen	11	i	1, 2
Factors 13, 22	G	S. poona	13, 22	z	1, 6

Reproduced by permission of Dr F. Kauffmann, formerly Director of the International Salmonella Centre. The complete tables, revised regularly, are too large for inclusion here. They are obtainable from Salmonella Centres.
*Identical serologically but differ biochemically.

The Kauffmann–White Scheme (Table 21.5) is brought up to date every few years by the International Salmonella Centre.

Shigella antigens

The genus *Shigella* is divided into four antigenic and biochemical subgroups (Table 21.6).

Table 21.6 *Shigella* classification (see also Table 28.6)

Subgroup	Species	Serotypes
A: Mannitol not fermented	*S. dysenteriae*	1–10 all distinct
B: Mannitol fermented	*S. flexneri*	1–6 all related; 1–4 divided into subserotypes
C: Mannitol fermented	*S. boydii*	1–15 all distinct
D: Mannitol fermented, lactose fermented late	*S. sonnei*	

Subgroup A

S. dysenteriae
Contains ten serotypes that have distinct antigens and do not ferment mannitol.

Subgroup B

S. flexneri
Contains six serotypes (I–VI) that can be divided into subserotypes according to their possession of some group factors designated 3,4; 4; 6; 7; and 7,8 (Table 21.7).

Table 21.7 Flexner subserotypes determined with Wellcome reagents sera

Subserotype	Agglutination with serotype	Abbreviated antigenic formula
1	1a	1:4
1 and 3	1b	1:6
2	2a	II:3, 4
2 and X (7, 8)	2b	II:7, 8
3 and X (7, 8)	3a	III:6, 7, 8
3 and Y (3, 4)	3b	III:3, 4, 6
3	3c	III:6
4	4a	IV:3, 4
4 and 3	4b	IV:6
5	5	V:7, 8
6	6	VI:−

The X variant is an organism that has lost its type antigen and is left with the group factors 7,8.

The Y variant is an organism that has lost its type antigen and is left with the group factors 3,4.

Subgroup C

S. boydii
Contains 15 serotypes with distinct antigens; all ferment mannitol.

Subgroup D

S. sonnei
Contains only one, distinct serotype; this ferments mannitol. It may occur in phase I or in phase II, sometimes referred to as 'smooth' and 'rough'. The change from phase I to phase II is a loss of variation and phase II organisms are often reluctant to agglutinate or give a very fine, slow agglutination. Phase II organisms are rarely encountered in clinical work but the sera supplied commercially agglutinate both phases.

Choice of media

Liquid media
These are generally inhibitory to coliforms but less so to salmonellas. Selenite F media are commonly used. If not overheated, they permit *S. sonnei* and *S. flexneri* 6 to grow but do not enrich the culture as they do salmonellas. Mannitol selenite and cystine selenite broths are preferred by some workers for the isolation of salmonellas from foods. Tetrathionate media enrich only salmonellas. Shigellas do not grow. *S. typhi* grows in 0.8% selenite.
    Cultures from these media are plated on solid media.

Solid media
Many media exist for the isolation of salmonellas and shigellas but in our experience the Hynes modification of deoxycholate citrate agar (DCA) and Xylose Lysine Deoxycholate agar (XLD) give the best results for faeces. For the isolation of salmonellas from foods, XLD and Brilliant Green agar (BGA) are superior to DCA. Bismuth sulphite agar (Wilson and Blair medium) is an excellent medium for salmonellas but there are many versions and some are variable in their selectivity.

Colony appearance

Deoxycholate citrate agar (DCA)
On the ideal DCA medium, colonies of salmonellas and shigellas are not sticky. Salmonella colonies are creamy brown, 2–3 mm in diameter at 24 h and usually have black or brown centres. Unfortunately, colonies of proteus are similar. Shigella colonies are smaller, usually slightly pink and do not have black centres. Colonies of *S. typhi* are similar. *S. cholerae-suis* grows poorly on this medium. White, opaque colonies are not significant. Some *Klebsiella* strains grow.

Xylose lysine decarboxylase agar (XLD)
Salmonella colonies are red with black centres, 3–5 mm in diameter. Shigella colonies are red, 2–4 mm in diameter.

Bismuth sulphite agar
Salmonella colonies are black, 2–3 mm across and usually surrounded by a halo showing a metallic sheen due to the diffusion of sulphide. The black coloration of the colony is due to the release of hydrogen sulphide by the organism and the formation of bismuth sulphide. This is a very variable medium.

Brilliant green agar (BGA)
Salmonella colonies are pale pink with a pink halo. Other organisms have yellow-green colonies with a yellow halo.

Isolation

Faeces, rectal swabs and urines
Plate faeces and rectal swabs on DCA and XLD agars. Inoculate selenite F broth
with a portion of stool about the size of a pea or with 0.5–1.0 ml of liquid faeces.
Break off rectal swabs in the broth. Add about 10 ml of urine to an equal volume of
double-strength selenite F broth.

Incubate plates and broths at 37°C for 18–24 h. Plate broth cultures on DCA,
XLD or BGA and incubate at 37°C for 18 h.

Drains and sewers
Fold three or four pieces of cotton gauze (approx. 15 × 20 cm) into pads of
10 × 4 cm. Tie with string and enclose in small-mesh chicken wire to prevent
sabotage by rats. Autoclave in bulk and transfer to individual plastic bags or glass
jars. Suspend in the drain by a length of wire and leave for several days. Squeeze
out the fluid into a jar and add an equal volume of double-strength selenite F broth.
Place the pad in another jar containing 200 ml of single-strength selenite F broth.
Incubate both jars at 43°C for 24 h and proceed as for faeces. This is a modification
of the method originally described by Moore (1948) (see also Vassiliades *et al.*
(1982) and Harvey and Price (1978)).

Foods, feedingstuffs and fertilizers
The method of choice depends on the nature of the sample and on the total
bacterial count. In general, raw foods such as comminuted meats have very high
total counts and salmonellas are likely to be present in large numbers. On the other
hand, heat-treated and deep-frozen foods should have lower counts and if any
salmonellas are present they may suffer from heat shock or cold shock and require
resuscitation and pre-enrichment.

Pre-enrichment   Make a 1:10 suspension or solution of the sample in buffered
peptone water. Use 25 g of food if possible. For dried milk make the suspension in
0.002% brilliant green in sterile water. Use a stomacher if necessary. Incubate at
37°C for 16–20 h.

Enrichment   Add 10 ml of the pre-enrichment to 100 ml of Muller–Kauffman
tetrathionate broth (MK-TB), mixing well so that the calcium carbonate stabilizes
the pH.

Incubate at 43°C for up to 48 h.

Selective plating   At 18–24 h and at 42–48 h subculture from the surface of the
MK-TB to brilliant green phenol red agar (BGA). Incubate at 37°C for 22–24 h.

Salmonella colonies are pink, smooth and low convex. Subculture at least two
colonies for identification.

Additional procedures   Add 0.1 ml of the pre-enrichment to 10 ml of Rappaport–Vasiliades (RV) enrichment medium. Subculture at intervals as with MK-TB.
This method may increase isolation rates, as will additional cultures on DCA and
bismuth sulphite agar.

For full reviews of the isolation of salmonellas from foods see Van Leusden *et al.*
(1982), Fricker (1987) and BSI (1982).

Salmonellas may also be detected in foods by the enzyme-linked immunosorbent assay method (ELISA). See Clayden *et al.* (1987).

Waters
See p.239.

Biochemical screening tests

Method A
Pick representative colonies on a urea slope as early as possible in the morning. Incubate in a water-bath at 35–37°C. In the afternoon, reject any urea positive cultures and from the others inoculate the following: Triple Sugar Iron (TSI) medium by spreading on slope and stabbing butt; broth for indole test; lysine decarboxylase medium. Incubate overnight at 37°C (see Table 21.8) and proceed to serological identification if indicated.

Table 21.8 Biochemical reactions of *Salmonella*, *Shigella*, etc.

	TSI[a]		H_2S	Indole	ONPG	Lysine decarboxylase
	Butt	*Slope*				
S. typhi	A	−	+[b]	−	−	+
S. paratyphi A	AG	−	−	−	−	−
Other salmonellas	AG	−	+/−	−	−	+
Shigella	A	−	−	+/−	+/−	−
Citrobacter	AG	−	+	−	+	−

A = acid; AG = acid and gas.
[a]In TSI medium acid in the butt indicates the fermentation of glucose in the slope that of lactose and/or sucrose.
[b]H_2S production by *S. typhi* may be minimal.

Method B
Use one of the kit systems. Follow the manufacturer's directions and proceed to serological identification if indicated.

Serological identification of salmonellas

Note on commercial agglutination sera
These are obtainable from several companies who vary in the range of sera they offer and in the packaging, e.g. of polyvalent sera.

Test the suspected organism by slide agglutination with Polyvalent O sera. If positive, do further slide agglutinations with the relevant single-factor O sera; only one should be positive. If in doubt, test against 1:500 acriflavine; if this is positive, the organisms are unlikely to be salmonellas.

Subculture into glucose broth and incubate in a water-bath at 37°C for 3–4 h, when there will be sufficient growth for tube H agglutinations. Tube H agglutinations are more reliable than slide H agglutinations because cultures on solid media may not be motile. The best results are obtained when there are about 6–8 × 10^8 organisms/ml. Dilute, if necessary, with formol saline. Add 0.5 ml of formalin to 5 ml of broth culture and allow to stand for at least 10 min to kill the organisms.

Place one drop of Polyvalent (phase 1 and 2) Salmonella H serum in a 75 × 9 mm tube. Add 0.5 ml of the formolized suspension and place in a

Table 21.9 Spicer–Edwards Rapid Salmonella Diagnostic (RSD) H agglutination sera (BBL and Difco)

H antigen	Agglutination with serum			
	A 1	B 2	C 3	D (BBL) 4 (Difco)
a	+	+	+	−
b	+	+	−	+
c	+	+	−	−
d	+	−	+	+
eh	+	−	+	−
G complex	+	−	−	+
i	+	−	−	−
k	−	+	+	+
r	−	+	−	+
y	−	+	−	−
z_4 complex	−	−	+	−
z_{10}	−	−	−	+
z_{29}	−	+	+	−

The G complex (Difco) and the g complex (BBL) contain all those antigenic groups which include f,g,m,p,q,s,t, and u.

The z_4 complex (Difco) and the z_4 complex (BBL) contain all those antigenic groups which include z_4.

water-bath at 52°C for 30 min. If agglutination occurs, the organism is probably a salmonella.

Identify the H antigens as follows. Do tube agglutinations against the Rapid Salmonella Diagnosis (RSD) sera, *S. typhimurium* (*i*) if it is not included in the RSD set, phase 2 complex *1–7*, EN (e,n) complex and L (l) complex. If agglutination occurs (see Tables 21.9 and 21.10). Check the result, if possible by doing tube agglutinations with the indicated monospecific serum. If the test indicates EN (e,n), G or L (l) complexes, send the cultures to a Public Health or Communicable Diseases Laboratory.

Table 21.10 Spicer–Edwards Rapid Salmonella Diagnostic (RSD) agglutination sera (Wellcome Reagents)

H antigen	Agglutination with serum		
	1	2	3
b	+	+	−
d	+	−	+
E complex	+	+	+
G complex	−	−	+
k	−	+	+
L complex	−	+	−
r	+	−	−

The E complex contains e,h; e,n,x; and e,n,z_{15}.

The G complex contains all those antigenic groups which include f,g,m,p,q,s,t and u.

The L complex contains l,v and l,w.

If agglutination occurs in the phase 2 *1–7* serum, test each of the single factors, 2, 5, 6 and 7. Note that with some commercial sera, factor *1* is not absorbed and it may be necessary to dilute this out by using one drop of 1:10 or weaker serum.

Changing the H phase
If agglutination is obtained with only one phase, the organism must be induced to change to the other phase.

Cut a 50 × 20 mm ditch in a well-dried nutrient agar plate. Soak a strip of previously sterilized filter-paper (36 × 7 mm) in the H serum by which the organism is agglutinated and place this strip across the ditch at right-angles. At one end, place one drop of 0.5% thioglycollic acid to neutralize any preservative in the serum. At the other end, place filter-paper disc, about 7 mm in diameter, so that half of it is on the serum strip and the other half on the agar. Inoculate the opposite end of the strip with a young broth culture of the organism and incubate overnight. Remove the disc with sterile forceps, place it in glucose broth and incubate it in a water-bath for about 4 h, when there should be enough growth to repeat agglutination tests to find the alternative phase. Organisms in the original phase demonstrated will have been agglutinated on the strip. Organisms in the alternate phase will not be agglutinated and will travel across the strip.

The phase may be changed with a Craigie tube. Place 0.1 ml of serum and 0.1 ml of 0.5% thioglycollic acid in the inner tube of a Craigie tube and inoculate the inner tube with the culture. Incubate overnight and subculture from the outer tube into glucose broth.

Some organisms, e.g. *S. typhi* and *S. montevideo*, have only one phase. These should be sent to a reference expert.

Antigenic formulae
List the O, phase I and phase II antigens in that order and consult the Kauffmann–White tables (Table 21.5) for the identity of the organism. In the table, the antigenic formula is given in full, but as we use single factor sera for identification it is usually written thus:

S. typhimurium	4, *i*, 2
S. typhi	9, *d*, –
S. newport	6, 8, *e,h*, 2

(The − sign indicates the organism is monophasic)

The Vi antigen
If this antigen is present and prevents the organisms agglutinating with O group sera, make a thick suspension in saline from an agar slope and boil it in a water-bath for 1 h. This destroys the *Vi* and permits the O agglutination.

Salmonella subgenera

The genus *Salmonella* has been divided into four subgenera, numbered I, II, III and IV. Subgenus I is the most important because it contains the majority of the human and animal pathogens. Serotypes ('species') are named. Subgenus II contains serotypes commonly found in reptiles but rarely found in man. Some are named; others known only by their antigenic formula. Subgenus III contains the organisms previously described as the Arizona group. Subgenus IV contains rare serotypes, only some of which are named.

Identification of subgenus is by biochemical tests but as this is of limited practical value it is not pursued here (see Ewing 1986).

Salmonella species and serotypes

Salmonella typhi and in some countries *S. paratyphi* A, B and C (see national lists), are Risk/Hazard group 3 pathogens and should be handled in Biosafety/ Containment Level 3 laboratories. There still seems to be some confusion about the status of species of salmonellas: some have been named as species (e.g. *S. typhi*) others as serotypes (e.g. *S. enteritidis* ser. *paratyphi* B). For practical purposes the binomials used in routine laboratory practice are to be preferred.

Salmonella typhi
Produces acid but no gas from glucose and mannitol. It may produce acid from dulcitol but fails to grow in citrate media and does not liquefy gelatin. It decarboxylates lysine but not ornithine. It is malonate and ONPG negative. Freshly isolated strains may not be motile at first. This organism causes enteric fever.

Other serotypes
Produce acid and usually gas from glucose, mannitol and dulcitol (anaerogenic strains occur) and rare strains ferment lactose. They grow in citrate and do not liquefy gelatin. They decarboxylate lysine and ornithine (except *S. paratyphi A*) and are malonate and ONPG negative. *S. paratyphi A, B* and *C* may also cause enteric fever. Strains with the antigenic formula 4; *b; 1, 2* may be *S. paratyphi B*, usually associated with enteric fever, or *S. java* which is usually associated with food poisoning. Many other serotypes cause food poisoning.

Arizona organisms
These are now included with salmonellas. They have the antigenic formula 23; z_4, z_{23}, z_{26}. There are a number of subserotypes. They fail to ferment dulcitol, may ferment lactose late, slowly liquefy gelatin, decarboxylate lysine and ornithine and are malonate and ONPG positive.

Reference laboratories

It is not possible to identify all of the 1400 or more salmonella species or serotypes with the commercial sera available and in the smaller laboratory. The following organisms should be sent to a Reference or Communicable Diseases Laboratory.
    Salmonellas placed by RSD sera into Groups E, G, or L.
    Salmonellas agglutinated by Polyvalent H or monospecific H sera but not by phase II sera.
    Salmonellas agglutinated by Polyvalent H but not by phase I or phase II sera.
    Salmonellas giving the antigenic formula 4; *b; 1, 2*, which may be either *S. paratyphi B* or *S. java*.
    S. typhi, S. paratyphi B, S. tyhpimurium, S. thompson, S. virchow, S. hadar and *S. enteritidis* should be sent for phage typing.
    Note that *Brevibacterium spp.*, which are sometimes founds in foods, possess Salmonella antigens. These organisms are Gram positive.

Identification of shigellas

Agglutinating sera for these organisms are obtainable commercially.
    The dysentery bacilli do not grow in citrate media, do not decarboxylate lysine,

may or may not ferment mannitol and may or may not produce indole. The ONPG test is variable. Other reactions are given in Tables 21.8 and 21.11.

Mannitol non-fermenters
Test by slide agglutination against polyvalent *S. dysenteriae* antiserum. If positive, test with individual type sera (1–10). If negative, test against *S. flexneri* 6 antiserum (some strains of *S. flexneri* 6 fail to ferment mannitol).

Mannitol fermenters
Test indole-negative strains against *S. sonnei* (phase I and II) serum. It is rarely necessary to use phase I and II sera separately. Test indole-negative strains that do not agglutinate, and indole-positive strains against Polyvalent Flexner and Polyvalent Boyd sera. If either is positive, test with individual type sera (Flexner, 1–6; Boyd, 1–15).

   If the correct biochemical reactions are obtained but there is no slide agglutination, boil a suspension of the organism for 1 h to destroy possible masking surface (K) antigens and re-test.

   Strains that cannot be identified locally, i.e. give the correct biochemical reactions but are not agglutinated by available sera, should be sent to a Reference Laboratory.

   Note that the motile organism *Plesiomonas shigelloides* (p.256) may be agglutinated by *S. sonnei* serum.

Species of *Shigella* (Table 21.11)

S. dysenteriae (S. shiga)
Forms acid but no gas from glucose but does not ferment mannitol. Types 1, 3, 4, 5, 6, 9 and 10 are indole negative. Types 2, 7 and 8 are indole positive and are also known as *S. schmitzii (S. ambigua)*. *S. dysenteriae* is responsible for the classic bacillary dysentery of the Far East.

S. flexneri
Serotypes 1–5 form acid but no gas from glucose and mannitol and are indole positive. Some type 6 strains (Newcastle strains) may not ferment mannitol and may be indole negative. A bubble of gas may be formed in the glucose tube. This serotype is not inhibited by selenite media. Serotypes 1–6 are widely distributed, especially in the Mediterranean area, and often cause outbreaks in mental and geriatric hospitals, nurseries and schools in temperate climes.

S. flexneri subserotypes
Wellcome Reagents do not absorb group 6 factor from their type 3 serum and they also provide X and Y variant sera containing group factors 7, 8 and 3, 4. These sera may therefore be used to determine the Flexner subserotypes for epidemiological purposes (Table 21.7).

S. sonnei
Forms acid but no gas from glucose and mannitol, is indole negative and decarboxylates ornithine. It causes the commonest and mildest form of dysentery, which mostly affects babies and young children and spreads rapidly through schools

Table 21.11 Shigellas

Species	Acid from					Indole	Lysine decar-boxylase	Ornithine decarboxy-lase
	Glucose	Lactose	Sucrose	Dulcitol	Mannitol			
S. dysenteriae	+	–	–	–	–	–/+	–	–
S. flexneri 1–5	+	–	–	–	+	v	–	–
S. flexneri 6 (Boyd 88)	+(G)	–	–	–	+/–	–	–	–
S. sonnei	+	(+)	(+)	–	+	–	–	+
S. boydii 1–15	+	–	–	–	+	+/–	–	–

+(G), a small bubble of gas may be formed; –/+, most strains negative; +/–, most strains positive.
None of these organisms grow in citrate or ferment salicin; none produce H₂S, liquefy gelatin or are motile.

and nurseries. It may be typed for epidemiological purposes by colicine production (Reference Laboratory). It survives in some selenite media.

S. boydii

Forms acid but no gas from glucose and mannitol; indole production is variable. The 15 serotypes of this species are widely distributed but not common and cause mild dysentery.

'Inactive escherichia'

These are anaerogenic, non-motile and late-lactose fermenters. The group includes enteroinvasive strains of *E. coli* (EIEC) which give shigella-like reactions and may be agglutinated by shigella antisera. They occupy an intermediate position between typical *Escherichia* and *Shigella* strains.

Serratia marcescens

This is motile, produces acid or acid plus a small amount of gas from glucose, ferments lactose late or not at all and liquefies gelatin. The VP reaction is positive, it is lysine decarboxylase variable and the DNase test is positive. A red or pink pigment may be produced on agar at room temperature. Old cultures smell of trimethylamine (Tables 21.1 and 21.4). Originally thought to be a saprophyte and hence used in aerosol and filter pore-size tests, this species is now considered to be important in hospital-acquired infections, e.g. meningitis, endocarditis, septicaemia and urinary tract infections.

Hafnia alvei

This gives reliable biochemical reactions only at 20–22°C. It gives acid and gas in glucose and mannitol and does not hydrolyse urea. It is MR negative, VP variable, indole negative, grows in citrate but does not liquefy gelatin (Tables 21.1 and 21.4). It is a soil and water organism that is sometimes found, in equivocal circumstances, in human material, (e.g. faeces).

Proteus, Providencia and Morganella

These organisms are widely distributed. Some strains are easily recognized by their ability to 'swarm' or spread over the surface of agar media in a series of successive waves. This is inconvenient in diagnostic bacteriology as the swarming obscures the presence of other organisms. Swarming does not take place on salt-free agar, e.g. cystine lactose electrolyte deficient (CLED) medium. Swarming is also inhibited on agar media containing 50 mg/l *p*-nitrophenyl glycerol (PNPG) (Senior, 1977), 10 g/l chloral hydrate, 1 g/l sodium azide or 6% agar and bile salts, e.g. MacConkey and DCA media. Non-motile strains are not uncommon.

Isolation and identification

Plate on nutrient agar or blood agar to observe swarming, and on CLED and chloral hydrate blood agar to inhibit it and to reveal other organisms if present. Incubate overnight at 37°C.

Subculture to test for phenylalanine deaminase (PPA test) and urease (PathoTec paper strip tests are convenient for both of these). Test PPA positives for fermentation of lactose, maltose, mannitol and dulcitol; for indole production, growth in citrate, gelatin liquefaction, hydrogen sulphide formation and ornithine decarboxylase. Proteus species are PPA and urease positive; Providence are PPA positive but urease negative (see Table 21.12 for species differentiation).

Table 21.12 Biochemical properties of *Proteus* and *Providencia*

	Acid from		Indole	Citrate	Gelatin liquefaction	Urease	H$_2$S	Ornithine decarboxylase
	Mannitol	Maltose						
P. vulgaris	−	+	+	v	+	+	+	−
P. mirabilis	−	−	−	+^{a}	+	+	+	+
P. rettgeri	+	−	+	+	−	+	−	−
Providencia	+	−	+	+	−	−	−	−
M. morganii	−	−	+	−	−	+	−	+

^{a}May be delayed

Proteus species

Proteus vulgaris
Common in soil and vegetation and in the animal intestine. It frequently contaminates food, leading to spoilage and decomposition, but is rarely found in pure culture. It is frequently found in hospital infections.

P. mirabilis
Associated with putrid and decaying animal and vegetable matter. It is a doubtful agent of gastroenteritis but is found in hospital infections, e.g. in urinary tract infections and suppurative lesions.

Morganella morganii
Is also a commensal but may be associated with infantile (summer) diarrhoea. Found in hospital infections.

Providencia rettgeri

This differs from *Proteus* spp. in failing to produce hydrogen sulphide and in fermenting mannitol. It is associated with fowl typhoid and similar diseases of poultry and with human infections.

Other providencia species
Are found in urinary tract infections and are also occasionally isolated from faeces but are not associated with diarrhoeal diseases.

References

BSI (1982) *BS 5763: Part 4. Standard Methods for the Microbiological Examination of Food and Animal Feeding Stuffs. Detection of Salmonella*, British Standards Institution, London
Clayden, J. A., Alcock, S. J. and Stringer, M. F. (1987) Enzyme-linked immunosorbent assays for the detection of *Salmonella* in foods. In *Immunological Techniques in Microbiology*, (eds J. M. Grange,

A. Fox and N. L. Morgan), Society for Applied Bacteriology Technical Series No. 24, Blackwells, Oxford, pp. 217–229

Cowan, S. T. (1974) *Cowan and Steel's Manual for the Identification of Medical Bacteria*, 2nd edn, University Press, Cambridge

Donovan, T. J. (1966) A Klebsiella screening medium. *Journal of Medical Laboratory Technology*, **23**, 194–196

Ewing, W. H. (ed) (1986) *Edwards and Ewing's Identification of the Enterobacteriaceae*, Elsevier, New York

Fricker, C. R. (1987) The isolation of salmonellas and campylobacters. *Journal of Applied Bacteriology*, **63**, 90–111

Gorbach, S. L. (1986) *Infectious Diarrhoea*, Blackwells, London

Gorbach, S. L. (1987) Bacterial diarrhoea and its treatment. *Lancet*, **ii**, 1378–1382

Harvey, R. W. S. and Price, T. H. (1982) Salmonella isolation techniques alternative to the standard method. In *Isolation and Identification Methods for Food Poisoning Organisms* (eds E. Corry, D. Roberts and F. A. Skinner), Society for Applied Bacteriology Technical Series No. 17, Academic Press, London

Moore, B. (1948) The detection of paratyphoid B carriers in towns by means of sewage examination. *Monthly Bulletin of the Ministry of Health*, **6**, 241–251

Senior, B. W. (1977) *p*-Nitrophenyl glycerol – a superior antiswarming agent for isolating and identifying pathogens from clinical material. *Journal of Medical Microbiology*, **11**, 59–61

Shinebaum, R. and Cooke, M. E. (1985) Klebsiellas. In *Isolation and Identification of Micro-organisms of Medical and Veterinary Importance* (eds C. H. Collins and J. M. Grange), Society for Applied Bacteriology Technical Series No. 21, Academic Press, London, pp. 35–42

Van Leusden, F. M., Van Schothorst, M. and Beckers, H. J. (1982) The standard salmonella identification method. In *Isolation and Identification Methods for Food Poisoning Organisms* (eds E. Corry, D. Roberts and F. A. Skinner), Society for Applied Bacteriology Technical Series No. 17, Academic Press, London, pp. 35–50

Vassiliades, P., Trichopoulos, D., Kalandidi, A. and Xirouchaki, E. (1978) Isolation of salmonellas from sewage with a new enrichment method. *Journal of Applied Bacteriology*, **44**, 233–239

Brucella, haemophilus, gardnerella, moraxella, bordetella, actinobacillus, yersinia, pasteurella, francisella, campylobacter, legionella, bartonella and mobiluncus

Brucella

The genus Brucella contains Risk/Hazard Group 3 pathogens. All manipulations that might produce aerosols should be done in microbiological safety cabinets in Biosafety/Containment Level 3 laboratories.

There are three important and several other species of small, regular Gram-negative bacilli in this genus. They are non-motile, reduce nitrates to nitrites, but carbohydrates are not metabolized when normal cultural methods are used. There are several biotypes within each species.

Isolation

Human disease
Do blood cultures, using the Castenada double phase method (p.142) in which the solid and liquid media are in the same bottle. This not only lessens the risk of contamination during repeated subculture but also minimizes the hazards of infection of laboratory workers. Use Brucella agar plus supplements or tryptone glucose media and, to suppress contaminants, add to each 100 ml of medium: bacitracin, 10 units; polymyxin B, 4 units; cycloheximide, 0.001 mg. Inoculate four bottles, each with 5 ml of blood. Replace the bottle caps with cotton wool plugs and incubate at 37°C in an atmosphere of 10% carbon dioxide (candle jars are inadequate).

Examine weekly for growth on the solid phase. If none is seen, tilt the bottle to flood the agar with the liquid, restore it to a vertical position and reincubate. Do not discard until 6 weeks have elapsed.

Treat bone marrow and liver biopsies in the same way. Subculture on serum glucose agar or chocolate blood agar.

Animal material
Plate uterine and cervical swabbings or homogenized fetal tissue on Farrell's selective medium (Farrell and Robertson, 1972). This is serum glucose agar containing (per ml); bacitracin, 25 units; polymyxin, 5 units; nalidixic acid, 5μg; vancomycin, 5μg; cycloheximide, 100 mg; amphotericin B, 10μg (Oxoid Brucella Supplement).

Milk
Dip a throat swab in the gravity cream and inoculate Farrell's medium and the fluid enrichment medium of Brodie and Sinton (1975). This is tryptone soya broth containing 5% horse serum and these antibiotics (per ml): bacitracin, 20 units; polymyxin 5 units; nalidixic acid, 5 µg; vancomycin, 20 µg; nystatin, 100 units; cycloheximide 100 µg; amphotericin B, 4 µg; cycloserine, 312.5 µg.

Incubate under 10% carbon dioxide (loosen the caps) and subculture in duplicate on serum glucose agar and Farrell's medium. Incubate one culture in air and the other in 10% carbon dioxide for 10 days.

Identification

Colonies on primary media are small, flat or slightly raised and translucent. Subculture on slopes of glucose tryptone agar with a moistened lead acetate paper in the upper part of the tube to test for hydrogen sulphide production.

Do dye inhibition tests. In Reference laboratories three concentrations of thionin (1:25 000; 1:50 000 and 1:100 000) and two of basic fuchsin (1:50 000 and 1:100 000) are used to identify biotypes.

For diagnostic purposes inoculate tubes of Brucella medium or glucose tryptone serum agar containing (1) 1:50 000 thionin, and (2) 1:50 000 basic fuchsin (National Aniline Division, Allied Chemical and Dye Corp.).

Inoculate tubes or plates with a small loopful of a 24-h culture.

It is advisable to control *Brucella* identification with reference strains, obtainable from Type Culture Collections or CDCs: *B. abortus* 554, *B. melitensis* 16M and *B. suis* 1330 (see Table 22.1).

Table 22.1 *Brucella* species

	Growth in		Needs CO_2	H_2S
	Fuchsin	Thionin		
B. melitensis	+	+	−	−
B. abortus	+	−	+	+ (most strains)
B. suis (American)	−	+	−	++
B. suis (other biotypes)	−	+	−	−

Do slide agglutination tests with available commercial sera. Absorbed mono-specific sera are not yet available commercially. Reference laboratories make their own and also use bacteriophage identification and typing methods, and metabolic tests (see Robertson *et al.*, 1980).

Species of *Brucella*

Brucella melitensis
Grows on blood agar and on glucose tryptone serum agar aerobically in 3–4 days. Does not need carbon dioxide to initiate growth. Does not produce hydrogen sulphide and is not inhibited by fuchsin or thionin in the concentrations used in the commercial media. This organism causes brucellosis in humans, the Mediterranean or Malta fever or undulant fever. The reservoirs are sheep and goats and infection occurs by drinking goat milk.

B. abortus

There are eight biotypes. Most require 5–10% carbon dioxide to initiate growth, produce hydrogen sulphide and are inhibited by thionin but not by fuchsin. Some biotypes resemble *B. melitensis*. They cause contagious abortion in cattle. Drinking infected milk can result in indulant fever in humans. Veterinarians and stockmen are frequently infected from aerosols released during birth or abortion of infected animals. Laboratory infections are usually acquired from aerosols released by faulty techniques.

B. suis

The four biotypes of this species do not require carbon dioxide for primary growth. The American biotype produces abundant hydrogen sulphide: the others do not. Biotypes 1 and 2 are inhibited by fuchsin and all four are inhibited by thionin. They cause contagious abortion in pigs and may infect humans, reindeer, hares and geese.

The Ring test

Purchase the stained antigen from a veterinary or commercial laboratory. It is a suspension of *Brucella abortus* cells stained with haematoxylin or another dye. Store the milk samples overnight at 4°C before testing.

To 1 ml of well mixed raw milk in a narrow tube (75 × 9 mm), add one drop (0.03 ml) of the stained antigen. Mix immediately by inverting several times and allow to stand for 1 h at 37°C.

If the milk contains antibodies, these will agglutinate the antigen and the stained aggregates of bacilli will rise with the cream, giving a blue cream line above a white column of milk. Weak positives give a blue cream line and a blue colour in the milk. Absence of antibodies is shown by a white cream line above blue milk. It may be necessary to add known negative cream to a low-fat milk.

False positives may be obtained with milk collected at the beginning and end of lactation, probably due to leakage of serum antibodies into the milk.

The Ring test does not give satisfactory results with pasteurized milk or with goat milk.

Whey agglutination test

This is a useful test when applied to milk from individual cows but is of questionable value for testing bulk milk.

Centrifuge quarter milk and remove the cream. Add a few drops of rennin to the skimmed milk and incubate at 37°C for about 6 h. When coagulated centrifuge and set up doubling dilutions of the whey from 1:10 to 1:2560, using 1-ml amounts in 75 × 9 mm tubes. Add one drop of standard concentrated *B. abortus* suspension and place in a water-bath at 37°C for 24 h. Read the agglutination titre. More than 1:40 is evidence of udder infection unless the animal has been vaccinated recently.

Serological diagnosis

Apart from the standard agglutination test (above) there are others that are outside the scope of this book. For information on the mercaptoethanol test, antihuman globulin and complement fixation tests see Robertson *et al.* (1980).

Useful information about the laboratory diagnosis of brucellosis is also given in the WHO Monograph No. 5 (1981) and Corbell and Hendry (1985).

Haemophilus

For laboratory culture these organisms require either or both of two factors that are present in blood: X factor, which is haematin, and V factor, which is diphospho-pyridine nucleotide and can be replaced by co-enzymes I or II. The X factor is heat stable; the V factor is heat labile. V factor is synthesized by *Staphylococcus aureus*. The bacilli are small (1.5 μm × 0.3 μm, non-motile, usually regular, fail to grow on ordinary media and reduce nitrate to nitrite. There are several species.

Isolation and identification

Plate sputum, spinal fluid, eye swabs, etc., on chocolate agar, which is the best medium for primary isolation because V factor may be inactivated by enzymes present in fresh blood. Incubate at 37°C for 18–24 h. Colonies may be haemolytic or non-haemolytic, are 1 mm in diameter on blood, larger on chocolate agar, grey and translucent. Bacilli are usually small and regular but filamentous forms may be seen (especially in spinal fluids and cultures from them).

'Satellitism', i.e. large colonies around *S. aureus* colonies and around colonies of other organisms synthesizing V factor, may be observed on mixed, primary cultures on blood agar.

Test for X and V factor requirements as follows.

Pick several colonies and spread on nutrient agar to make a lawn. Place X, V and X + V factor discs on the medium and incubate at 37°C overnight. Observe growth around the discs, i.e. if the organisms require both X and V or X or V or neither (see Table 22.2). To test sugar reactions, add Fildes' extract to glucose, maltose, lactose and mannitol fermentation broths. These reactions are not entirely reliable.

Table 22.2 *Haemophilus* **species**

	Haemolysis	Factors required			Growth enhanced by 10% CO_2
		X + V	V only	X only	
H. influenzae[a]	−	+			−
H. parainfluenzae	−		+		−
H. haemolyticus	+	+			−
H. parahaemolyticus	+		+		−
H. haemoglobinophilus[b]	−			+	−
H. ducreyi	±			+	−
H. aphrophilus	−			±	+

[a] *H. aegyptius*   [b] *H. canis*

Agglutination and capsular swelling tests

Do slide agglutination tests using commercial sera (e.g. Wellcome) to confirm identity. These are not satisfactory with capsulated strains as agglutination can occur with more than one serum. To test for capsule swelling with homologous sera use a very light suspension of the organism in saline, coloured with filtered methylene blue solution. Mix a drop of this with a drop of serum on a slide and cover with a cover-slip. Examine with an oil immersion lens with reduced light. Swollen capsules should be obvious, compared with non-capsulated organisms. More information about these tests is given by Turk (1982).

Species of *Haemophilus*

Haemophilus influenzae

Requires both X and V factors, produces acid from glucose but not lactose or maltose and varies in indole production. It occurs naturally in the nasopharynx and is associated with upper respiratory tract infections, including sinusitis and life-threatening acute epiglottitis. It also causes pneumonia, particularly post-influenzal. In eye infections ('pink eye'), it is known as the Koch–Weekes bacillus. Is also found in the normal vagina and is one of the causative organisms of purulent meningitis and purulent otitis media. There are several serological types and growth 'phases'. *H. aegyptius* is probably this species.

H. parainfluenzae

This differs only in not requiring X factor. It produces porphyrins from 6-aminolaevulinic acid. Normally present in the throat, but may be pathogenic.

H. haemolyticus and parahaemolyticus

Haemolytic organisms with the same properties as *H. influenzae* and *H. parainfluenzae*, respectively.

H. haemoglobinophilus (H. canis)

Requires X but not V factor. It is found in preputial infections in dogs and in the respiratory tract of humans.

H. ducreyi

Ducrey's bacillus is associated with chancroid or soft sore. In direct films of clinical material the organisms appear in a 'school of fish' arrangement. They are difficult to grow but may be obtained in pure culture by withdrawing pus from a bubo and inoculating inspissated whole rabbit blood slopes or 30% rabbit blood nutrient agar or on Mueller Hinton medium containing Isovitalex. Culture at 30–34°C but no higher, in 10% carbon dioxide. Colonies at 48 h are green, grey or brown, intact and easily pushed along the surface of the media. It requires X but not V factor.

H. aphrophilus

Colonies on chocolate agar are small (0.5 mm) at 24 h, smooth and translucent. Better growth is obtained in a 10% carbon dioxide atmosphere. Some strains require X factor. There is no growth on MacConkey agar. Acid is produced from glucose, maltose, lactose and sucrose but not from mannitol. Fermentation media should be enriched with Fildes' extract. It is oxidase and catalase negative.

Human infections, including endocarditis, have been reported.

This organism closely resembles *Actinobacillus actinomycetemcomitans* (p.288) but it does not produce acid from lactose and sucrose (see Tables 22.2 and 22.5).

Gardnerella

The species *Gardnerella vaginalis* was previously known as *Haemophilus vaginalis* or *Corynebacterium vaginale*.

Direct examination
Giemsa-stained films made from vaginal secretions usually show many squamous epithelial cells covered with large numbers of organisms ('clue cells'). Gram-stained films show large numbers of Gram-indifferent bacilli rather than the usual Gram-positive lactobacilli.

Culture
Place specimens in transport medium (Stuart's or Amies'). Plate on Columbia blood agar plus supplement (Oxoid) and incubate under 5–10% carbon dioxide. Colonies at 48 h are very small (1 mm), glistening 'dew-drops' and β-haemolytic in human blood agar. Gram-stained films show thin, poorly stained bacilli (unlike the solidly stained diphtheroids). They are Gram positive when young but become Gram negative later.

Make lawns on two plates of Columbia blood agar. On one place a drop of 3% hydrogen peroxide. On the other place discs of trimethoprim (5 μg) and metronidazole (50 μg). Incubate 48 h. Gardnerellas are inhibited by the peroxide and are sensitive to trimethoprim and metronidazole. Lactobacilli and diphtheroids are resistant. Neither X nor V factor is required. There is no growth on MacConkey agar and apart from acid from glucose (add Fildes' extract), biochemical tests seem to be all negative.

Subculture on blood agar and apply discs of metronidazole (50 μg) and sulphonamide (100 μg) (Oxoid). *G. vaginalis* is sensitive to metronidazole but resistant to sulphonamide.

This organism is found in the human vagina and may be associated with non-specific vaginitis (1985). See Easmon and Ison (1985).

Moraxella

The bacilli are plump (2 μm × 1 μm), often in pairs end to end, and non-motile. They are oxidase positive and do not attack sugars. They are indole negative, do not produce hydrogen sulphide and are sensitive to penicillin. Some species require enriched media.

Plate exudate, conjunctival fluid, etc., on blood agar and incubate at 37°C overnight. Subculture colonies of plump, oxidase positive Gram-negative bacilli on blood agar and on nutrient agar (not enriched), on gelatin agar (use the plate method) and on Loeffler medium and incubate at 37°C. Do the nitrate reduction test (see Table 22.3).

Table 22.3 Species of *Moraxella*

	Growth on nutrient agar	Gelatin liquefaction	Nitrate reduction	Catalase	Urease	PPA
M. lacunata[a]	−	+	+	+	−	−
M. nonliquefaciens	v	−	−	+	−	−
M. bovis	+	+	−	v	−	−
M. osloensis	+	−	v	+	−	−
M. kingii[b]	+	v	v	−	−	−
M. phenylpyruvica	+	−	+	+	+	+

PPA, phenylalanine test
[a]Also known as *Moraxella liquefaciens*
[b]*Kingella kingii*

Species of Moraxella

Moraxella lacunata

Colonies on blood agar are small and may be haemolytic. There is no growth on non-enriched media. Colonies on Loeffler medium are not visible but are indicated by pits of liquefaction ('lacunae'). Gelatin is liquefied slowly. Nitrates are reduced.

This organism is associated with angular conjunctivitis and is known as the Morax–Axenfeld bacillus.

M. liquefaciens

This is similar to *M. lacunata* and may be a biotype of that species. It liquefies gelatin rapidly.

M. nonliquefaciens

This is also similar to *M. lacunata* but fails to liquefy gelatin and to reduce nitrates.

M. bovis

Requires enriched medium, liquefies gelatin but does not reduce nitrates. It causes pink eye in cattle but has not been reported in human disease.

M. osloensis (M. duplex, Mima polymorpha var. oxidans)

Enriched medium is not required. Gelatin is not liquefied; nitrates may be reduced. It is found on the skin and in the eyes and the respiratory tract of humans but its pathogenicity is uncertain.

M. kingii (now Kingella kingii)

Colonies are haemolytic and may be mistaken for haemolytic streptococci or haemolytic haemophilus. It does not require enriched media. This is the only member of the group which is catalase negative. It has been found in joint lesions and in the respiratory tract.

M. phenylpyruvica

This does not require enriched media. It is strongly urea positive and reduces phenylalanine to phenylpyruvic acid (PPA positive) – the only species in this group to do this. It has been isolated from assorted human material but its pathogenicity is uncertain.

M. lwoffii

As this is oxidase negative, it is included in *Acinetobacter* (p.245).

Bordetella

The bacilli in this genus are small, 1.5 μm × 0.3 μm and regular. Nitrates are not reduced and carbohydrates are not attacked.

Isolation and identification

Pernasal swabs are better than cough plates, but swabs are best conveyed in one of the commercial transport media. Plate on Bordetella, Bordet-Gengou or Charcoal agar medium containing cephalexin 40 mg/l (Oxoid Bordetella supplement).

Incubate under conditions of high humidity and at 37°C for 3 days. Examine daily and identify by slide agglutination, using commercially available sera. FA reagents are also available.

Test for growth and pigment formation on nutrient agar, urease, nitratase and motility (see Table 22.4).

Table 22.4 *Bordetella* **species**

	Growth on nutrient agar	Brown coloration	Urease	Nitrate reduction	Motility
B. pertussis	−	−	−	−	−
B. parapertussis	+	+	+	−	−
B. bronchiseptica	+	−	+	+	+

Species of *Bordetella*

Bordetella pertussis
Growth on one of the above media has been described as looking like a 'streak of aluminium paint'. Colonies are small (about 1 mm in diameter) and pearly grey. This organism cannot grow in primary culture without heated blood, but may adapt to growth on nutrient agar on subculture. *B. pertussis* causes whooping cough. For detailed information about the bacteriological diagnosis of this disease (see Smith, 1988).

B. parapertussis
Growth on blood agar may take 48 h and a brown pigment is formed under the colonies. It grows on nutrient and on MacConkey agar. On Bordetella or similar media, the pearly colonies ('aluminium paint') develop earlier than those of *B. pertussis*. It does not change Hugh and Leifson medium and does not metabolize carbohydrates. It is nitrate negative and urease positive.

This organism is one of the causative organisms of whooping cough.

B. bronchiseptica
This organism has been in the genera *Brucella* and *Haemophilus*. It forms small smooth colonies, occasionally haemolytic on blood agar, grows best as 37°C and is motile, urease positive and grows in citrate medium.

It causes bronchopneumonia in dogs, often associated with distemper, broncho-pneumonia in rodents and snuffles in rabbits. It has been associated with whooping cough.

Actinobacillus

This genus contains non-motile Gram-negative rods that are fermentative but do not produce gas and vary in their oxidase and catalase reactions.

Isolation and identification

Plate pus, which may contain small white granules, or macerated tissue on blood agar and incubate at 37°C under a 10% carbon dioxide atmosphere. Subculture small flat colonies on to MacConkey agar, in Hugh and Leifson medium and test for fermentation of glucose, lactose and sucrose in enriched fermentation media (see Table 22.5).

Table 22.5 *Actinobacillus*, *Haemophilus aphrophilus* and *Cardiobacterium*

	Growth on MacConkey agar	Acid from		Oxidase	Catalase
		Lactose	Sucrose		
A. lignieresii	+	late	+	+	+
A. equuli	+	early	+	v	v
A. actinomycetemcomitans	−	−	−	−	+
H. aphrophilus	−	+	+	−	−
Cardiobacterium hominis	−	−	+	+	−

Species of *Actinobacillus*

Actinobacillus lignieresii
This species grows on MacConkey agar, is fermentative and produces acid only from glucose, lactose (late) and sucrose.

It is oxidase and catalase positive. It is associated with woody tongue in cattle and human infections have been reported.

A. equuli
This species resembles *A. ligniersii* but ferments lactose rapidly. The oxidase and catalase tests give variable reactions. It is associated with joint ill and sleepy disease of foals, and is also known as *Shigella equirulis* or *S. equulis*.

A. actinomycetemcomitans
This organism requires carbon dioxide on primary isolation and some strains require the X factor. Use an enriched medium and place an X factor disc on the heavy part of the inoculum. There is no growth on MacConkey agar and the oxidase test is negative. Acid is produced from glucose (fermentatively) but not from lactose or sucrose. It is catalase positive.

This organism is difficult to distinguish from *H. aphrophilus* (see Table 22.5). It is sometimes associated with infections by *Actinomyces israelii* and it has been recovered from blood cultures of patients with endocarditis.

Cardiobacterium hominis
This species is included here for convenience.

It is a facultative anaerobe and requires an enriched medium and high humidity. Growth is best under carbon dioxide. Colonies on blood agar are minute at 24 h and about 1 mm in diameter after 48 h and are convex, glossy and butyrous. There is no growth on MacConkey medium, the oxidase test is positive and glucose and sucrose, but not lactose, are fermented without gas production. The catalase, urea and nitrate reduction tests are negative. Hydrogen sulphide is produced but gelatin is not liquefied (see Table 22.5).

This organism has been isolated from blood cultures of patients with endocarditis but may also be found in the upper respiratory tract.

Yersinia

Yersinias cause plague, pseudotuberculosis and gastroenteritis in humans and animals. They are now classified among the enterobacteria but are retained here for pragmatic reasons. They are catalase positive, fermentative in Hugh and Leifson medium (slow reaction), reduce nitrates to nitrites and fail to liquefy gelatin.

The plague bacillus, Yersinia pestis, is in Risk/Hazard Group 3 and all work should be done in a microbiological safety cabinet in a Biosafety/Containment Level 3 laboratory. After isolation and presumptive identification cultures should be sent to a Reference laboratory for any further tests.

Isolation and identification

Yersinia pestis
In suspected bubonic plague examine pus from buboes; in pneumonic plague, sputum; and in rats the heart blood, enlarged lymph nodes and spleen. Before examining rats immerse them in disinfectant for several hours to kill fleas which might be infected.

Make Gram- and methylene blue-stained films and look for small oval Gram-negative bacilli with capsules. Plate on blood agar containing 0.025% sodium sulphite to reduce the oxygen tension, on 3% salt agar and on MacConkey agar. Incubate for 24 h.

Colonies of plague bacilli are flat or convex, greyish white and about 1 mm in diameter at 24 h. Subculture into nutrient broth and cover with liquid paraffin. Test for motility at 22°C, urease, indole, ornithine decarboxylase and inoculate sucrose, cellobiose, amygdalin and rhamnose media (see Table 22.6).

Films of colonies on salt agar show pear-shaped and globular forms. *Yersinia pestis* grows on MacConkey agar and usually produces stalactite growth in broth covered with paraffin. It is non-motile, is urease, indole and ornithine decarboxylase negative and does not produce acid from sucrose, cellobiose, amygdalin and or rhamnose.

It is the causative organism of plague rats, transmitted from rat to rat and from rat to humans by the rat flea. In humans plague is bubonic, pneumonic or septicaemic.

Materials and cultures should be sent to the appropriate Reference Laboratory or Communicable Diseases Centre. Reference laboratories use FA methods, bacteriophage, specific agglutination tests, precipitin tests and animal inoculations.

Yersinia enterocolitica
Culture faeces on Yersinia Selective medium (Cefsulodin–Irgasan–Novobiocin agar, CIN). Incubate at 30°C for 24–48 h and look for 'bullseye' colonies. Enrichment is not usually recommended for isolating this organism from faeces. Some strains grow on DCA as small non-lactose fermenting colonies, even after incubation at 37°C.

Make a 10% suspension of food in peptone water diluent. Keep at 4°C for 3 weeks, and subculture weekly to CIN.

Gram-stained films show small, coccoid bacilli, unlike the longer bacilli of other enterobacteria.

Test for growth on nutrient and MacConkey agar, indole, motility, urease, ornithine decarboxylase and acid production from sucrose, cellobiose, amygdalin, melibiose, rhamnose and raffinose. Kits may be used for all or most of these. Incubate at 30°C rather than at 37°C. See Table 22.6.

There are several serotypes and biotypes. Some are pathogenic for humans (O:3, O:5, O:8, O:9), others for animals. Serological tests for specific antibodies may be of value in diagnosis.

Y. enterocolitica causes enteritis in humans and animals. In humans there may also be lymphadenitis, septicaemia, arthritis and erythema nodosum. The reservoirs are pigs, cattle, poultry, rats, cats, dogs and chinchillas. The organism has been found in milk and milk products, water, oysters and mussels. It grows at 4°C and may therefore multiply in cold storage of food. Person to person spread occurs, mainly in families.

Y. fredericksenii, Y. kristensenii and *Y. intermedia* resemble *Y. enterocolitica* but their roles in disease are uncertain. For further information see Swaminathan *et al.* (1982) and Mair and Fox (1986).

Yersinia pseudotuberculosis

Homogenize lymph nodes, etc. in tryptone broth. Inoculate blood agar with some of the suspension. Retain the remainder at 4°C and subculture to blood agar every 2–3 days for up to 3 weeks.

Emulsify faeces in peptone water diluent and inoculate Yersinia Selective medium (CIN).

Colonies on blood agar are raised, sometimes umbonate, granular and about 1–2 mm in diameter. Rough variants occur. Growth on MacConkey agar is obvious but poor. *Y. pseudotuberculosis* is motile at 22°C, indole negative, urease positive and ornithine decarboxylase negative; acid is produced in glucose and maltose. See Table 22.6.

This organism causes pseudotuberculosis in rodents, especially in guinea pigs. Human infections occur and symptoms resemble those of infection with *Y. enterocolitica*. *Y. pseudotuberculosis* is rarely isolated from faeces, however, and serological tests for specific antibodies are useful.

Pasteurella

Pasteurella multocida

Bipolar staining and pleomorphism are less obvious than in other species. The bacilli are very small (about 1.5 μm × 0.3 μm) and non-motile. Colonies on nutrient agar are translucent, slightly raised, about 1 mm in diameter at 24 h; on blood agar they are slightly larger, more opaque and non-haemolytic. There is no growth on MacConkey agar. Acid is produced from glucose; this is the only species which may not produce acid from maltose and is indole positive. It is urease variable and ornithine decarboxylase positive (see Table 22.6).

This organism causes haemorrhagic septicaemia in domestic and wild animals and birds, e.g. fowl cholera, swine plague, transit fever. Humans may be infected by animal bites or by inhaling droplets from animal sneezes. It has also been isolated from septic fingers.

It is often given specific names according to its animal host, e.g. *avicida, aviseptica, suilla, suiseptica, bovicida, boviseptica, ovicida, oviseptica, cuniculocida, lepiseptica, muricida, muriseptica*. There is some host interspecificity: Cattle strains

Table 22.6 Species of *Yersinia* and *Pasteurella multocida*

	Growth on		Indole	Motility at 22°C	Urease	Ornithine decarboxylase	Acid from				Melibiose	Raffinose	Growth at 4°C
	Nutrient agar	MacConkey					Sucrose	Cellobiose	Amygdalin	Rhamnose			
Y. pestis	+	+	–	–	–	–	–	–	–	–	v	–	–
Y. enterocolitica	+	+	v	+	+	+	+	+	+	+	–	+	+
Y. pseudotuberculosis	+	+	–	+	+	–	–	–	–	+	+	v	–
Y. fredericksenii	+	+	v	+	+	+	+	+	+	+	–	–	+
Y. kristenseii	+	+	+	+	+	+	–	+	+	–	–	–	+
Y. intermedia	+	+	+	+	+	+	+	+	+	+	+	+	+
P. multocida	+	–	+	–	v	+	v						

are also pathogenic for mice but not for fowls; fowl cholera strains are pathogenic for cattle and mice; lamb septicaemia strains are not pathogenic for rodents.

Pasteurella uraea

P. uraea, found in the respiratory tract in humans, resembles *P. multocida* but is indole and ornithine decarboxylase negative. It gives a rapid positive urease test.

Francisella

Francisella tularensis is in Risk/Hazard Group 3. It is highly infectious and has caused many laboratory infections. It should be handled with great care in microbiological safety cabinets in Biosafety/Containment Level 3 laboratory.

Culture blood, exudate, pus or homogenized tissue on several slopes of blood agar (enriched), cystine glucose agar and inspissated egg yolk medium. Also inoculate tubes of plain nutrient agar. Incubate for 3–6 days and examine for very small drop-like colonies. If the material is heavily contaminated, add 100 µg/ml of cycloheximide or 200 units/ml of nystatin, and 2.5–5 µg/ml of neomycin.

Subculture colonies of small, swollen or pleomorphic Gram-negative bacilli on blood and nutrient agar, on MacConkey agar and in enriched nutrient broth. Test for acid production from glucose and maltose, motility at 22°C and urease activity.

The bacilli are very small, show bipolar staining and are non-motile. There is no growth on nutrient or MacConkey agar, poor growth on blood agar with very small, grey colonies; on inspissated egg, at 3 to 4 days, colonies are minute and drop-like. It is a strict aerobe. Acid is formed in glucose and maltose. Indole and urease tests are negative; there is no motility at 22°C.

The organism is responsible for a plague-like disease (tularaemia) in ground squirrels and other rodents in western USA and Scandinavia.

Cultures and materials in cases of suspected tularaemia should be sent to a Reference Laboratory. Fluorescent antibody procedures are used for rapid diagnosis.

Campylobacter

These small, curved, actively motile rods are microaerophilic, reduce nitrates to nitrites but do not attack carbohydrates. They have become very important in recent years as a cause of food poisoning (Chapter 12).

Isolation

Plate emulsions of faeces on blood agar containing (commercially available) campylobacter growth and antibiotic selective supplements. Incubate duplicate cultures at 37 and 42°C for 24–48 h in an atmosphere of approximately 5% oxygen, 10% carbon dioxide and 85% nitrogen, preferably using a gas generating kit, otherwise a candle jar.

Identification

Campylobacter colonies are about 1 mm in diameter at 24 h, grey, watery and flat. Other colonies because opaque. Examine Gram-stained films and inoculate chocolate agar for sensitivity to nalidixic acid disc (30 µg). Test for growth at 25°C and 42°C, production of hydrogen sulphide in cysteine medium (with lead acetate strips) and for growth in the presence of 1% glycine and 8% glucose in broth media and for hippurate hydrolysis (see Table 22.7).

Table 22.7 Species of Campylobacter

European name	North American name	Growth at 25°C	42°C	Resistance to nalidixic acid	H_2S	1% glycine	Hippurate hydrolysis
C. fetus venerealis	C. fetus fetus	+	−	+	−[a]	−	−
C. fetus fetus	C. fetus intestinalis	+	−	+	+	+	−
C. jejuni biotype 1	C. fetus jejuni	−	+	−	−	+	+
biotype 2	C. fetus jejuni	−	+	−	+	−	+
C. coli	C. fetus jejuni	−	+	−	+	+	−
C. laridis		v	+	+	+	+	−

[a]A subtype produces limited H_2S

Species of *Campylobacter*

The names in current use vary with the geographical location of their authors and do not always agree with the Approved Lists of Skerman *et al.* (1980). The text and tables therefore give alternative names.

Campylobacter venerealis (C. fetus subsp. fetus)
Is found in the genitalia of bulls, who act as carriers, and causes abortion and sterility in cows. It cannot multiply in the intestine and therefore is not known to cause human disease.

C. fetus fetus (C. fetus intestinalis)
Is found in sheep, when it may cause abortion. It can multiply in the intestine and is the usual cause of systemic campylobacteriosis, but rarely enteritis in humans.

C. jejuni (C. fetus subsp. jejuni)
Forms part of the normal flora of many domestic and wild animals and is the cause of most cases of campylobacter enteritis.

C. coli (C. fetus subsp. jejuni)
Is also found in animals. It multiplies in the small bowel and causes infectious abortion in sheep.

Other species
There are un-named nalidixic acid resistant thermophilic campylobacters (NARTC) and also a species, *C. sputorum*, which is not regarded as significant.

    For more information abut campylobacters see Newell (1982), Lander and Gill (1985), Humphrey (1986).

Legionella

Legionella pneumophila and related organisms are small Gram-negative bacilli which will not grow on ordinary, unenriched media. About 20 species have been described, some of which are responsible for severe often fatal respiratory tract infections, sometimes in explosive outbreaks. There are no major differences between the pneumonias caused by the different species. Some of the strains isolated from water have not so far been found in humans.

Isolation and identification

Examine bronchial secretions and lung tissue by FA methods. Homogenize the material in buffered saline or peptone water diluent and culture on blood agar containing (per ml): L-cystein HCl, 400 μg; ferric pyrophosphate, 250 μg; sodium selenate, 10 μg; colistin, 15 units; vancomycin, 5 μg, trimethoprim, 2.5 μg; amphotericin B 2.5 μg (Greaves, 1980). These are conveniently supplied as Legionella Growth and Legionella Selective Supplements by Oxoid.

Incubate under 5% carbon dioxide and humid conditions at 37°C and examine daily for 5 days and then at 7 and 14 days. Colonies of *Legionella* are up to 3 mm in diameter, circular, low convex, with moist, glistening surface and are grey- or greenish brown in colour. They are weakly catalase and oxidase positive. Confirm their identity with FA techniques.

Identification to species is important only for epidemiological purposes.

An immunofluorescence method is described by Smith (1987).

For further information about *Legionella* and legionellosis see Wright and Dennis (1985), Ager and Ticknell (1985), HSE (1987) and Harrison and Taylor (1988).

Bartonella bacilliformis

This small Gram-negative rod causes Oroya fever (Carrion fever; bartonellosis) and is transmitted by the sandfly *(Phlebotomus* spp.).

It is best observed in Giemsa-stained blood films and the bacilli are seen in and on red blood cells.

For blood culture use a leptospira medium and incubate at 25°C for up to 4 weeks. Subculture to lysed blood agar or semisolid leptospira medium. There are no commercial test kits as yet but mono- and polyclonal test sera will soon be available. It is best to send specimens or cultures to a reference laboratory.

Mobiluncus

An old species revived in 1980 and now of interest. It is found in vaginal secretions along with other agents of vaginitis but its significance is not fully documented.

It grows on the media used for vaginitis investigations and there appear to be two morphological variants: one long and Gram negative and the other short and Gram variable.

References

Ager, B. P. and Ticknell, J. A. (1985) *The Control of Microorganisms Responsible for Legionnaires' Disease and Humidifier Fever*, Science Reviews, Leeds

Bartlett, L. R., Macrae, A. D. and Macfarlane, J. T. (1988) *Legionella Infections*, Edward Arnold, London

Brodie, J. and Sinton, G. P. (1975) Fluid and solid media for the isolation of *Brucella abortus*. *Journal of Hygiene (Cambridge)*, **74**, 359–367

Corbell, M. J. and Hendry D. (1985) Brucellas. In *Isolation and Identification of Micro-organisms of Medical and Veterinary Importance* (eds C. H. Collins and J. M. Grange), Society for Applied Bacteriology Technical Series No 21, Academic Press, London

Easmon, C. S. F. and Ison, C. A. (1985) *Gardnerella vaginalis*. In *Isolation and Identification of Micro-organisms of Medical and Veterinary Importance* (eds C. H. Collins and J. M. Grange), Society for Applied Bacteriology Technical Series No. 21, Academic Press, London, pp. 115–122

Farrell, I. D. and Robertson, L. (1972) Comparison of various selective media and a new medium for the isolation of brucella from milk. *Journal of Applied Bacteriology*, **35**, 625–630

Greaves, P. W. (1980) New methods for the isolation of *Legionella pneumophila*. *Journal of Clinical Pathology*, **33**, 581–584

Harrison, T. G. and Taylor, A. G. (1988) *A Laboratory Manual for Legionella*, John Wiley, Chichester

HSE (1987) *Legionnaires' Disease*, Guidance Note EH 48: Health and Safety Executive, HMSO, London

Humphrey, T. J. (1986) Techniques for the optimum recovery of cold-injured *Campylobacter jejuni* from milk or water. *Journal of Applied Bacteriology*, **61**, 125–132

Lander, K. P. and Gill, K. P. W. (1985) Campylobacters. In *Isolation and Identification of Micro-organisms of Medical and Veterinary Importance* (eds C. H. Collins and J. M. Grange), Society for Applied Bacteriology Technical Series No. 21, Academic Press, London, pp. 123–142

Mair, N. S. and Fox, E. (1986) *Yersiniosis: Laboratory Diagnosis, Clinical Features and Epidemiology*, Public Health Laboratory Service, London

Newell, D. G. (ed.) (1982) *Campylobacter; Epidemiology, Pathogenesis and Biochemistry*, MTP Press, Lancaster

Robertson, L., Farrell, I. D., Hinchcliffe, P. M and Quaife, R. D. (1980) *Benchbook on Brucella*. Public Health Laboratory Service Monograph No. 14, HMSO, London

Skerman, V. D. B., McGowan, V. and Sneath, P. H. A. (eds) (1980) Approved lists of bacterial names. *International Journal of Systematic Bacteriology*, **30**, 225–420

Smith, J. W. G. (1984) Bacterial infection of the respiratory tract. In *Topley and Wilson's Principles of Bacteriology, Virology and Immunology*, 7th edn, Vol. 3, (eds. G. S. Wilson, A. A. Miles and M. T. Parker), Edward Arnold, London, pp. 391–406.

Smith, M. G. (1987) Immunofluorescent techniques for Legionnaires' disease. In *Immunological Techniques in Microbiology* (eds J. M. Grange, A. Fox and N. L. Morgan), Society for Applied Bacteriology Technical Series No. 24, Blackwells, Oxford, pp. 123–127

Swaminathan, B., Harmon, M. C. and Mehlman, I. J. (1982) A review: *Yersinia enterocolitica*. *Journal of Applied Bacteriology*, **52**, 151–183

Turk, D. C. (1982) *Haemophilus influenzae*. Public Health Laboratory Service Monograph No. 17, HMSO, London

WHO (1981) *A Guide to the Diagnosis, Treatment and Prevention of Human Brucellosis* (ed. S. S. Elberg), World Health Organization, Geneva

Wright, A. E. and Dennis, P. J. (1985) Legionellas. In *Isolation and Identification of Micro-organisms of Medical and Veterinary Importance* (eds C. H. Collins and J. M. Grange), Society for Applied Bacteriology Technical Series No 21, Academic Press, London, pp. 105–114

Chapter 23

Neisseria and branhamella

The Gram-negative cocci include the genera *Neisseria, Branhamella* and *Veillonella*. The first two are oxidase and catalase negative and aerobic. *Veillonella* are obligate anaerobes and are considered in Chapter 28.

Neisseria

These occur mostly as oval or kidney-shaped cocci arranged in pairs with their long axes parallel. They are non-motile, non-sporing, oxidase positive and reduce nitrates to nitrites. Most species are aerobic, facultatively anaerobic but obligate anaerobes are known. The two important pathogens are *N. gonorrhoeae* (the gonococcus) causative organism or gonorrhoea, and *N. meningitidis* (the meningococcus), one of the organisms causing meningitis.

Gonococcus

Examine Gram-stained films of exudates (urethra, cervix, conjunctivae, etc.) for intracellular Gram-negative diplococci. Exercise caution in reporting the presence of gonococci in vaginal and conjunctival material and in specimens from children.

Culture at once or use one of the transport media. Culture on deep plates (20–25 ml of medium) of New York City (NYC) or Thayer Martin (TM) media supplemented with enriched additives and antibiotics. Incubate all cultures 35–36°C (better than 37°C) in 5–10% CO_2 and 70% humidity. Examine at 24 and 48 h. Colonies of *N. gonorrhoeae* are transparent discs about 1 mm in diameter, later increasing in size and opacity, when the edge becomes irregular. Test suspicious colonies by the Gram film and oxidase test. Typical morphology and a positive oxidase test are presumptive evidence of the gonococcus in material from the male urethra, but not from other sites.

For rapid results test colonies by FA method, coagglutination (Phadebact) or rapid biochemical tests (Gonocheck II, Rapid IDNH).

Subculture oxidase positive colonies on chocolate agar and incubate overnight in a 10% carbon dioxide atmosphere. Immediate subculture into carbohydrate test media may not be satisfactory as the primary medium contains antibiotics; other organisms may grow. Repeat the oxidase test on the subculture and emulsify positive colonies in about 1 ml of serum broth. Use this to test for acid production from glucose, maltose, sucrose and lactose in Flynn and Waitkin's sugar-free medium. Some serum media may contain a maltase that may give a false-positive reaction with maltose. The best results are obtained with a semisolid medium, as gonococci do not like liquid media (Table 23.1). API QUADFERM is useful.

Confirm fermentation results by agglutination or coagglutination tests (Phadebact).

Exercise caution in reporting these organisms without adequate experience, especially if they are from children, eye swabs, anal swabs or other sites. They may be meningococci or other *Neisseria* which are genital commensals which do not cause sexually transmitted disease. Sugar reactions and colonial morphology are not entirely reliable. Certain other organisms, *Moraxella*, (p.285) and *Acinobacter calcoaceticus* (p.245) may resemble gonococci in direct films and on primary isolation. *Moraxella*, like *Neisseria*, is oxidase positive but *Acinetobacter* is oxidase negative.

For further information on gonococci: see the WHO Report (1978), Jephcott and Egglestone (1985), Jephcott (1987).

Table 23.1 *Neisseria* **and** *Branhamella*

	Growth on		DNase	ONPG	Acid from			
	Nutrient agar	Thayer Martin			Glucose	Maltose	Sucrose	Lactose
N. gonorrhoeae	−	+	−	−	+	−	−	−
N. meningitidis	−	+	−	−	+	+	−	−
N. lactamicus	+	+	−	+	+	+	−	+
Other Neisseria	v	−	−	v	v	v	v	v
Branhamella	+	−	+	−	−	−	−	−

v, variable

Meningococcus
Examine films of spinal fluid for intracellular Gram-negative diplococci. Inoculate blood agar and chocolate agar with spinal fluid, or its centrifuged deposit. Incubate, preferably in a 5% carbon dioxide atmosphere, at 37°C overnight.

Incubate the remaining CSF overnight and repeat culture on chocolate agar.

Meningococcal colonies are transparent, raised discs about 2 mm in diameter at 18–24 h. In spinal fluid culture, the growth will be pure, while in eye discharges and vaginal swabbings of young children, etc., other organisms will be present. Subculture oxidase positive colonies of Gram-negative diplococci in semisolid serum sugar media (glucose, maltose, sucrose). Meningococci give acid in glucose and maltose (see Table 23.1).

Agglutinating sera are available but should be used with caution. There are at least four serological groups. Slide agglutinations may be unreliable. It is best to prepare suspensions in formol saline for tube agglutination tests.

N. lactamicus (N. lactamica)
This is important in that it might be confused with pathogenic species because it is oxidase positive and is ONPG positive. It is fermentative and produces acid from lactose but this may be delayed. The pathogens do not change lactose and are oxidative.

Other species and similar organisms
These grow on nutrient agar. Colonies are variable: 1–3 mm, opaque, glossy or sticky, fragile or coherent; others may be rough and granular. They often show a lemon or deep yellow pigment.

Branhamella

The single species in the genus, *B. catarrhalis*, was known for many years as *N. catarrhalis*. Previously regarded as a commensal of the respiratory tract it is now considered to be an opportunist pathogen of the upper and lower respiratory system and is associated with lung abscesses. Middle ear infections are not uncommon. It grows well on blood agar, forming non-pigmented, non-haemolytic colonies. It is β-lactamase and DNase positive (Corkill and Makin, 1982; Johnson, 1983). Carbohydrates are not fermented, oxidase and catalase tests are negative (see Table 23.1). It also degrades tributyrin (a butylase strip test is available).

References

Corkill, J. E. and Makin, T. (1982) A selective medium for non-pathogenic aerobic Gram-negative cocci from the respiratory tract, with particular reference to *Branhamella catarrhalis*. *Medical Laboratory Sciences*, **39**, 3–10

Jephcott, A. E. (1987) Gonorrhoea. In *Sexually Transmitted Diseases*, Public Health Laboratory Service, London, pp. 24–40

Jephcott, A. E. and Egglestone, S. I. (1985) *Neisseria gonorrhoeae*. In *Isolation and Identification of Micro-organisms of Medical and Veterinary Importance* (eds C. H. Collins and J. M. Grange), Society for Applied Bacteriology Technical Series No. 21, Academic Press, London, pp. 143–160

Johnson, A. P. (1983) The pathogenic potential of commensal species of *Neisseria*. *Journal of Clinical Pathology*, **36**, 213–223

WHO (1978) *Neisseria gonorrhoeae and Gonococcus Infection*, World Health Organization, Geneva

Chapter 24

Staphylococcus and micrococcus

Staphylococci and micrococci are frequently isolated from pathological material and foods. Distinguishing between the two groups is important. Some staphylococci are known to be pathogens; some are doubtful or opportunist pathogens; others, and micrococci, appear to be harmless but are useful indicators of pollution. Staphylococci are fermentative, capable of producing acid from glucose anaerobically; micrococci are oxidative and produce acid from glucose only in the presence of oxygen.

As with other groups of microorganisms recent taxonomic and nomenclatural changes have not exactly facilitated the identification of the Gram-positive cocci which are isolated in routine laboratories from clinical material and foods. The term 'micrococci' is still loosely but widely used to include many that cannot be easily identified (see also under *Aerococcus*, p.312 and *Pediococcus*, p.314).

Isolation

Pathological material
Plate pus, urine, swabs, etc., on blood agar or blood agar containing 10 mg/l each of colistin and nalidixic acid to prevent the spread of proteus and inoculate salt meat broth. Incubate overnight at 37°C. Plate the salt meat broth on blood agar.

Foodstuffs
Prepare 10% suspensions in 0.1% peptone water in a Stomacher. If heat stress is suspected add 0.1 ml amounts to several tubes containing 10 ml of brain heart infusion broth. Incubate for 3–4 h and then subculture. Use one or more of the following media. Milk salt agar (MSA), phenolphthalein phosphate polymyxin agar (PPP), Baird-Parker medium (BP) or tellurite polymyxin egg yolk (TPEY). Incubate for 24–48 h. Test the PPP medium with ammonia for phosphatase-positive colonies (see below).

MSA agar relies on the high salt content to select staphylococci. In PPP medium, the polymyxin suppresses many other organisms. BP medium is very selective but may be overgrown by proteus. Add 50 µg/ml of sulphamethazine.

For enrichment or to assess the load of staphylococci, use salt meat broth or mannitol salt broth. Add 0.1 and 1.0 ml of the emulsion to 10 ml of broth and 10 ml to 50 ml. Incubate overnight and plate on one of the solid media. If staphylococci are present in large numbers in the food, there will be a heavy growth from the 0.1-ml inoculum; very small numbers may give growth only from the 10-ml sample.

If counts are required, use the Miles and Misra method (p.131) and one of the solid media.

There are kits for the detection of toxins.

Identification

Colonies of staphylococci and micrococci on ordinary media are golden brown, white, yellow or pink, opaque, domed 1–3 mm in diameter after 24 h on blood agar and are usually easily emulsified. There may be β-haemolysis on blood agar. Aerococci show α-haemolysis.

On Baird-Parker medium after 24 h, *Staphylococcus aureus* gives black, shiny, convex colonies, 1–1.5 mm in diameter; there is a narrow white margin and the colonies are surrounded by a zone of clearing 2–5 mm in diameter. This clearing may be evident only at 36 h.

Other staphylococci, micrococci, some enterococci, coryneforms and enterobacteria may grow and may produce black colonies but do not produce the clear zone. Some strains of *S. epidermidis* have a wide opaque zone surrounded by a narrow clear zone. Any grey or white colonies can be ignored. Most other organisms are inhibited (but not proteus: see above).

On TPEY medium, colonies of *S. aureus* are black or grey and give a zone of precipitation around and/or beneath the colonies.

Examine Gram-stained films. Do coagulase and DNase tests on Gram-positive cocci growing in clusters grown from clinical material. This is a short cut: strains positive by both tests are probably *S. aureus*.

Coagulase test

Possession of the enzyme coagulase which coagulates plasma is an almost exclusive property of *S. aureus*. There are two ways of performing this test:

(1) *Slide coagulase test*   Emulsify one or two colonies in a drop of water on a slide. If no clumping occurs in 10–20 s dip a straight wire into human or rabbit plasma (EDTA) and stir the bacterial suspension with it. *S. aureus* agglutinates, causing visible clumping in 10 s.

Use water instead of saline because some staphylococci are salt sensitive, particularly if they have been cultured in salt media. Avoid excess (e.g. a loopful) of plasma as this may give false positives. Check the plasma with a known coagulase positive staphylococcus.

(2) *Tube test*   Do this (a) to confirm the slide test, (b) if the slide test is negative. Add 0.2 ml of plasma to 0.8 ml of nutrient (not glucose) broth in a small tube. Inoculate with the suspected staphylococcus and incubate at 37°C in a water-bath. Examine at 3 h and if negative leave overnight at room temperature and examine again. Include known positive and negative controls. In the tube test, citrated plasma may be clotted by any organism that can utilize citrate, e.g. by faecal streptococci (but these are catalase negative), *Pseudomonas* and *Serratia*. It is advisable therefore to use EDTA plasma (available commercially) or oxalate or heparin plasma. Check Gram films of all tube coagulase positive organisms.

S. aureus produces a clot, gelling either the whole contents of the tube or forming a loose web of fibrin. Longer incubation may result in disappearance of the clot due to digestion (fibrinolysis).

The slide test detects 'bound' coagulase ('clumping factor'), which acts on fibrinogen directly; the tube test detects 'free' coagulase, which acts on fibrinogen in conjunction with other factors in the plasma. Either or both coagulases may be present.

Kits remove the potential hazard of using human plasma. Some combine tests for clumping factor and protein A. Others employ haemagglutination. Berke and Tilton (1986) compared kits with the standard coagulase technique.

DNase test

Inoculate DNase agar plates with a loop so that the growth is in plaques about 1 cm in diameter. Incubate at 37°C overnight. Flood the plate with 1 N hydrochloric acid. Clearing around the colonies indicates DNase activity. The hydrochloric acid reacts with unchanged deoxyribonucleic acid to give a cloudy precipitate. A few other bacteria, e.g. Branhamella, S. pyogenes and Serratia, may give a positive reaction.

If the organisms are coagulase and DNase negative or have been grown on a selective medium that distinguishes between staphylococci and micrococci by the OF test using Baird–Parker's method and one or more of the following tests.

Schleifer and Kloos' test

Use commercial purple agar base containing 10 ml glycerol and 0.4 mg erythromycin/litre. Staphylococci change the colour of the indicator (bromocresol purple) to yellow: micrococci do not grow.

Lysostaphin test

Prepare a solution of Lysostaphin (Sigma Chemicals) by dissolving 50 mg in 40 ml phosphate buffer (0.02 M). Adjust to pH 7.4 and add NaCl to give 1% (w/v). Dispense in 1-ml amounts and store at −60°C. To use, add 1 ml to 9 ml phosphate buffer and mix 5 drops of this with 5 drops of an overnight broth culture in a small test tube. Incubate at 35°C and examine for lysis at 30, 60 and 120 min. Include a control using phosphate buffer instead of Lysostaphin. Most staphylococci are lysed (Kloos and Schleifer, 1975).

Phosphatase test

Inoculate phenolphthalein phosphate agar and incubate overnight. Expose to ammonia vapour. Colonies of phosphatase positive staphylococci will turn pink.

S. aureus gives a positive test (but negative strains have been reported). Coagulase-negative staphylococci and micrococci are usually phosphatase negative (see Table 24.1).

Table 24.1 *Staphylococcus, Micrococcus* **and** *Aerococcus*

	OF	EK	SK	Lys	Coagulase	DNase	Phosphatase	Morphology
S. aureus	F	AN	+	+	+	+	+	Clusters
Staphylococcus spp	F	AN	+	+/−	−	−	−	Clusters
Micrococcus spp.	O	A	−	−	−	−	−	Clusters, Tetrads or packets
Aerococcus viridans[a]	F	AN	−	−	−		−	Clusters or pairs

OF, Hugh and Leifson; EK, Evans and Kloos; SK, Schleifer and Kloos; Lys, lysostaphin sensitivity; F, fermentative; O, oxidative; AN, anaerobic; A, aerobic.
[a]Includes *Gaffkya*

If identification to species is required (other than *S. aureus*) use the API Staph system or test for nitratase reduction, susceptibility to novobiocin (5 μg disc) and for fermentation of mannitol, trehalose and sucrose (Baird–Parker sugar medium, p.66). See under *Species* below and for further information Baird–Parker (1979), Kloos and Jorgensen (1988), Jones *et al.* (1990).

Staphylococci

There are about 30 species of *Staphylococcus* but some are of little interest. They may be divided into three groups: (1) coagulase positive; (2) coagulase negative and novobiocin susceptible; (3) coagulase negative and novobiocin resistant. Only known and opportunist pathogens are mentioned below.

Table 24.2 Properties of some *Staphylococcus* species

Species	Pigment	Coagulase	DNAse	Phosphatase	Novobiocin*	Mannitol	Trehalose	Sucrose
S. aureus	+	+	+	+	S	+	+	+
S. chromogenes	+	+	+	v	S	+	+	−
S. hyicus	−	+	+	v	S	v	+	+
S. intermedius	−	+	+	v	S	v	+	+
S. epidermidis	−	−	+	v	S	−	−	+
S. warneri	−	−	−	−	S	v	+	+
S. cohnii	−	−	−	−	R	+	+	−
S. saprophyticus	−	−	−	−	R	+	+	+

* 5-μg disc
S, susceptible; R, resistant; v, variable

Staphylococcus aureus
This species is coagulase and DNase positive, forms acid from lactose, maltose and mannitol, reduces nitrate, hydrolyses urea and reduces methylene blue. It is usually phosphatase positive but does not grow on ammonium phosphate agar.

Some strains are haemolytic on horse blood agar but the zone of haemolysis is relatively small compared with the diameter of the colony (differing from the haemolytic streptococcus).

Production of the golden yellow pigment is probably the most variable characteristic. Young cultures may show no pigment at all. The colour may develop if cultures are left for 1 or 2 days on the bench at room temperature.

Pigment production is enhanced by the presence in the medium of lactose or other carbohydrates and their breakdown products. It is best demonstrated on glycerol monoacetate agar.

In clinical laboratories, *S. aureus* is usually identified by *either* the coagulase *or* the DNase test. False-positive coagulase tests are possible with enterococci. *Pseudomonas* and *Serratia* if citrated plasma is used. *Serratia* may give a positive DNase test. Coagulase negative, DNase positive strains do occur. We believe, therefore, that both coagulase and DNase tests should be done.

The fermentation of mannitol is not reliable. Mannitol-fermenting, coagulase negative, strains occur.

S. aureus is a common cause of pyogenic infections and food poisoning (Chapter 12). Staphylococci are disseminated by common domestic and ward activities such as bedmaking, dressing or undressing. They are present in the nose, on the skin and in the hair of a large proportion of the population.

Cultures of methicillin-resistant staphylococci (MRSA) are usually a mixture of resistant and susceptible organisms. Growth of the resistant strains is favoured by incubation at 30°C for 48 h and by media containing 2% sodium chloride.

Air sampling, with scatter plates or slit samplers, is an interesting exercise.

In food poisoning and epidemiological investigations, *S. aureus* strains should be sent to a reference expert for bacteriophage typing.

Staphylococcus epidermidis (S. albus)
Is coagulase, DNase and phosphatase negative and may liquefy gelatin. Occasional strains ferment mannitol. It produces no pigment; the colonies are 'china white'. Antibiotic-resistant strains are not uncommon and are often isolated from clinical material, including urine, where they may be opportunist pathogens. (See Schleifer and Kloos, 1975.)

Staphylococcus epidermidis (sensu stricto) is associated with infections with implanted material, e.g. prosthetic heart valves and joints, ventricular shunts and cannulas. Strains resistant to many antibiotics occur. See Table 24.2.

S. saprophyticus
This causes urinary tract infections in young sexually active women. There is a diagnostic kit (Dermaci, Sweden).

Several other coagulase negative species occur. See Kloos and Jorgensen (1988).

Other species
S. chromogenes produces pigment like that of *S. aureus* but does not ferment sucrose. It is an opportunist pathogen and has been found in pigs and in cows' milk. *S. hyicus* is an opportunist pathogen and has been found in dermatitis of pigs, in poultry and cows' milk. *S. intermedius* is known to cause infections in dogs. *S. warneri* and *S. cohnii* are opportunist pathogens.

Staphylococcal toxins
There are kits for testing for the presence of these in foods and also for investigating toxic shock syndrome which is associated with the use of tampons.

Micrococci

These are Gram-positive, oxidase negative, catalase positive cocci that differ from the staphylococci in that they utilize glucose oxidatively or do not produce enough acid to change the colour of the indicator in the medium. They are common saprophytes of air, water and soil and are often found in foods.

The classification is at present confused: the genus contains the organisms formerly called *Gaffkya* and *Sarcina*, tetrad- and packet-forming cocci. These morphological characteristics vary with cultural conditions and are not considered constant enough for taxonomic purposes. The genus *Sarcina* now includes only anaerobic cocci.

At present, it does not seem advisable to describe newly isolated strains by any of the specific names that abound in earlier textbooks, but two 'old' organisms are:

(1) *M. luteus*   Forms yellow colonies, is biochemically inactive and differs from other micrococci in being sensitive to novobiocin.
(2) *M. roseus*   Forms pink colonies.

References

Baird-Parker, A. C. (1979) Methods for identifying staphylococci and micrococci. In *Identification Methods for Microbiologists,* 2nd edn (eds F. A. Skinner and D. W. Lovelock), Society for Applied Bacteriology Technical Series No. 14, Academic Press, London, pp. 201–209

Berke, A. and Tilton, R. C. (1981) Evaluation of rapid coagulase methods for the identification of *Staphylococcus aureus. Journal of Clinical Microbiology,* **23,** 916–919

Kloos, W. E. and Jorgensen, J. H. (1988) Staphylococci. In *Manual of Clinical Microbiology,* 4th edn (eds E. H. Lennette, A. Balows, W. J. Hauser and H. J. Shadomy), Association of American Microbiologists, Washington, pp. 143–153

Jones, D., Board, R. G. and Sussman, M. (1990) *Staphylococci. Journal of Applied Applied Bacteriology* Symposium No. 19, 69, 1S–188S

Kloos, W. E. (1990) Systematics and the natural history of staphylococci. In *Staphylococci.* Eds Jones, D. *et al. Journal of Applied Applied Bacteriology* Symposium No. 19, 69, 25S–38S

Kloos, W. E. and Schleifer, K. H. (1975) Simplified scheme for routine identification of human staphylococci. *Journal of Clinical Microbiology,* **1,** 82–88

Schleifer, K. H. and Kloos, W. E. (1975) Isolation and characterization of staphylococci from human skin. *International Journal of Systematic Bacteriology,* **25,** 50–61

Chapter 25

Streptococcus, enterococcus, aerococcus, leuconostoc and pediococcus

This group includes organisms of medical, dental and veterinary importance as well as starters used in the food and dairy industries, spoilage agents and saprophytes. The cells divide in one plane or in two planes at right angles to one another.

Streptococcus and Enterococcus

Gram-positive cocci that always divide in the same plane, forming pairs or chains; the individual cells may be oval or lanceolate. They are Gram-positive, non-sporing, non-motile and some are capsulated. Most strains are aerobic but it is best to culture clinical material anaerobically. Important streptococci will grow and many other organisms will be suppressed. The catalase test is negative.

Isolation

From clinical material
Blood agar is the usual primary medium. If the material is known to contain many other organisms place a 30 µg neomycin disc on the heavy part of the inoculum. Alternatively use blood agar containing 10 µg/ml colistin and 5 µg/ml oxolinic acid (COBA medium; Petts, 1984) which is very selective for streptococci. (Crystal violet blood agar, formerly recommended, is not very satisfactory because of batch variation in the dye affects colony size and amount of haemolysis.)
Islam's (1977) medium facilitates the recognition of Group B streptococci, e.g. in antenatal screening. Incubate anaerobically at 37°C.

Dental plaque
Plate on blood agar and trypticase yeast extract cystine agar or mitis salivarius agar.

Bovine mastitis
Use Edwards' medium. The crystal violet and thallous sulphate inhibit most saprophytic organisms; aesculin-fermenting saprophytic streptococci give black colonies and mastitis streptococci pale grey colonies.

Foods
Prepare 10% suspensions in 0.1% peptone water using a stomacher. For cold-

stressed organisms inoculate tryptone yeast glucose broth. For heat stressed cocci inoculate glucose broth or tryptone soy broth. Incubate for 3–4 h and plate on one of the azide or thallous acetate media (Mead's or Slanetz and Bartley's). On these media most other organisms are inhibited and 2,3,5-triphenyltetrazolium chloride (TTC) is reduced by enterococci, giving red colonies. Other streptococci give white or pink colonies. On Mead's medium, human enterococci ferment the sorbitol and decompose tyrosine. Typical colonies are maroon in colour and are surrounded by clear zones. See Chapter 18 for isolation from water.

Dairy products
Use yeast glucose agar for mesophiles and yeast lactose agar for thermophiles.

Water, mineral waters and sewage
Membrane filtration gives best results with one of the membrane enterococcus agars (see also p.58).

Air
For β-haemolytic streptococci, in hospital cross-infection investigations use crystal violet agar containing 1:500 000 crystal violet (satisfactory for this purpose) with slit samplers. For evidence of vitiation use mitis salivarius agar.

Identification of streptococci

Colonies on blood agar are usually small, 1–2 mm in diameter and convex with an entire edge. The whole colony can sometimes be pushed along the surface of the medium. Colonies may be 'glossy', 'matt' or 'mucoid'. Growth in broth is often granular, with a deposit at the bottom of the tube.

The primary classification is made on the basis of alteration of haemolysis on horse blood agar.

α-Haemolytic or 'viridans' streptococci produce a small, greenish zone around the colonies. This is best observed on chocolate blood agar.

α′ (Alpha prime)-haemolytic streptococci are surrounded by an area of haemolysis which superficially resembles that of β-haemolytic streptococci (below) but with a hazy outline and unaltered red blood cells within the haemolysed area.

β-Haemolytic streptococci give small colonies surrounded by a much larger, clear haemolysed zone in which all the red cells have been destroyed.

Some streptococci show no haemolysis. Minute β-haemolytic colonies may be S. milleri (anginosus).

Haemolysis on blood agar is only a rough guide to pathogenicity. The β-haemolytic streptococci include those strains which are pathogenic for humans and animals but the type of haemolysis may depend on conditions of incubation and the medium used as a base for the blood agar. Some α-haemolytic streptococci show β-haemolysis on Columbia-based blood agar.

Some saprophytic streptococci are β-haemolytic; so are organisms in other genera having similar colonial morphology. Haemolytic Haemophilus spp. are often reported as haemolytic streptococci in throat swabs because films are not made. C. pyogenes is also haemolytic.

Clinical strains
β-Haemolytic streptococci from clinical material are usually identified by rapid

coagglutination methods (see below) or as *S. pyogenes* (group A) by their sensitivity to bacitracin (see below). Pneumococci are usually identified by their sensitivity to optochin and their lysis in bile. Other streptococci require further tests. Coleman and Ball (1984) give a system for identifying streptococci in clinical laboratories. See Beighton (1985) and Manning and Hogg (1987) for information about oral streptococci.

Streptococcal antigens

Species and strains of streptococci are usually identified by their serological group and type. There are 15 Lancefield groups (A–P, excluding I) characterized by a series of carbohydrate antigens contained in the cell wall. Rabbits immunized with known strains of each group produce serum which will react specifically *in vitro* by a precipitin reaction with an extract of the homologous organism. The carbohydrate is known as the C substance. The streptococci within *Group A* and a few other groups can be divided into serological Griffith types (1–30) by means of two classes of protein antigens, M and T.

M is a type-specific antigen near the surface of the organism which can be removed by trypsinization and is present in matt and mucoid types of colonies but not in glossy types. It is demonstrated by a precipitin test.

T is not type-specific, may be present with or independent of the M antigen, and is demonstrated by agglutination tests with appropriate antisera.

Grouping is usually carried out in the laboratory where the organisms are isolated, using sera and control extracts obtained commercially. Only sera for Groups A, B, C, D and G need be used for routine purposes.

Workers should be aware that the groups are not always species-specific. Some group D streptococci and enterococci possess the G antigen; *S. milleri* may group as A, C, F or G and, when they are haemolytic they may be confused with pyogenic streptococci. Some viridans streptococci may also group as A or C and under certain conditions may appear to be β-haemolytic.

Typing of haemolytic (Group A, etc.) strains is necessary for epidemiological purposes. This is best done by Reference laboratories.

Serological grouping of streptococci

There are two approaches. Coagglutination which requires commercially available kits, is rapid and gives results within 1 h of obtaining a satisfactory growth on the primary culture. Precipitin tests take longer but allow a larger number of groups to be identified.

Coagglutination: latex agglutination

At least three kits are in common use. The streptococcal antibody is attached to staphylococci or latex particles. These antibody-coated particles are agglutinated when mixed on a slide with suspensions or extracts of streptococci of the same group. In the Phadebact (Pharmacia) method colonies from the primary plates are mixed with the reagents on a slide. In the Streptex (Wellcome) and Oxoid systems the antigens are extracted before mixing with the latex reagent on a slide. Groups A, B, C, D, F and G may be identified but it is recommended that biochemical tests are done on Group D streptococci to identify enterococci and 'viridans' streptococci.

Precipitin tube methods

These have largely been superseded by the slide (kit) methods but the following methods are still in use.

Centrifuge 50 ml of an overnight culture of the streptococcus in 0.1% glucose broth or a suspension prepared by scraping the overnight growth from a heavily inoculated blood agar plate.

For Lancefield method suspend the deposit in 0.4 ml of 0.2 N hydrochloric acid and place in boiling water-bath for 10 min. Cool, add 1 drop of 0.02% phenol red and loopsful of 0.5 N sodium hydroxide solution until the colour changes to faint pink. Centrifuge; the supernatant is the extract.

For Fuller's formamide method suspend the deposit in 0.1 ml of formamide and place in an oil-bath at 160°C for 15 min. Cool and add 0.25 ml of acid alcohol (95.5 parts ethanol:52 parts N hydrochloric acid). Mix and centrifuge. Remove the supernatant fluid and add to it 0.5 ml of acetone. Mix, centrifuge and discard the supernatant. To the deposit add 0.4 ml of saline, 1 drop of 0.02% phenol red and loopsful of 0.2 N sodium hydroxide solution until neutral. This is the extract. This is not a very good method for Group D streptococci.

In Maxted's method the C substance is extracted from streptococcal suspension by incubation with an enzyme prepared from a strain of *Streptomyces*.

To prepare this enzyme, obtain *Streptomyces* sp. No. 787 from the National Collection of Type Cultures and grow it for several days at 37°C on buffered yeast extract agar. The medium is best sloped in flat bottles (120-ml 'medical flats'). When there is good growth, place the cultures in a bowl containing broken pieces of solid carbon dioxide. After 24 h allow the medium to thaw and remove the fluid; this contains the enzyme. Bottle it in small amounts and store in a refrigerator.

To prepare the streptococcal extract, scrape the growth from a heavily inoculated 24-h blood agar culture of the organisms in 0.5 ml of enzyme solution in a small tube and place in a water-bath at 37°C for 2 h. Centrifuge and use the extract to do precipitin tests as described below.

Prepare capillary tubes from pasteur pipettes. Dip the narrow end in the grouping serum so that a column a few millimetres long enters the tube. Place it in a block of Plasticine and, with a very fine pasteur pipette, layer extract on the serum so that the two do not mix but a clear interface is preserved. Some practice is necessary in controlling the pipette. If an air bubble develops between the two liquids, introduce a very fine wire, when an interface is usually produced.

A positive result is indicated by a white precipitate that develops at the interface. It is sometimes necessary to dilute the extract 1:2 or 1:5 with saline to obtain a good precipitate.

The bacitracin disc method

Inoculate a blood agar plate heavily and place a commercial bacitracin disc on the surface and incubate overnight. A zone of inhibition appears around the disc if the streptococci are Group A. This test is not wholly reliable and is declining in popularity except as a screening method.

Biochemical and other tests

These are used in clinical laboratories for identifying α- and non-haemolytic streptococci, enterococci and other Group D streptococci. They are useful in oral (dental) bacteriology and in food bacteriology.

Table 25.1 Groups A, B, C, G, D and 'viridans' streptococci

Group	Haemolysis	Hippurate hydrolysis	Growth in 6% NaCl medium	Bile-aesculin
A	β	–	–	–
B	β none	+	+/–	–
D (enterococci)	α/β/none	+[a]	+	+
D (not enterococci)	α	–	–	–

[a] usually, see Table 25.4

The most useful tests may be selected according to the known source of the material. Nevertheless, identification of streptococci is far from easy. The API 20 Strep and Rapid ID Strep systems are useful for identifying streptococci.

See Tables 25.2, 25.3 and 25.4.

Table 25.2 Oral streptococci

	Acid from		Arginine	Aesculin	VP
	Mannitol	Sorbitol			
S. mutans	+	+/–	–	+/–	+
S. sanguis	–	–	+	+	–
S. mitior	–	–	–	–	–
S. milleri	–	–	+	+	+
S. salivarius	–	–	–	+	–

From Beighton (1985) (Reproduced by permission of the author and editors of the *Journal of Applied Bacteriology*)

Table 25.3 *Enterococcus faecalis* biotypes

	Acid from		Hippurate hydrolsis	β-galactosidase
	Arabinose	Sorbitol		
Biotype 1	–	+	+	–
2	–	+	–	–
3	–	+	+	+

Table 25.4 *Enterococcus faecium* biotypes

	Acid from				Amygdalin	Hippurate hydrolysis	β-galactosidase
	Arabinose	Sorbitol	Inulin	Raffinose			
Biotype 1	+	–	–	–	–	+/–	–
2	+	–	–	+	+	+	+
3	+	–	+	+	+	–	+

Fluorescent antibody tests

These are useful for Group A streptococci but not so reliable for other groups.

Groups and species of streptococci

NB: There are important differences between the British and American taxonomies, particularly about *S. milleri (S. anginosus)*. See Facklam and Carey (1985).

Group A

These are β-haemolytic, are the so-called haemolytic streptococci of scarlet fever, tonsillitis, puerperal sepsis and other infections of humans, and are known as *S. pyogenes*. Some strains are capsulated and form large (3-mm) colonies like water drops on the surface of the medium. '*S. mucosus*' and '*S. epidemicus*' were names once used to describe these mucoid strains, which have been associated with milk-borne outbreaks. Capsule formation is, however, not uncommon when streptococci are grown in milk.

Group B

The β-haemolytic streptococci in this group correspond to *S. agalactiae*, the causative organism of chronic bovine mastitis. This is an important cause of neonatal meningitis and septicaemia. It is found in the female genital tract and gut. It has been reported as a cause of urinary tract infections. It is usually β-haemolytic but non-haemolytic variants occur. It grows on bile salt media.

Group C

The most important β-haemolytic member of this group is *S. equi*, which causes strangles in horse. Others include *S. equisimilis*, responsible for some human infections and *S. zooepidemicus* for outbreaks among animals. *S. dysgalactiae*, which is α-haemolytic, is associated with acute bovine mastitis (but is less common than *S. agalactiae*).

Group D

The so-called faecal streptococci or enterococci are now placed in a separate genus, *Enterococcus* (Schleifer and Klipper-Balz, 1983; Collins *et al.*, 1984). Some of these strains are β-, some α- and some non-haemolytic. *E. faecalis* and *E. faecium* are the enterococci, commensals in human and animal intestines and are used as an indicator of 'faecal' pollution in sanitary bacteriology. Most of the organisms in this group are able to resist 60°C for 30 min, all grow at 45°C and in the presence of 40% bile and hydrolyse aesculin, but these properties are not exclusive to the group.

E. faecalis is common in the human and poultry gut but rare in other animal intestines. This is the only enterococcus which reduces 2,3,5-triphenyltetrazolium chloride (TTC), produces acid from sorbitol, decomposes tyrosine and grows in the presence of 0.03% potassium tellurite and at 10°C.

E. faecium, present in the intestines of pigs and other animals, has caused spoilage of canned ham, which is pasteurized ('commercially sterilized') (see p.201). TTC is not reduced; there is no growth on medium containing 0.03% potassium tellurite. *E. faecium* ferments arabinose; *E. durans* does not. Most enterococci reduce litmus milk. Other Group D streptococci do not (use a heavy inoculum). See Tables 25.3 and 25.4 for the biotypes of *E. faecalis* and *E. faecium*.

E. bovis and *E. equinus* are also found in the animal intestine and in milk and are difficult to tell apart. The Sims test allows *S. bovis* to be distinguished from other streptococci; inoculate Rogosa medium stabs in small screw-capped bottles in which little air space is left and cap tightly. *S. bovis* will grow in about 4 days if carbon dioxide is not allowed to escape. Other streptococci do not grow.

These organisms do not reduce TTC, are inhibited by 0.03% potassium tellurite. *E. bovis* ferments raffinose; *E. equinus* does not.

The D antigen is deeper in the cell than are the antigens of other streptococci. For this reason, extracts prepared by acid hydrolysates of the streptococci and the antisera prepared with them do not give good precipitation. The commercial antisera are satisfactory but Fuller's method should be used for grouping. Most, but not all, will group with coagglutination reagents.

Group D streptococci can be confused with aerococci but the latter do not grow at 45°C and do not hydrolyse arginine. Both cause greening of bacon.

Group N

These are non-haemolytic and non-pathogenic and are of importance in the dairy industry. *S. lactis* and *S. cremoris* are the most common. They are used in the ripening and curing of Cheddar-type cheese and in the preparation of cultured buttermilk and the manufacture of sauerkraut. Both organisms are susceptible to a phage which rapidly destroys them, interfering with the commercial processes.

Other groups and species

Group G is known to contain some human pathogens and some strains in Group L and M are pathogenic for animals. Groups H and K contain oral streptococci. Some species react with several group specific sera and others have not yet been grouped. See Table 25.2.

S. salivarius

These are commensals in the human upper respiratory tract and are therefore useful indicators in air hygiene and ventilation investigations. They are rapidly identified by their ability to produce a levan when grown on media containing 5% sucrose. The colonies are large and mucoid.

'Viridans' streptococci

This name is often given in error to any streptococcus that shows α-haemolysis. It should be restricted to *S. mitior*, which is found in the mouth. It may react with sera of Groups O, K or M. *S. uberis*, which also gives α-haemolysis, is a saprophyte found in soil, often gaining access to milk. This may react with sera of Groups E, C, D or P.

S. mutans

Appears to play a major role in dental caries, but does not appear to belong to any group.

S. milleri (S. anginosus)

This is also found in the mouth, in dental abscesses and other deeper abscesses. It may require 5% carbon dioxide for growth and some cultures smell of caramel. It is difficult to group but some strains may react with sera of Groups A, C, G or F.

S. mitior

See 'Viridans' streptococci above.

S. sanguis

Is found in the mouth and in dental plaque and has been reported in heart valve disease. It usually reacts with Groups H or K sera.

See Beighton (1985) for more information about oral streptococci.

S. thermophilus

Resembles Group D streptococci biochemically but has no D antigen. The optimum temperature for growth is 50°C. It is used (along with *Lactobacillus bulgaricus*) in the manufacture of yoghurt.

S. pneumoniae

The pneumococcus is a causative organism of lobar pneumonia, otitis media and meningitis in humans and of various other infections in humans and animals, including (rarely) mastitis. In pus and sputum, the organism appears as a capsulated diplococcus but usually grows on laboratory media in chains and then shows no capsule. There are several serological types, but serological identification is rarely attempted nowadays. One type produces highly mucoid colonies but most strains give flat 'draughtsman'-type colonies 1–2 mm in diameter with a greenish haemolysis.

Pneumococci often resemble 'viridans' streptococci on culture. The following tests allow rapid differentiation.

(1) Pneumococci are sensitive to optochin (ethylhydrocupreine hydrochloride). Optochin discs are supplied by most media manufacturers.

A disc placed on a plate inoculated with pneumococci will give a zone of inhibition of at least 10 mm when incubated aerobically (not in carbon dioxide).
(2) Pneumococci lysed by bile; other streptococci are not. Add 0.2 ml of 10% sodium deoxycholate in saline to 5 ml of an overnight broth culture (do not use glucose broth). Incubate at 37°C. Clearing should be complete in 30 min.

Pneumococci are alone among the streptococci in fermenting inulin and are not heat resistant as are many 'viridans' streptococci. These two criteria should not be used alone for differentiation.

Aerococcus

These are Gram-positive, oxidase negative, fermentative cocci that are usually in clusters, pairs, tetrads or short chains.

Isolation

Use blood agar for food investigations and in slit samplers for aerobiology. Incubate at 30–35°C.

Identification

Aerococci give green α-haemolysis on blood agar and a weak or negative catalase test. It is not easy to distinguish them from 'viridans' streptococci. They differ from *S. mitior* in growing at pH 9.6 in 6.6% sodium chloride broth and in being resistant to 60°C for 30 min. They grow on bile salt media but differ from Group D streptococci in not possessing the D antigen, in failing to grow at 45°C and in not hydrolysing arginine.

Aerococcus viridans

Is the only named species. It is a common airborne contaminant and is also found in curing brines. It appears to be the same organism as *Gaffkya homari* which causes an infection in lobsters; they develop a pink discoloration on the ventral surface and the blood loses its characteristic bluish green colour, becoming pink.

Leuconostoc

This genus contains several species of microaerophilic streptococci of economic importance. Some produce a dextran slime on frozen vegetables.

Isolation

Culture material on semi-solid yeast glucose agar at pH 6.7–7.0. Incubate at 20–25°C under 5% carbon dioxide.

Identification

Inoculate MRS sugars: glucose, lactose, sucrose, mannose, arabinose and xylose and look for dextran slime on solid medium containing glucose (see Table 25.5).

Table 25.5 *Leuconostoc* species

	Acid from						Slime
	Arabinose	*Xylose*	*Glucose*	*Mannose*	*Lactose*	*Sucrose*	
L. cremoris	−	−	+	−	+	−	−
L. dextranicum	−	v	+	v	+	+	+
L. lactis	−	−	+	v	+	+	−
L. mesenteroides	+	v	+	+	v	+	++
L. paramesenter-oides	v	v	+	+	v	+	−

Species of *Leuconostoc*

Leuconostoc mesenteroides
This is the most important species. It produces gas and a large amount of dextran slime on media and vegetable matter containing glucose. Colonies on agar media without the sugar are small and grey. Growth is very poor on media without yeast extract. It takes part in sauerkraut fermentation and silage production and is responsible for slime disease of pickles (unless *Lactobacillus plantarum* is encouraged by high salinity). It is also responsible for 'slimy sugar' or 'sugar sickness'. It spoils frozen peas and fruit juices.

L. paramesenteroides
Is similar to *L. mesenteroides* but does not produce a dextran slime. Widely distributed on vegetation and in milk and dairy products.

L. dextranicum
Produces rather less slime than *L. mesenteroides* and is less active biochemically. Widely distributed in fermenting vegetables and in milk and dairy products.

L. cremoris
Is rare in nature but extensively used as a starter in dairy products.

L. lactis
Not common. Found in milk and dairy products.

Pediococcus

Pediococci are non-capsulated microaerophilic cocci that occur singly and in pairs. They are nutritionally exacting and are of economic importance in the brewing, fermentation and food processing industries.

Isolation

Use enriched media and wort agar for beer spoilage pediococci. Tomato juice media at pH 5.5 is best. Incubate at temperatures according to species sought (see Table 25.6) under 5–10% carbon dioxide.

Identification

If it is necessary to proceed to species identification inoculate MRS sugar media: lactose, sucrose, galactose, salicin, maltose, mannitol, arabinose and xylose. Incubate at appropriate temperature. Gas is not produced by an species (see Table 25.6).

Table 25.6 *Pediococcus* species

	Growth at		Acid from							
	37°C	45°C	Lactose	Sucrose	Maltose	Mannitol	Galactose	Salicin	Arabinose	Xylose
P. damnosus	−	−	−	−	+	−	−	−	−	−
P. acidi-lactici	+	+	v	v	−	−	+	−	−	−
P. pentosaceus	+	+	+	v	+	−	+	+	+	+
P. urinae-equi	+	−	+	+	+	−	+	+	v	v
P. halophilus	+	−	v	+	+	+	+	+	v	v

Species of *Pediococcus*

Pediococcus damnosus (P. cerevisiae)
Is found in yeasts and wort. Contamination of beer results in a cloudy, sour product with a peculiar odour – 'sarcina sickness'. *P. damnosus* is resistant to the antibacterial agents in hops.

P. acidi-lactici
Is found in sauerkraut, wort and fermented cereal mashes. Is sensitive to the antibacterial activity of hops.

P. pentosaceus
Is also found in sauerkraut as well as in pickles, silage and cereal mashes but not in hopped beer.

P. urinae-equi
Is a contaminant of brewers' yeasts. (The name indicates that it was originally isolated from urine of a horse.)

References

Beighton, D. (1985) *Streptococcus mutans* and other streptococci from the oral cavity. In *Isolation and Identification of Micro-organisms of Medical and Veterinary Importance* (eds C. H. Collins and J. M. Grange), Society for Applied Bacteriology Technical Series No. 21, Academic Press, London, pp. 177–190

Coleman, G. and Ball, C. (1984) Identification of streptococci in a medical laboratory. *Journal of Applied Bacteriology*, **57**, 1–4

Collins, M. D., Jones, D., Farrow, J. E. A., Kilpper-Balz, R. and Scheifer, K. H. (1984) *Enteroccocus avium* nom.rev., comb. nov.; *E. casseliflavus* nom. rev. comb. nov.; *E. durans* non. rev., comb. nov.; *E. gallinarum* comb. nov. and *E. malodoratus* sp. nov. *International Journal of Systematic Bacteriology*, **34**, 220–223

Facklam, R. R. and Carey, R. B. (1985) Streptococci and aerococci. In *Manual of Clinical Microbiology*, 4th edn (eds A. Lennette, W. J. Hauser and H. J. Shadomy), Association of American Microbiologists, Washington DC, pp. 154–175

Islam, A. K. M. S. (1977) Rapid recognition of Group B streptococci. *Lancet*, **i**, 256–257

Manning, J. E. and Hogg, S. D. (1987) A short scheme for the identification of 'viridans' streptococci isolated from the human mouth. *Letters in Applied Microbiology*, **4**, 17–19

Petts, D. N. (1984) Colistin-oxolinic acid blood agar: a new selective medium for streptococci. *Journal of Clinical Microbiology*, **19**, 4–7

Schleifer, K. H. and Kilpper-Balz, R. (1983) Transfer of *Streptococcus faecalis* and *Streptococcus faecium* to the genus *Enterococcus* nom. rev. as *Enterococcus faecalis* comb. nov. and *Enterococcus faecium* comb. rev. *International Journal of Systemic Bacteriology*, **34**, 31–34

Corynebacterium, microbacterium, brochothrix, propionibacterium, brevibacterium, erysipelothrix, listeria, lactobacillus and bifidobacterium

Corynebacterium

These organisms are frequently club shaped, thin in the middle with swollen ends that contain metachromatic granules. These are best seen when Albert's or Neisser's stain is used. The stained bacilli may also be barred, with a 'palisade' or 'Chinese letter' arrangement, due to a snapping action in cell division.

This section is restricted to corynebacteria of medical and veterinary importance. Other 'corynebacteria' or 'coryneforms' are discussed under their generic headings.

The diphtheria group

Diphtheria is now a very rare disease and may be clinically atypical. *If an organism is suspected of being a diphtheria bacillus because of its colonial and microscopical appearance, inform the physician at once and before proceeding to identify the bacillus.* It is in the best possible interests of the patient that this is done. No harm will result if the final result is negative, but a delay of 24 h or more to confirm a positive case may have serious consequences, especially in an unimmunized child.

A culture should be sent immediately to the nearest reference expert. A case of diphtheria will usually generate large numbers of swabs from contacts. For the logistics of mass swabbing and the examination of large numbers of swabs in the shortest possible time see Collins and Dulake (1983).

Isolation and identification of the diphtheria organism and related species

Plate throat, nasal swabs, etc., on Hoyle's or other tellurite media and on blood agar. Incubate at 37°C and examine at 24 and 48 h.

For colonial characteristics, see species descriptions.

Make Gram-stained films of suspicious colonies. *C. diphtheriae* tends to decolorize easily and appears Gram-negative compared with diphtheroids which stain solidly Gram positive. When staining, make a control film of staphylococci treated in the same way. This apparent reaction is a useful characteristic.

Subculture a colony from the tellurite to Loeffler serum as soon as possible. Incubate at 37°C for 4–6 h. The best results are obtained with Loeffler medium containing an appreciable amount of water of condensation and when the tube has a cotton wool plug or the cap is left loose. Examine Gram- and Albert-stained films for typical beaded bacilli which may show metachromic granules at their extremities. Inoculate Robinson's serum water sugars (glucose, sucrose and starch) but do not wait for the result before doing the plate toxigenicity test.

Note: Robinson's sugar medium is buffered and contains horse serum. Unbuffered horse serum medium may give false-positive results because of fermentation of the small amount of glycogen it contains, which is metabolized by many corynebacteria. Similarly, media that contain unheated rabbit serum may give false-positive starch fermentation reactions. Natural amylases hydrolyse starch, forming glucose, which is fermented by the organism. Misidentification of *C. diphtheriae* by inexperienced workers using various other fermentation tests is not uncommon. The biochemical reactions are given in Table 26.1. The API CORYNE kit is useful.

Table 26.1 Corynebacteria from human sources

	Acid from				
	Glucose	*Sucrose*	*Starch*	*β-Haemolysis*	*Urea*
C. diphtheriae gravis	+	−	+	−	−
C. diphtheriae mitis	+	−	−	+	−
C. diphtheriae intermedius	+	−	−	−	−
C. ulcerans	+	−	+	−	+
C. pseudodiphtheriticum	−	−		−	−
C. xerosis	+	+	+	−	−
JK diphtheroids	+	−	v	−	−

Plate toxigenicity test
Elek's method, modified (Davies, 1974): inoculate a moist slope of Loeffler's medium with a colony from the primary plate as early as possible in the morning. Also inoculate Loeffler slopes with the stock control cultures maintained on Dorset egg medium. NCTC No. 10648 is toxigenic and NCTC 10356 is non-toxigenic. Incubate at 37°C for 4–6 h.

Melt two 15-ml tubes of Elek's medium and cool to 50°C. Add 3 ml of sterile horse serum to each and pour into plastic or very clear glass petri dishes. Place immediately in the still liquid medium in each plate a strip of filter-paper (60 × 15 mm) which has been soaked in diphtheria antitoxin. With the antitoxin available until 1982 we found that 750 units/ml gave satisfactory results. This material is no longer available and new products vary. Laboratories should titrate each new batch against stock cultures, using concentrations between 500 and 1000 units/ml. This should be done in anticipation, not when a diphtheria investigation is imminent. Place the strip along a diameter of the plate. Dry the plates. These may be stored in a refrigerator for several days.

On each plate, streak the 4–6 h Loeffler cultures at right-angles to the paper strip so that the unknown strain is between the toxigenic and non-toxigenic controls as in Figure 26.1, approximately 10-mm apart. Incubate at 37°C for 18–24 h. If no lines are visible re-incubate a further 12 h. Examine by transmitted light against a black background. If the unknown strain is toxigenic its precipitation lines should join those of the toxigenic control, i.e. a reaction of identity. Strains that produce little toxin may not show lines but turn the lines of adjacent toxigenic strains.

An alternative method is that of Jameson (1965). Prepare the Loeffler cultures as above. Pour the plates but do not add the antitoxin strip. Dry the plates and place along the diameter of each a filter-paper strip (75 × 5 mm) which has been soaked

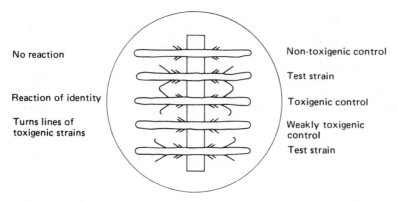

No reaction

Reaction of identity

Turns lines of
toxigenic strains

Non-toxigenic control

Test strain

Toxigenic control

Weakly toxigenic
control

Test strain

Figure 26.1 Elek plate toxigenicity test

in diphtheria antitoxin (for concentration see above). Place the plates over a
cardboard template marked out as in Figure 26.2 and heavily inoculate them from
the Loeffler cultures in the shape of the arrowheads. Place the unknown strain
between the two controls. Incubate at 37°C for 18–24 h and examine for reactions
of identify.

To be of value to the clinician, the plate toxigenicity test must give clear results in
18–24 h. If the toxigenic control does not show lines at 24 h, the medium or serum is
unsatisfactory. The test requires a certain amount of experience and the medium
and serum used must be tested frequently. For reasons not yet appreciated, certain
batches of medium and serum give poor results. To test the medium and serum, use
the control strains noted above and also NCTC 3894, which is weakly toxigenic.
The strength of the antitoxin used is critical.

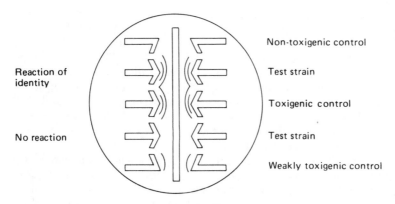

Reaction of
identity

No reaction

Non-toxigenic control

Test strain

Toxigenic control

Test strain

Weakly toxigenic control

Figure 26.2 Jameson's method for plate toxigenicity test

Report the organism as toxigenic or non-toxigenic *C. diphtheriae* cultural type
gravis, mitis or *intermedius*.

Note that in any outbreak of diphtheria due to a toxigenic strain, a proportion of
contacts may be carrying non-toxigenic *C. diphtheriae*.

Species of corynebacteria from human sources

The following general descriptions of colony appearance and microscopic morphology of diphtheria bacilli and diphtheroids is a guide only. The appearance of the colony varies considerably with the medium and it is advisable to check this periodically with stock (NCTC and ATCC) organisms and, if possible, strains from recent cases of the disease. Even then, strains may be encountered that present new or unusual colony appearances.

C. diphtheriae

Three biotypes, *gravis, mitis* and *intermedius*, were originally given these names because of their respective association with the severe, mild and intermediate clinical manifestations of the disease. These names have become attached to the colonial forms and related to starch fermentation. These properties do not always agree and it is more important to carry out the toxigenicity test than to attempt to interpret the colonial and biochemical properties, which should be used only to establish that the organism belongs to the species. The following colony appearances relate to tellurite medium of the Hoyle type.

C. diphtheriae, gravis type

At 18–24 h, the colony is 1–2 mm in diameter, pearl grey with a darker centre; the edge is slightly crenated. It fractures easily when touched with a wire but is not buttery. At 48 h, it has enlarged to 3–4 mm and is much darker grey with the edge markedly crenated and with radial striations. This is the 'daisy head' type of the colony. The organism does not emulsify in saline. Films stained with methylene blue show short, often barred or bearded, bacilli arranged in irregular 'palisade' form or 'Chinese letters'. Well decolorized Gram films compared on the same slide with staphylococci or diphtheroids appear relatively Gram negative. Glucose and starch are fermented, but not sucrose. There is no haemolysis on blood agar.

C. diphtheriae, mitis type

At 18–24 h, colonies are 1–2 mm across and are dark grey, smooth and shining. The colonies fracture when touched but are much more buttery than the gravis type. At 48 h, they are larger (2–3 mm diameter), less shiny and darker in the centre, which may be raised, giving the 'poached egg' appearance. The organisms emulsify easily in saline. Films show long, thin bacilli, usually beaded and with granules, which may not be very Gram positive (see above). Glucose is fermented but not starch or sucrose. Colonies on blood agar are haemolytic.

C. diphtheriae, intermedius type

Colonies at 18–24 h are small (1 mm diameter), flat with sharp edges, black and not shining or glossy. At 48 h, there is little change. They are easily emulsified and films show beaded bacilli, intermediate in size between gravis and mitis types. Glucose is fermented but not starch or sucrose. Colonies on blood agar are non-haemolytic.

C. ulcerans

This organism resembles very closely the *gravis* type of diphtheria bacillus and is often mistaken for it. The colony tends to be more granular in the centre. Glucose and starch are fermented but not sucrose. *C. ulcerans* splits urea; *C. diphtheriae* does not. *C. ulcerans* also grow quite well at 25–30°C. It gives a reaction of incomplete identity with the plate toxigenicity test and *C. diphtheriae* antitoxin. This organism does not cause diphtheria and is not readily communicable, but can cause ulcerated tonsils.

C. pseudodiphtheriticum (C. hofmanii)

Hoffman's bacillus at 18–24 h gives colonies 2–3 mm in diameter that are domed, shiny and dark grey or black. They are buttery (occasionally sticky), easily emulsified and films show intensely Gram-positive bacilli that are smaller and much more regular in shape, size and arrangement than are diphtheria bacilli. Carbohydrates are not fermented. These organisms are common and harmless commensals.

C. xerosis

At 18–24 h, colonies are 1–2 mm in diameter, flat, grey and rough, with serrated edges. At 48 h, they are 2–3 mm in diameter. Films show beaded bacilli, very Gram positive. This is a dubious species, a commensal of the conjunctivae and often found in the female genitalia. It is not a pathogen but resembles very closely the diphtheria bacillus in microscopical (but not colonial) morphology. It is, however, much more strongly Gram positive than *C. diphtheriae* and produces acid from glucose, usually sucrose but not starch.

C. haemolyticum

Also known as *Arcanobacterium haemolyticum*, this may be mistaken for *C. diphtheriae* in films made from Loeffler's medium (but not on tellurite media). Colonies on blood agar are haemolytic, resembling those of streptococci. Glucose and maltose are fermented but sucrose fermentation is variable.

'JK' diphtheroids (C. jeikeium)

These are non-haemolytic, catalase positive, nitrate, gelatin and urease negative and ferment glucose but not sucrose. Maltose fermentation is variable. They have been isolated from blood cultures and a variety of 'sterile' body fluids. They are resistant to many antibiotics but are susceptible to vancomycin (see Riley *et al.* 1975 and Jackman *et al.*, 1987).

Corynebacteria from animal sources

Isolation and identification

Some of these have been removed to other genera but are retained here for convenience. Culture, pus, etc., on blood agar and tellurite medium, incubate at 37°C and examine at 24 and 48 h for colonies of beaded Gram-positive bacilli. Some corynebacteria are haemolytic on blood agar. Small grey-black colonies are usually found on tellurite medium.

Subculture on Loeffler medium and examine Albert-stained films at 6–18 h for clubbed or beaded bacilli which may have metachromatic granules. From the Loeffler inoculate gelatin medium, nutrient agar, Robinson's serum water glucose, sucrose, mannitol, maltose, lactose and urea medium. Incubate at 37°C for 24–48 h (see Table 26.2). The API CORYNE kit is useful.

Table 26.2 'Corynebacteria' from animal sources

	Liquefaction of gelatin and Loeffler	Acid from					β-Haemolysis	Urea
		Glucose	Sucrose	Mannitol	Maltose	Lactose		
C. pyogenes	+	+	+/−	−	+	+	+	−
C. kutscheri	−	+	+	+	+	−	−	+/−
C. pseudotuberculosis	v	+	v	−	+	v	+	+
C. equi	−	−	−	−	−	−	−	+
C. renale	−	+	−	−	v	v	+/−	+

Species of 'corynebacteria' infecting animals

C. pyogenes

Now considered to belong to *Actinomyces*. The colonies on blood agar are pin-point and haemolytic and may be mistaken for haemolytic streptococci except for their slow rate of growth. They are often not apparent for 36–48 h. Films show very small bacilli that may not easily be recognized as corynebacteria. The catalase test is negative (other corynebacteria mentioned here are positive). It does not grow on nutrient agar but Loeffler's serum cultures are liquefied. This organism causes suppurative lesions including mastitis in domestic and wild animals.

C. kutscheri (C. murium)

Colonies on blood agar are larger than *C. pyogenes*, flat, grey and non-haemolytic. The bacilli are slender, often filamentous. Loeffler medium is not liquefied. It is responsible for fatal septicaemia in mice and rats.

C. pseudotuberculosis (C. ovis)

Colonies are small, usually grey but may be yellowish and are haemolytic, especially in anerobic cultures. Loeffler serum may be liquefied. The bacilli are slender and clubbed, very like *C. diphtheriae*. It is responsible for epizootics in rodents and is a nuisance in laboratory animal houses.

This is the Preisz–Nocard bacillus which causes pseudotuberculosis in sheep, cattle and swine and lymphangitis in horses.

C. equi

This is now regarded as *Rhodococcus equi* (see also p.369). It grows well on nutrient agar, forming large, moist, pink colonies. The bacilli are very pleomorphic, varying from coccoid to large clubbed forms with granules. It is non-haemolytic, does not liquefy Loeffler medium and does not form acid from carbohydrates. It is responsible for serious pyaemia in young horses.

C. renale

Causes pyelitis and cystitis in cattle.

Microbacterium

The 'coryneforms' in this genus are found mostly in dairy products and in association with animals. They are relatively heat resistant (70°C for 15 min).

Microbacterium lacticum
This is thermoresistant and a strict aerobe. It is non-motile, grows slowly at 30°C on milk agar, forms acid from glucose and lactose, is catalase positive, hydrolyses starch, is lipolytic and does not liquefy gelatin in 7 days. A pale yellow pigment may be formed. These organisms may cause spoilage in milk and indicate unsatisfactory dairy hygiene. They are also found in deep litter in hen houses.

Brochothrix

Brochothrix thermosphacta (formerly Microbacterium thermosphactum)
Is found in souring sausages of the British fresh type and in meat. It grows aerobically and anaerobically on Gardner's STAA medium and is highly pleomorphic; coccoid or filamentous forms may be seen. It is Gram-variable and grows best at 22°C. It will grow at 1°C but not at 37°C and does not survive heating at 63°C for 3 min. It is non-motile, catalase positive and produces acid from glucose, maltose and cellobiose and it grows in the presence of 6.5% sodium chloride (see Gardner (1966) and Roberts *et al.* (1981)).

Propionibacterium

These are pleomorphic coryneforms varying from coccoid to branched forms. They are aerobic but may be aerotolerant, non-motile and grow poorly in the absence of carbohydrates. They are catalase positive and do not usually liquefy gelatin (but see *Propionibacterium acnes*, below). The optimum temperature is 30°C. There are several species of interest to food microbiologists. They occur naturally in the bovine stomach and hence in rennet and are responsible for flavours and 'eyes' in Swiss cheese. A method for presumptive isolation is given on p.229. It is difficult to identify species.

There is one species of medical interest: *P. acnes.*

Propionibacterium acnes
The bacilli are small, almost coccoid, or about 2.0×0.5 μm, and may show unstained bands. The best growth is obtained anaerobically. The colonies on blood agar are either small, flat, grey-white and buttery, or are larger, heaped up and more granular. Both colony forms are β-haemolytic. Acid is produced from glucose, mannose, trehalose and fructose but not from lactose, maltose and sucrose. It is indole positive, nitratase positive, catalase positive and liquefies gelatin slowly. Slide agglutination is useful.

It is a commensal on human skin, in hair follicles and in sweat glands. It appears to be involved in the pathogenesis of acne. Oral and systemic infections have been suspected.

Brevibacterium

This includes the organisms previously known as *Kurthia* and *Zopfius*. They grow on ordinary media, aerobically and anaerobically, are motile at 20–22°C but not at 35°C and fail to attack carbohydrates. They are non-haemolytic, are indole and VP negative, attack urea slowly and some strains liquefy gelatin. Some strains are thermoduric.

They are found in milk and dairy products, decomposing farmyard material and faeces and have been isolated from human material.

Erysipelothrix

This genus contains small (2 × 0.3 μm) Gram-positive rods that are found in and on healthy animals and on the scales of fish. Humans may be infected through abrasions when handling such animals, particularly pigs and fish. The resulting skin lesions are called erysipeloid. Epizootic septicaemia of mice has also been reported. There is one species, *Erysipelothrix insidiosa*, also known as *E. rhusiopathiae*. Mouse strains have been called *E. muriseptica*.

Isolation and identification

Culture tissue or fluid from the edge of a lesion. Emulsified biopsy material is better than swabs of lesions. Plate on blood agar and on crystal violet or azide blood agar and incubate at 37°C in 5–10% carbon dioxide for 24–48 h.

Colonies are either small (0.5–1.0 mm) and 'dew-drop' or larger and granular. Young cultures are Gram-variable; older cultures show filaments. Test for motility with a Craigie tube. Test for fermentation of glucose, maltose and mannitol using enriched medium, e.g. broth containing Fildes' extract. Inoculate aesculin broth. Inoculate three tubes of enriched agar and incubate one each at 4, 22 and 37°C. Do a catalase test and test for neomycin sensitivity by the disc method.

Erysipelothrix insidiosa
Is non-motile, produces acid from glucose but not from maltose and mannitol, does not hydrolyse aesculin and is catalase negative and is resistant to neomycin. It grows at 22°C but not at 4°C (see Table 26.3).

Listeria

These small (2 × 0.5 μm) Gram-positive rods are found in grass, silage, soil, sewage and water. The cause various infections in mammals, including septicaemia (with monocytosis), encephalitis and abortion in farm animals and necrotic hepatitis in poultry. Human infections include meningitis (especially in neonates) and septicaemia. Several milk- and cheese-borne outbreaks have been reported. The organisms survive in cold storage and can multiply at low temperatures. If large numbers are present some may survive pasteurization. There are at least eight species, but the most important is *Listeria monocytogenes*.

Isolation from clinical material

Plate CSF deposits and blood cultures on blood agar and on (Oxford) Listeria Selective Agar (Curtis *et al.* 1989) which is less inhibitory than McBride's medium. Macerate tissues and plate on blood agar containing 0.004% nalidixic acid to discourage other organisms. Incubate at 37° for 24–48 h.

Inoculate two tubes each of tryptone soy broth and Listeria Enrichment Broth. Incubate one of each at 30°C and the others at 4°C. Subculture the 30°C broths after 24 and 48 h and the 4°C broths at 7, 14 and 28 days to Listeria Selective Agar and McBride's medium.

Isolation from food

Cold enrichment (4°C) is too slow for food bacteriology. Emulsify duplicate 25 g samples in 225 ml of Listeria Enrichment Broth or similar medium and incubate at 30°C. After incubation for 24 and 48 h subculture to Listeria Selective Agar and McBride's medium (i.e. without blood) and to Columbia agar containing 50 mg/l ceftazidime and 10 mg/l acriflavine (Bannerman and Bille, 1988).

Identification

Colonies on blood agar are small (1 mm), smooth and surrounded by a narrow zone of β-haemolysis. On Oxford Listeria Selective Agar colonies are usually about 1 mm in diam. black, surrounded by a black halo and have a sunken centre (Curtis *et al.* 1989). On modified McBride's agar they are smaller, 'dew-drops', and are blue and finely textured when viewed by reflected light using the Henry (1933) illumination technique. On the ceftazidime–acriflavine agar they are yellow and smooth.

L. monocytogenes shows a tumbling motility in broth cultures grown at 20–25°C. It is catalase positive, hydrolyses aesculin and produces acid but no gas from glucose, maltose and salicin (see Table 26.3).

Table 26.3 *Erysipelothrix* and *Listeria*

| | Growth at 4°C | Motility | Acid from | | | Aesculin hydrolysis | Catalase | Haemolysis |
			Glucose	Maltose	Mannitol			
Erysipelothrix insidiosa[a]	−	−	+	−	−	−	−	α
Listeria monocytogenes[b]	+	+[c]	+	+	−	+	+	β

[a]Resistant to neomycin; [b]Sensitive to neomycin; [c]At 22°C

For identification of other species of listerias and their epidemiology see McLauchlin (1987) and for further information see Gitter (1985) and Watkins (1985).

Lactobacillus

Lactobacilli are important in the food, dairy and brewing industries and are commensals in the human and animal bodies.

Isolation

Plate in duplicate on MRS, Rogosa and similar media at pH 5.0–5.8. LS Differential medium is particularly useful in the examination of yoghurt. Use Raka–Ray agar for beer (see Oxoid Manual). Many strains grow poorly, if at all, at pH 7.0, but plate human pathological material on blood agar (see below). Incubate one set of plates aerobically and the other in a 5:95 carbon dioxide–hydrogen atmosphere. Incubate cultures from food at 28–30°C and from human or animal material at 35–37°C for 48–72 h. Examine for very small, white colonies. Few other organisms, apart from moulds, will grow on the acid media.

For provisional identification of *L. acidophilus*, which is the commonest species in human material, place on the blood agar plates with:

(1) a sulphonamide sensitivity disc which has been dipped in saturated sucrose solution, and
(2) a penicillin (5 unit) disc.

L. acidophilus gives a green haemolysis. Growth is stimulated by sucrose; it is resistant to sulphonamides and sensitive to penicillin.

Identification

Subculture if necessary to obtain pure growth and heavily inoculate the modified MRS broth containing the following sugars: glucose (plus Durham's tube), lactose, sucrose, salicin, mannitol, sorbose and xylose in MRS broth containing 2% glucose and in which the ammonium citrate has been replaced by 0.3% arginine (for arginine hydrolysis test), and in MRS broth containing 4% sodium chloride. Incubate 28–30°C for 3–4 days. Inoculate tubes of MRS broth and incubate at 15°C and at 45°C (see Table 26.4). The API system is useful.

Culture on solid medium low in carbohydrate to test for catalase (high-carbohydrate medium may give false positives). The catalase test is negative (corynebacteria and listeria, which may be confused with lactobacilli, are catalase positive).

Table 26.4 Some species of lactobacilli

	Growth at		Acid from						NH_3 from arginine	Growth in 4% NaCl broth
	15°C	45°C	lac-tose	su-crose	sali-cin	mann-itol	sor-bose	xy-lose		
L. acidophilus	−	+	+	+	+	−	−	−	−	−
L. bulgaricus	−	+	+	−	−	−	−	−	−	−
L. casei	+	v	+/−	+/−	+	+	+	−	−	+
L. plantarum	+	v	+	+	+	+	−	−	−	+
L. delbruckii	−	+	−	+	−	−	−	−	−	−
L. leichmannii	−	+	+/−	+	+	−	−	−	+/−	−
L. brevis	+	−	+/−	+	+/−	+/−	−	+	+	+
L. fermenti	−	+	+	+/−	−	−	−	+/−	+	−

Identification of lactobacilli is not easy; reactions vary according to technique and the taxonomic position of some species is confused.

To grow the heterofermentative lactobacilli which cause greening of cured meats, add 0.1 g of thiamine hydrochloride to each litre of medium, or use APT medium. For further information see Carr *et al.* (1975).

Species of lactobacilli

Lactobacillus acidophilus and L. casei
These are widely distributed and found in milk and dairy products and as

commensals in the alimentary tract of mammals. They are used to make 'acido-philus milk' and are associated with dental caries in humans as a secondary invader. *L. acidophilus* is probably the organism described as the Boas–Oppler bacillus, observed in cases of carcinoma of the stomach in humans, and also Doderlein's bacillus, found in the human vagina.

Colonies of this organism on agar media are described as either 'feathery' or 'crab-like'.

L. bulgaricus
Is also associated with milk and dairy products but is thermophilic and not found in the mammalian intestine. It is used in the manufacture of yoghurt, along with *S. thermophilus*, and cream and to produce gas holes in hard cheese. The natural fermentation of silage is in part due to this organism. Colonies on agar medium are small and grey.

L. casei and L. plantarum
Are difficult to separate except by paper chromatography of the end-products of carbohydrate fermentation. *L. casei* is used in making whey and occurs naturally in milk and cheese. *L. plantarum* is widely distributed on plants, alive and dead, and is one of the organisms responsible for fermentation in pickles and is used in the manufacture of sauerkraut. Both organisms give very poor growth on agar.

L. delbruckii
Is found in vegetation and in grain, and is used as a 'starter' to initiate acid conditions in yeast fermentation of grains. This organism grows on agar medium as small, flat colonies with crenated edges.

L. leichmannii
Is found in most dairy products. It is used in the commercial production of lactic acid and grows on agar as small white colonies.

L. bifidus
See *Bifidobacterium* (below).

Counting lactobacilli

Enumeration of lactobacilli is a useful guide to hygiene and sterilization procedures in dairy industries. Suitable dilutions of bottle or churn rinsings or of swabbings steeped in 0.1% peptone water are prepared (see Plate counts, Chapter 9) in Rogosa medium or one of the other media recommended for lactobacilli. Plates are incubated at 37°C for 3 days or at 30°C for 5 days in an atmosphere of 5% carbon dioxide.

Dental surgeons and oral hygienists sometimes request lactobacillus counts on saliva. The patient chews paraffin wax to encourage salivation, then 10 ml of saliva are collected and diluted for plate counts. Usually dilutions of 1:10, 1:100 and 1:1000 are sufficient.

Counts of 10^5 lactobacilli/ml of saliva are thought to indicate the likelihood of caries. There is a simple colorimetric test in which 0.2 ml of saliva is added to medium, as a shake tube culture, and a colour change after incubation suggests a predisposition to caries.

Bifidobacterium

Bifidobacterium are small Gram-positive rods resembling coryneforms which predominate in the faeces of breast-fed infants, but they are also commensal in the adult bowels, mouth and vagina. They are anaerobic and may need to be differentiated from anaerobic corynebacteria. Bifidobacteria are catalase negative, nitratase negative and do not produce gas from glucose. The anaerobic corynebacteria are catalase and nitratase positive and do produce gas from glucose.

There is one pathogen, *B. eriksonii*, formerly *Actinomyces eriksonii*, associated with mixed infections in the upper respiratory tract.

References

Bannerman, E. S. and Billie, J. (1988) A new selective medium for isolating *Listeria* spp. from heavily contaminated material. *Applied and Environmental Microbiology*, **54**, 165–167

Carr, J. G., Cutting, C. V. and White, G. C. (1975) *Lactic Acid Bacteria*, Academic Press, London

Collins, C. H. and Dulake, C. (1983) Diphtheria: the logistics of mass swabbing. *Journal of Infection*, **6**, 227–230

Curtis, G. W. D., Mitchell, R. G., King, A. F. and Griffin, E. J. (1989) A selective differential medium for the isolation of *Listeria monocytogenes*. *Letters in Applied Microbiology*, **8**, 95–98

Davies, J. R. (1974) Identification of diphtheria bacilli. In *Laboratory Methods, I*, Public Health Laboratory Service Monograph No. 4, HMSO, London

Gardner, G. A. (1966) A selective medium for the enumeration of *Microbacterium thermosphactum* in meat and meat products. *Journal of Applied Bacteriology*, **29**, 455–457

Gitter, M. (1985) Listeriosis in farm animals in Great Britain. In *Isolation and Identification of Micro-organisms of Medical and Veterinary Importance* (eds C. H. Collins and J. M. Grange), Society for Applied Bacteriology Technical Series No. 21, Academic Press, London, pp. 191–200

Henry, B. S. (1933) Dissociation in the genus *Brucella*. *Journal of Infectious Diseases*, **52**, 374–402

Jackman, P. J. H., Pitcher, D. G., Pelcznska, S. and Borman, P. (1987) Classification of corynebacteria associated with endocarditis (Group JK) as *Corynebacterium jeikeium* sp. nov. *Systematic and Applied Microbiology*, **9**, 83–90

Jameson, J. E. (1965) A modified Elek test for toxigenic *Corynebacterium diphtheriae*. *Monthly Bulletin of the Ministry of Health*, **24**, 55–58

McLauchlin, J. (1987) *Listeria monocytogenes:* recent advances in the taxonomy and epidemiology of listeriosis in humans. *Journal of Applied Bacteriology*, **63**, 1–11

Riley, P. S., Hollis, D. G., Utter, G. B. Weaver, R. E. and Baker, C. N. (1975) Characterization and identification of 95 diphtheroid (group JK) cultures isolated from clinical specimens. *Journal of Clinical Microbiology*, **9**, 418–422

Roberts, T. A., Hobbs, G., Christian, J. H. B. and Skovgaard, N. (eds) (1981) *Psychrotrophic Micro-organisms in Spoilage and Pathogenicity*, Academic Press, London

Watkins, J. (1985) *Listeria monocytogenes. In Isolation and Identification of Micro-organisms of Medical and Veterinary Importance* (eds C. H. Collins and J. M. Grange), Society for Applied Bacteriology Technical Series No.21, Academic Press, London, pp. 20–206

Chapter 27

Bacillus

This is a large genus of Gram-positive spore-bearing bacilli that are aerobic, facultatively anerobic and catalase positive. Many species are normally present in soil and in decaying animal and vegetable matter. One species, *Bacillus anthracis*, is responsible for anthrax in man and animals. *B. cereus* causes food poisoning. This and several other species are now known to cause disease in humans and animals (see Logan, 1988). Some are responsible for food spoilage.

The bacilli are large (up to 10×1 μm) and commonly adhere in chains. Most species are motile; some form motile colonies. Some species are capsulated, some produce a sticky levan on media that contain sucrose. Spores may be round or oval, central or subterminal.

Physiological characters vary. There are strict aerobes and species which are facultative anerobes. A few are thermophiles. All are catalase positive. In old cultures, Gram-negative forms may be seen and some species are best described as 'Gram-indifferent'.

Isolation of *B. anthracis* from pathological material

Veterinarians often diagnose anthrax in animals on clinical grounds and by the examination of blood films stained with polychrome methylene blue. These show chains of large, square-ended bacilli with the remains of their capsules forming pink-coloured debris between the ends of adjacent organisms. This is M'Fadyean's reaction.

Culture animal blood, spleen substance or other tissue ground in a Griffith's tube or macerator. In humans, culture material from the cutaneous lesions, faeces, urine and sputum (pulmonary anthrax is now very rare).

Plate on blood agar and on PLET medium which is brain heart infusion agar containing (per ml) polymyxin, 30 units; lysozyme, 40 μg; EDTA, 300 μg; thallous acetate, 40 μg and incubate overnight. Pick woolly or waxy 'medusa head' colonies of Gram-positive rods into Craigie tubes to test for motility and on chloral hydrate blood agar. It may be difficult to get pure growths without several subcultures. For identification, see below.

Isolation of *B. anthracis* from hairs, hides, feedingstuffs and fertilizers

The sample should be in a 200–300 ml screw-capped jar, and should occupy about 4 fluid ounces. Add sufficient warm 0.1% peptone solution to cover it. Shake, stand

328

at 37°C for 2 h, decant and heat the fluid at 60°C for 30 min. Place 0.1, 1.0 and 2 ml of the uncentrifuged fluid and the deposit after centrifuging in petri dishes. Add to each plate either 0.5 ml of 5% egg albumen or 0.25 ml of 0.01% dibromopropamidine isethionate, pour on 15 ml of yeast extract agar at 50°C, mix well, allow to set and leave on the bench until 5 p.m. Incubate at 37°C until 9 a.m. the next day. Longer incubation produces atypical colonies. The lysozyme in the egg albumen and the dibromopropamidine compound reduce the numbers of other organisms. Polymyxin B (20 units/ml) in the agar is also useful for heavily contaminated material. Also plate 0.25 ml amounts of deposit on several plates of PLET medium (p.328).

Examine under a low-power binocular (plate) microscope. Ignore surface colonies and look for deep colonies that resemble dahlia tubers or Chinese artichokes. Pick into Craigie tubes and plate on chloral hydrate agar.

Identification of B. anthracis

Bacillus anthracis is non-motile. The Craigie tube is the best way of demonstrating this. B. cereus, which is commonly mistaken for B. anthracis, is motile.

The growth of B. anthracis on agar media has a characteristic woolly or waxy nature when touched with a wire (the 'tenacity test'). Inoculate ammonium salt sugar medium containing 1% salicin. Do not use media that contain peptones; sufficient ammonia may be produced by some species to mask acid production. Culture on nutrient agar containing 10 units/ml of penicillin, nutrient agar containing 0.3% 2-phenylethanol and in nutrient broth incubated at 45°C.

Test for the 'string of pearls' appearance: grow in nutrient broth containing 0.5 units of penicillin per ml for 3–6 h at 37°C and examine a wet preparation. Strings of spherical bodies suggest B. anthracis.

B. anthracis fails to ferment salicin, is sensitive to penicillin, is inhibited by 2-phenylethanol and does not grow at 45°C. B. cereus ferments salicin, is resistant to penicillin and 2-phenylethanol and grows at 45°C.

The biochemical properties are given in Table 27.1. The API method is useful but the best way to identify B. anthracis is by immunofluorescence. Sera are available commercially. Unfortunately the specific W phage, which we used

Table 27.1 Properties of some mesophilic aerobic spore bearers

Species	Lecithinase	Acid from		Hydrolysis of starch	Anaerobic growth	Growth at 60°C	VP
		Glucose	Xylose				
B. subtilis	−	A	A	+	−	+	+
B. pumilus	−	A	A	−	−	+	+
B. coagulans	−	+	v	+	+	+	v
B. megaterium	−	A	v	+	−	−	−
B. cereus	+	A	−	+	+	−	+
B. anthracis	+	A	−	+	+	−	+
B. polymyxa	−	AG	AG	+	+	−	+
B. macerans	−	AG	AG	+	+	+	−
B. circulans	−	A	A	+	v	+	−
B. laterosporus	+	A	−	−	+	+	−
B. brevis	−	A	−	−	−	+	−
B. licheniformis	−	AG	AG	+	+	+	+

A, acid; AG, acid and gas; v, variable; all in ammonium salt sugar medium in 1% sodium glucose phosphate broth successfully for many years at the Anthrax Reference Laboratory in London, is no longer generally available.

B. anthracis is regarded by some bacteriologists as a non-motile variant of *B. cereus (B. cereus* var. *anthracis).* The 'medusa head' surface colonies of the two organisms are similar. It causes anthrax in animals, a fatal septicaemia. Anthrax is transmissible to humans, producing a localized cutaneous necrosis ('malignant pustule') or pneumonia (woolsorters' disease) due to inhalation of spores.

Imported bone meal, meat meal and other fertilizers and sometimes feeding-stuffs may be infected. These are often prepared from the remains of animals that died of anthrax. In most developed countries, an animal dying of anthrax must be buried in quicklime under the supervision of a veterinarian and below the depth at which earthworms are active.

There are regulations governing the importation of hairs and hides from foreign countries where anthrax is endemic.

For further information on the isolation of *B. anthracis* see Carman *et al.* (1985) and Turnbull (1990).

Food poisoning

Bacillus cereus
May be associated with food poisoning (Chapter 12). One vehicle is fried rice.

Make 10% suspensions of the food in 0.1% peptone water, using a Stomacher. Inoculate glucose tryptone agar, chloral hydrate blood agar (to suppress some other organisms) and, if available, the Bacillus Cereus Selective agar. Incubate at 30°C for 24–48 h.

B. cereus produces large, flat, irregular, 'ground glass' colonies on glucose tryptone agar, showing acid reaction. On the selective agar the colonies are turquoise blue and are surrounded by a blue halo of precipitated egg.

Test colonies for the egg yolk reaction, acid production from glucose and xylose (use ammonium salt sugars), VP (use 1% NaCl glucose phosphate broth) gelatin liquefaction, citrate utilization, hydrolysis of starch and ability to grow anaerobically (see Table 27.1). The API 20B system is useful.

NB Food poisoning due to other *Bacillus* spp., e.g. *B. subtilis* and *B. licheniformis*, has been reported (Gilbert *et al.*, 1981; Logan, 1988).

Rapid staining identification
Place air-dried and fixed films made from young colonies over boiling water and stain with 5% malachite green for 2 min. Wash and blot dry. Stain with 0.3% sudan black in 70% alcohol for 15 min. Wash with xylol and blot dry. Counterstain with 0.5% safranin for 20 s. Wash and dry. *B. cereus* cells stain red and contain black-stained lipid granules. The spores do not swell the cells and stain green.

Food spoilage

Examine Gram-stained films for spore-bearing Gram-positive bacilli. Make 10% suspensions of food in 0.1% peptone water, and pasteurize some of the suspension at 75–80°C for 10 min to kill vegetative forms.

Inoculate glucose tryptone agar and 5% egg yolk agar media with both unheated and pasteurized material.

Incubate at 25–30°C overnight. If the food was canned, incubate replicate cultures at 60°C. Identify colonies of Gram-positive bacilli.

Cold-tolerant spore-formers are important spoilage organisms. Incubate at 5°C for 7 days.

Test colonies of Gram-positive bacilli for egg yolk reaction, acid from glucose and xylose in ammonium salt sugars, VP (use 1% NaCl glucose phosphate broth), gelatin liquefaction, citrate utilization, starch hydrolysis and ability to grow anaerobically (see Table 27.1). Or use the API 20B system.

Species of *Bacillus*

Bacillus cereus   General properties are described above. Phospholipinase activity produces large haloes around colonies on egg yolk agar. With most other species, activity is usually limited to beneath the colony. Methylene blue milk is rapidly decolorized. This organism is responsible for 'bitty cream', particularly in warm weather, and causes milk and ice-cream to fail the methylene blue test. It may cause food poisoning (see above). Bacteraemia and pneumonia associated with *B. cereus* have been reported. It is a common contaminant on shell eggs. It does not ferment mannitol, unlike *B. megaterium*, which it may otherwise resemble.

B. subtilis   Wrinkled or smooth and folded colonies, 4–5 mm across on glucose tryptone agar with yellow halo (acid production). Strongly alkaline reaction in Crossley medium but no gas or blackening. No zone around colonies on egg yolk agar. Responsible for spoilage in dried milk and in some fruit and vegetable products.

B. mesentericus   Now considered to be identical, in Europe, with *B. subtilis*, but American strains are thought to be *B. pumilus*. Both cause 'ropy bread'. Auto-enzymes from *B. subtilis* are used in certain laundry products. These may be toxic to factory operatives.

B. stearothermophilus   This and associated thermophiles form large colonies (4 mm in diameter) on glucose tryptone agar with a yellow halo due to acid production at 60°C but grow poorly at 37°C and not at 20°C. This organism is responsible for 'flat-sour' (i.e. acid but no gas) spoilage in canned foods. For more detailed investigation see Chapter 15.

B. mycoides   A variant of *B. cereus* with rhizoid colonial form.

B. anthracoides   Another variant of *B. cereus*. Could be confused with *B. anthracis* because of surface colony appearance, but is motile.

B. licheniformis   May cause food poisoning. It does not produce lecithinase.

B. megaterium   A very large bacillus. Widely distributed. May cause *B. cereus*-like food poisoning.

B. pumilus   Found on plants. Common contaminant of culture media.

B. coagulans   Widely distributed. Found in canned foods.

B. anthracis   See p.328.

B. polymyxa   Widely distributed in soil and decaying vegetables (possesses a pectinase). Source of the antibiotic polymyxin.

B. macerans   Common in soil. Found in rotting flax.

B. circulans   Motile colonies with non-motile variants. Found in soil.

B. laterosporus   Soil organism. Source of the antibiotics tyrothricin and gramicidin.

B. popilliae   Insect pathogen.

B. lentimorphus   Insect pathogen.

B. thuringensis   Insect pathogen, used in America in pest control.

B. sphaericus   Large spores, resembles *C. tetani* in appearance. Forms motile colonies. Soil organism.

B. rotans   Probably synonym of *B. sphaericus*. Motile colonies.

B. pasteuri   Found in decomposing urine. Grows only in media containing peptone and urea.

References

Carman, J. A., Hambleton, P. and Melling, J. (1985) *Bacillus anthracis*. In *Isolation and Identification of Micro-organisms of Medical and Veterinary Importance* (eds C. H. Collins and J. M. Grange), Society for Applied Bacteriology Technical Series No. 21, Academic Press, London, pp. 207–214

Gilbert, R. J., Turnbull, P. C. B., Parry, J. M. and Kramer, J. M. (1981) *Bacillus cereus* and other *Bacillus* species: their part in food poisoning and other clinical infections. In *The Aerobic Endospore-forming Bacteria* (eds R. C. W. Berkeley and M. Goodfellow), Society for General Microbiology Special Publications Series No. 4, Academic Press, London, pp. 297–314

Logan, N. A. (1988) *Bacillus* species of medical and veterinary importance. *Journal of Medical Microbiology*, **25**, 157–165

Turnbull, P. C. B. (ed.) (1990) Proceedings of an International Workshop on Anthrax, Winchester, 1989. Salisbury Medical Bulletin No 68 Special Supplement. PHLS Porton Down, Salisbury

Chapter 28

Anaerobes: clostridium, bacteroides and obligate anaerobic cocci

Clostridium

This is a large genus of Gram-positive spore-bearing anaerobes (a few are aerotolerant) that are catalase negative.

They are normally present in soil; some are responsible for human and animal disease; others are associated with food spoilage

Clostridia are classified according to the shape and position of the spores (see Table 28.1) and by their physiological characteristics. They may be either predominantly saccharolytic or proteolytic in their energy-yielding activities.

Saccharolytic species decompose sugars to form butyric and acetic acids and alcohols. The meat in Robertson's medium is reddened and gas is produced.

Proteolytic species attack amino acids. Meat in Robertson's medium is blackened and decomposed, giving the culture a foul odour.

Most species are mesophiles; there are a few important thermophiles and some psychrophiles and psychrotrophs.

The clostridia are conveniently considered under four headings: food poisoning; tetanus and gas gangrene; pseudomembranous colitis; and food spoilage.

Table 28.1 Morphology and colonial appearances of some species of *Clostridium*

	Spores	Bacilli	Haemolysis	Colony appearance on blood agar
C. botulinum	OC or S	normal	+	large, fimbriate, transparent
C. perfringens	a	large thick	+	flat, circular, regular
C. tetani	RT	normal	+	small, grey, fimbriate, translucent
C. novyi	OS	large	+	flat, spreading, transparent
C. septicum	OS	normal	+	irregular, transparent
C. fallax	OS	thick	−	large, irregular, opaque
C. sordellii	OC or S	large thick	+	small, crenated
C. bifermentans	OC or S	large thick	+	small, circular, transparent
C. histolyticum	OS	normal	−	small, regular, transparent
C. sporogenes	OS	thin	+	medusa head, fimbriate, opaque
C. tertium	OT	long thin	−	small, regular, transparent
C. cochlearium		thin	−	circular, transparent
C. butyricum	OC	normal	−	white, circular, irregular
C. nigrificans		normal		black
C. thermosaccharolyticum		normal		granular, feather edges
C. difficile	OS	normal	−	grey, translucent, irregular

Spores: O, oval; R, round; S, subterminal; C, central; T, terminal
a not usually seen

Food poisoning

Botulism

This is the least common but most often fatal kind of food poisoning (Chapter 12) and is caused by *Clostridium botulinum*. This is a strict anaerobe and requires a neutral pH and absence of competition to grow in food (e.g. canned, underprocessed). It is a difficult organism to isolate in pure culture.

Emulsify the suspected material in 0.1% peptone water and inoculate several tubes of Robertson's cooked meat medium which has been heated to drive off air and then cooled to room temperature. Heat some of these tubes at 75–80°C for 30 min and cool. Incubate both heated and unheated tubes at 35°C for 3–5 days. Plate them on pre-reduced blood agar and egg yolk agar and incubate under strict anaerobic conditions at 35°C for 3–5 days.

If the material is heavily contaminated, heat some of the emulsion as described above, make several serial dilutions of heated and unheated material and add 1-ml amounts to 15–20 ml of glucose nutrient agar melted and at 50°C in 152 × 16 mm tubes (Burri tubes). Mix, cool and incubate. Cut the tube near the sites of suspected colonies, aspirate these with a pasteur pipette and plate on blood agar or on egg yolk agar containing (per ml): cycloserine, 250 μg; sulphamethoxazole, 76 μg: trimethoprim, 4μg (Dezfullian *et al.*, 1981).

Clostridium botulinum gives a positive egg yolk and Nagler (half-antitoxin plate) reaction (see below). Types A, B and F liquefy gelatin, blacken and digest cooked meat medium and produce hydrogen sulphide. Types C, D and E do none of these. All strains produce acid from glucose, fructose and maltose and are indole negative. See p.340 for methods and Tables 28.1 and 28.2.

It is best to use fluorescent antibody methods and to send suspected cultures to a Reference laboratory

Type A *C. botulinum* is usually associated with meat. Type E is found in fish and fish products and estuarine mud and is psychrotrophic.

Perfringens (welchii) food poisoning

Clostridium perfringens is a common cause of food poisoning (Chapter 12).

Faeces

Make a 1:10 suspension of faeces in nutrient broth and add 1 ml to each of two tubes of Robertson's cooked meat medium. Heat one tube at 80°C for 10 min and then cool. Incubate both tubes at 37°C overnight and plate on blood agar with and without neomycin and on egg yolk agar. Place a metronidazole disc on the streaked-out inoculum. Incubate at 37°C anaerobically overnight.

Emsulsify faeces in ethanol (industrial grade) to give a 50% suspension. Mix well and stand for 1 h. Inoculate media and incubate as above. See also Chapter 12.

The aerobic plates act as a control to enable anaerobic colonies to be distinguished.

Food

Examine Gram-stained films of the material (usually meat or meat dishes) for short, plump, swollen bacilli, which may be present in large numbers. Spores are rarely seen. Make 10% emulsions in 0.1% peptone water using a Stomacher. Plate on Willis and Hobbs medium, tryptose sulphite cycloserine agar (TSC) and

Table 28.2 Cultural and biochemical properties of some species of clostridia

	RCM				Purple milk	Acid from				Indole	Gelatin liquefaction	Lecithinase
	Colour	Digestion	Odour	Gas		Glucose	Sucrose	Lactose	Salicin			
C. botulinum	black	+	−	+	D	+	>	−	−	−	+	−
C. perfringens	black	+	+	+	CD	+	+	+	>	−	+	+
C. tetani	black	+	−	−	C	−	−	−	−	+	+	−
C. novyi	red	−	−	+	GC	+	−	−	−	>	+	>
C. septicum	red	−	−	+	AC	+	−	+	+	−	+	−
C. fallax	red	−	−	+	AC	+	+	+	+	−	+	−
C. sordellii[a]	black	+	+	+	CD	+	−	−	−	+	+	+
C. bifermentans[a]	black	+	+	−	CD	+	−	−	>	+	+	+
C. histolyticum	black	+	+	−	D	−	−	−	−	−	+	−
C. sporogenes	black	+	+	+	D	+	−	−	>	−	+	−
C. tertium	black	+	−	+	AC	+	+	+	+	−	−	−
C. cochlearium	red	−	−	+		−	−	+	+	−	−	−
C. butyricum		−	−	+	ACG	+	+	+	−	−	−	−
C. nigrificans		−	−	−		−	−	−	+	−	+	−
C. thermosaccharolyticum		−	−	+		+	+	+	+	−	−	−

RCM: Robertson's cooked meat medium
Purple milk: A, acid; C, clot; D, digestion; G, gas
[a] C. sordellii also splits urea; C. bifermentans does not.

neomycin blood agar. Do pour plates in oleandomycin–polymyxin–sulphadiazine-
–perfringens agar (OPSP). Incubate duplicate cultures aerobically and anaerobically at 37°C overnight.

Add 2–3 ml of the emulsion to several tubes of Robertson's cooked meat
medium. Heat some tubes at 80°C for 1 h. Incubate overnight and plate out both
heated cultures on TSC, neomycin blood agar and Willis and Hobbs medium.

There is a kit (Oxoid) for detecting enterotoxin.

Identification

Heat-resistant strains of *C. perfringens* yield very slightly haemolytic colonies
(other strains usually show much more haemolysis) which are moist, raised and
smooth, about 2 mm in diameter. Heat-sensitive strains do not grow from the
pasteurized cultures. Spores are not usually seen in laboratory cultures of this
organism. Heat-sensitive strains may also cause food poisoning.

It is usually sufficient to identify *C. perfringens* by the Nagler half-antitoxin test
(see below) but other properties are given in Tables 28.1 and 28.2. No great
reliance should be placed on the 'stormy fermentation' of purple milk. General
cultural methods are given on p.340.

Nagler half-antitoxin plate test
Dip a cotton wool swab in standard antitoxin and spread on one-half of a plate of
Willis and Hobbs medium or on nutrient agar containing 10% egg yolk. Dry the
plates and inoculate the organism across both halves of the plate. Incubate
anaerobically overnight at 37°C.

The lecithinases of *C. perfringens* and *C. novyi* produce white haloes round
colonies on the untreated half of the plate. This activity is inhibited by the half
treated with antitoxin. The lecithinases of *C. bifermentans* and *C. sordellii* are
partially inhibited by *C. perfringens* antitoxin, but they may be distinguished on
Willis and Hobbs medium. A diffuse pink halo appears round colonies of *C.
perfringens* due to fermentation of the lactose. *C. bifermentans* and *C. sordellii* give
white haloes because they do not ferment lactose.

On Willis and Hobbs medium, some clostridia show a 'pearly layer' due to
lipolysis. This is a useful differential criterion.

Tetanus and gas gangrene

Tetanus is caused by *C. tetani*. Clostridia associated with gas gangrene include *C.
novyi (oedematiens), C. perfringens, C. septicum* and *C. sordellii. C. histolyticum* is
a possible pathogen and some other species, such as *C. sporogenes, C. tertium* and
C. bifermentans may also be found in wounds but are not known to be pathogenic.

Examine Gram-stained and FA films of wound exudates. Bacilli with swollen
terminal spores ('drumstick') suggest *C. tetani* but are not diagnostic. Thick,
rectangular, box-like bacilli suggest *C. perfringens*. Very large bacilli might be *C.
novyi* and thin, almost filamentous bacilli may be seen.

Inoculate two plates of blood agar, one for anaerobic and the other for aerobic
culture, and also Willis and Hobbs or Lowbury and Lilly medium. One of the
commercial media may be used as well as or instead of the latter. Inoculate two
tubes of Robertson's cooked meat medium. Heat one at 80°C for 30 min and cool.

Incubate blood agar and all the other media anaerobically at 37°C for 24 h. Plate out the Robertson's medium on blood agar and other clostridial media and incubate anaerobically with 10% carbon dioxide.

Identification

Test suspected colonies (see Table 28.1) with half-antitoxin plates (see above), inoculate Crossley milk medium, glucose, lactose, sucrose, maltose and salicin peptone waters and note nature of growth and appearance in the cooked meat medium (see Tables 28.1 and 28.2).

Pseudomembranous colitis

This is caused by *C. difficile*.

Treat faeces by the ethanol shock method (p.334). Plate on blood agar containing cefoxitin, 8 μg/ml, and cycloserine 250 μg/ml or on cycloserine cefoxitin fructose agar (Clostridium difficile Selective Medium) and inoculate cooked meat medium. Or plate on Columbia agar containing these antibiotics and 5% egg yolk so that lecithinase-producing colonies may be ignored *(C. difficile* is lecithinase negative). Incubate anaerobically for 48 h.

Colonies of *C. difficile* are 2–5 mm in diameter, irregular and opaque. Confirm by latex agglutination (Mercia; Diagnostics Ltd). There may be cross reaction with *C. bifermentans*, *C. sordellii*, and also *C. glycolicum*.

Examine blood agar cultures under long wave UV light. *C. difficile* colonies (and those of some other organisms) fluoresce yellow-green.

Final identification of *C. difficile* is not easy. GLC of volatile fatty acids and *p*-cresol may be necessary. The API ZYM may be useful.

Species of *Clostridium* of medical importance

C. botulinum
Strict anaerobe; requires also a neutral pH and absence of competition to grow in food (e.g. canned, underprocessed). Free spores may be seen. Cultural characteristics are variable within and between strains. There are six antigenic types (A–F). Types A, B and F are proteolytic, causing clearing on Willis and Hobbs medium; types C, D and E are not. Causative organism of botulism, one of the least common but most fatal kind of food poisoning. Type A is usually associated with meat and Type E (a psychrotroph) with fish products. Found in soil and marine mud.

C. perfringens (C. welchii)
This highly aerotolerant anaerobe may grow in broth, e.g. in MacConkey broth inoculated with water. Spores are rarely seen in culture (a diagnostic feature) but can be obtained on Ellner's medium and medium with added bile, bicarbonate and quinoline (Phillips, 1986). There are six antigenic types (A–F). Type A is associated with gas gangrene and with food poisoning and antibiotic-associated diarrhoea. Causes lamb dysentery, sheep 'struck' and pulpy kidney in lambs. Found in soil, water and animal intestines.

C. tetani

Strict anaerobe, easily dies on exposure to air. Swarms over the medium, but this may be difficult to see. It is, however, an important diagnostic characteristic. Found in soil, especially animal manured, and in the animal intestine. Causative organism of tetanus.

C. novyi (C. oedematiens)

A strict anaerobe which dies rapidly in air. There are four antigenic types (A–D). Types A and B cause gas gangrene, type D bacillary haemoglobinuria in cattle and 'black disease' of sheep.

C. septicum and chauvoei

Strict anaerobes   Once considered to be a single species, these are now known to be antigenically distinct. *C. septicum* causes gas gangrene in humans, braxy and blackleg in sheep. *C. chauvoei* is not pathogenic for humans but causes quarter evil, blackleg and 'symptomatic anthrax' of cattle and sheep. Both can be identified by fluorescent antibody methods.

C. bifermentans and C. sordellii

Separate species, distinguished biochemically but conveniently considered together because both have a toxin (lecithinase) similar to that of *C. perfringens* and give the same reaction as that organism on half-antitoxin plates with *C. perfringens* type A antiserum. *C. sordellii* is urease positive, *C. bifermentans* urease negative, *C. bifermentans* is not pathogenic but *C. sordellii* is associated with wound infections and gas gangrene. Both are found in soil.

C. difficile

May cause a wide spectrum of disease ranging from mild antibiotic-associated diarrhoea to full blown pseudomembraneous colitis which may be fatal.

Species of clostridia of doubtful or no significance in human pathological material

C. histolyticum   Not a strict anaerobe. Filamentous forms may grow as surface colonies on blood agar aerobically. Associated with gas gangrene but usually in mixed infection with other clostridia. Found in soil and intestines of man and animals.

C. sporogenes   Not pathogenic but is a common contaminant. Widely distributed; found in soil and animal intestines.

C. tertium   Aerotolerant and non-pathogenic.

C. tetanomorphum   Of note because its sporing forms resemble the drumsticks of *C. tetani*. It is not pathogenic.

C. cochlearium   The only common clostridium that is biochemically inert. It is not pathogenic.

C. fallax

A strict anaerobe, associated with gas gangrene. Spore are not very resistant to heat and the organism may be killed in differential heating methods. Found in soil.

Food spoilage clostridia

Clostridium species are known to be responsible for spoilage of a wide variety of foods, including milk, meat products, fresh water fish and vegetables.

Prepare 10% emulsions of the product in 0.1% peptone water using a Stomacher or blender. Heat some of this emulsion at 75–80°C for 30 min (pasteurized sample). Inoculate several tubes of Robertson's cooked meat medium and/or liquid reinforced clostridium medium with pasteurized and unpasteurized emulsion and incubate pairs at various temperatures, e.g. 5–7°C for psychrophiles, 22 and 37°C for mesophiles and 55°C for thermophiles. Inoculate melted reinforced clostridial agar (RCM) in deep tubes with dilutions of the emulsions. Allow to set. Incubate at the desired temperature and look for black colonies, or blackening of the media which suggests clostridia. Inoculate several tubes of Crossley's Milk medium each with 2 ml of suspension and incubate at the required temperatures. This gives a useful guide to the identity of the clostridias.

Appearance in Crossley's milk medium

C. putrifaciens, C. sporogenes, C. oedematiens, C. histolyticum are indicated by slightly alkaline reaction, gas, soft curd subsequently digested leaving clear brown liquid, a black sediment and a foul smell.

C. sphenoides gives a slightly acid reaction, soft curd, whey and some gas.

C. butyricum produces acid, firm clot and gas.

C. perfringens usually gives a stormy clot.

C. tertium usually gives a stormy clot.

Note that aerobic spore bearers may give similar reactions, so plate out and incubate both aerobically and anaerobically.

Thermophiles associated with canned food spoilage

Two kinds of clostridial spoilage occur due to underprocessing: 'hard swell' and 'sulphur stinkers'.

Examine Gram-stained films for Gram-positive spore-bearers. Inoculate reinforced clostridial medium or glucose tryptone agar and iron sulphite medium in deep tube cultures by adding about 1-ml amounts of dilutions of a 10% emulsion of the food material in peptone water diluent to 152 × 16 mm tubes containing 15–20 ml of medium melted and at 50°C. Either solid or semisolid media may be used. Incubate duplicate sets of tubes for up to 3 days at 60 and 25°C.

C. thermosaccharolyticum
Produces white lenticular colonies in all three media and changes the colour of glucose tryptone agar from purple to yellow. This organism is responsible for 'hard swell'.

C. nigrificans
Produces black colonies, particularly in media that contain iron. This causes 'sulphur stinkers'.

Neither organism grows appreciably at 25°C.

Counting clostridia

Emulsify 10 g of the food in 90 ml of a 0.1% peptone water in a homogenizer. Divide into two portions and heat one at 75°C for 30 min.

Do MPN tests (10-, 1- and 0.1-ml amounts) using the five or three tube method, on each portion in liquid differential reinforced clostridial medium (DRCM). Clostridia turn this medium black ('black tube method'). The unheated portion gives the total count, the heated portion the spore count.

For the pour plate method add 0.1-ml amounts of serial dilutions of the emulsion to melted OPSP agar. Incubate anaerobically for 24 h and count the large black colonies.

It may be necessary to dilute the inoculum 1:10 or 1:100 if many clostridia are present. If the load of other organisms is heavy, add 75 units/ml of polymyxin to the medium used for the unheated count.

To recover individual colonies, Burri tubes may be used. These are open at both ends and are closed with rubber bungs. Use solid DRCM and, after incubation and counting, remove both stoppers and extrude the agar cylinder. Cut it with a sterile knife and aspirate colonies with a pasteur pipette. These tubes need not be incubated in an anaerobic jar. After inoculation, cover the medium with a layer of melted paraffin wax about 2 cm deep.

Miller–Prickett tubes (flattened test-tubes) are also useful. Pipette dilutions into the tubes and then add 15 ml of melted medium at 50°C to each tube. Seal with paraffin wax as above.

Membrane filters can be used for liquid samples. Roll up the filters and place in test-tubes; cover with melted medium.

DRCM is suitable for most counts but for *C. nigrificans* iron sulphite medium is good.

Select incubation temperatures according to the species to be counted.

General identification procedure

Subculture each kind of colony of anaerobic Gram-positive bacillus in the following media: Robertson's cooked medium; purple milk; peptone water for indole production; gelatin medium; glucose, lactose, sucrose and salicin peptone waters. These media should be in cotton wool-plugged test-tubes, not in screw-capped bottles. Omit the indicator from these as it may be decolorized during anaerobiosis. Test for acid production after 24–48 h with bromocresol purple. It may be necessary to enrich some media with Fildes' extract and an iron nail in each tube assists anaerobiosis. Gas production in sugar media is not very helpful and Durham's tubes can be omitted. The API 20A system for anaerobes is very helpful.

Do half-antitoxin (Nagler) plates as above using *C. perfringens* type A and *C. novyi* antitoxin (see Tables 28.1 and 28.2).

Food spoilage species

(1) Thermophiles
    C. nigrificans   Sulphur stinkers
    C. thermosaccharolyticum   Hard swell

(2) Mesophiles

C. butyricum	Cheese disorders. Butter and milk products
C. sporogenes *C. sphenoides** *C. novyi* *C. perfringens*	Meat and dairy products

(3) Psychrophile

*C. putrefaciens**	Bone taint, off-odours

*Not described in this book.

For further information on clostridia see Willis (1977), Holdeman *et al.* (1977) and Phillips *et al.* (1985).

Bacteroides, Fusobacterium, Eikenella

These Gram-negative anaerobic bacilli are found in the alimentary and genitourinary tracts of humans and other animals and in necrotic lesions, often in association with other organisms. They are pleomorphic, often difficult to grow and die easily in cultures. Identification is often difficult. Reference experts use gas-liquid chromatography. They are generally classified into the *Bacteroides fragilis* group, the *Bacteroides melaninogenicus* group and the *Fusobacterium necrophorum* group. *Eikenella* are not strict anaerobes and will grow in 10% carbon dioxide. The organism known as *E. corrodens* may be confused with another corroder – *Bacteroides ureolyticus*.

In pathological material they are not uncommonly mixed with other anaerobes and capnophilic organisms. In general, primary infections above the diaphragm are due to *Fusobacterium*, but *Bacteroides* may occur anywhere and are frequently recovered from blood cultures.

Isolation

Examine Gram films of pathological material. Diagnosis of Vincent's angina can be made from films counterstained with dilute fuchsin, if fusiform bacilli can be seen in large numbers, associated with spirochaetes (*Borrelia vincenti*). In pus from abdominal, brain, lung or genital tract lesions, large numbers of Gram-negative bacilli of various shapes and sizes, with pointed or rounded ends suggest *Bacter oides* infection. Such material often has a foul odour. Infections involving *B. melaninogenicus* are rapidly diagnosed by viewing the pus under long wave UV (365 nm) light. This organism produces porphyrins which give a bright red fluorescence.

Culture pus and other material anaerobically, aerobically and under 5–10% carbon dioxide. For anaerobic cultures use both non-selective and enriched selective media, e.g. with haemin, menadione, sodium pyruvate, cysteine HCl. Or use Fastidious anaerobe or Wilkins–Chalgren agar. The usual selective agents are neomycin 75 μg/ml, nalidixic acid 10 μg/ml and vancomycin 2.5 μg/ml. Various combinations may be used to restrict contaminating facultative flora from certain specimens, e.g. neomycin and vancomycin for pus from the upper respiratory tract.

Place a 5 µg metronidazole disc on the streaked-out inoculum for the rapid recognition of obligate anaerobes, all of which are sensitive. Incubate at 37°C for 48 h.

Compare growths on selective and non-selective media and the anaerobic with aerobic cultures.

Identification

Full identification of these organisms is difficult and rarely necessary. All that is required in a clinical laboratory is confirmation that anaerobes are involved. A simple report that '...mixed anaerobes are present – sensitive to metronidazole...' usually suffices.

Identification is more important when anaerobes are present in blood cultures and other normally sterile sites. These simple presumptive tests are useful.

Potassium hydroxide test   The Gram stain is important but may give equivocal results. Emulsify a few colonies in a drop of 3% KOH on a slide. Bacteroides tend to become 'stringy' and form long strands when the loop is moved away from the drop.

UV fluorescence   Examine under long wave UV light. Red fluorescence suggests the *B. melaninogenicus* group and develops much earlier than the characteristic black pigment. Yellow fluorescent colonies of Gram-negative rods are probably fusobacteria (Brazier, 1986).

Phosphomycin test   Place a disc soaked in phosphomycin, 300 µg/ml, on a seeded plate and incubate. Fusobacteria are usually sensitive and bacteroides resistant (Bennett and Duerden, 1985).

Carbohydrate fermentation   See Phillips (1976) and Holdeman and Moore (1977).

Antibiogram and commercial identification kits
Antibiogram patterns give a low level of accuracy and offer only a limited range of identifications. Commercial kits are an improvement but results based on a generated number may be misleading. Much depends on the data base on which the scheme is based. The new API 32 ATB kit is one of the most useful.

Anaerobic cocci

These may occur in cultures from a wide variety of human and animal material.

Peptostreptococci and peptococci

It is difficult to separate these genera. Cocci occur singly, in pairs, clumps or short chains. They grow on blood agar at 37°C and are non-haemolytic but grow better on Wilkins–Chalgren agar containing a supplement for non-sporing anaerobes.

Peptostreptococci are sensitive to novobiocin (5 µg disc) but peptococci are resistant.

Peptostreptococci may be pathogenic for humans and animals, associated with gangrenous lesions. They are commensals in the intestine and have been isolated from the vagina in health and in puerpural fever. They produce large amounts of hydrogen sulphide from high-protein media, e.g. blood broth and a putrid odour in ordinary media. The API system is useful for identifying these organisms.

The organism formerly known as *Gaffkya tetragenus* is now *Peptostreptococcus tetradius*.

This genus *Peptococcus* now contains only one species, *P. niger*.

Sarcina

Cocci in packets of eight. Obligate anaerobes, requiring incubation at 20–30°C and carbohydrate media. Found in air, water and soil.

Sarcina ventriculi is a large coccus arranged in clumps of four to eight. It is found in the human gut, particularly in vegetarians.

Veillonella

These are obligate anaerobes and occur as irregular masses of very small Gram-negative cocci, cultured from various parts of the respiratory and alimentary tracts and genitalia. They do not appear to cause disease.

Growth may be improved by adding 1% sodium pyruvate and 0.1% potassium nitrate to the basal medium.

Taxonomists now recognize seven species. *Veillonella atypica*, *V. dispar* and *V. parvula* are found in humans; *V. caviae*, *V. criceti*, *V. ratti* and *V. rodentium* in rodents.

Fluorescence under UV light is generally weak, fades rapidly and is medium-dependent. It works only on brain infusion agars except for *V. criceti* which fluoresces on a range of media (Brazier and Riley, 1988). Phenotypic differentiation is based on catalase, fermentation of fructose and the origin of the strain. For further information see Mays *et al.* (1982).

References

Bennett, K. W. and Duerden, B. I. (1985) Identification of fusobacteria in a routine diagnostic laboratory. *Journal of Applied Bacteriology*, **59**, 171–181

Brazier, J. S. (1986) Yellow fluorescence of fusobacteria. *Letters in Applied Microbiology*, **2**, 125–126

Brazier, J. S. and Riley, T. V. (1988) UV red fluorescence of *Veillonella* spp. *Journal of Clinical Microbiology*, **26**, 383–384

Dezfullian, M., McCroskey, C. L., Hatheway, C. L. and Dowell, V. R. (1981) Isolation of *Clostridium botulinum* from human faeces. *Journal of Clinical Microbiology*, **13**, 526–531

Holdeman, L. V., Cato, E. P. and Moore, W. E. C. (eds) (1977) *Anaerobic Laboratory Manual*, 4th edn, Virginia State University, Blacksburg, VA

Mays, T. D., Holdeman, L. V., Moore, W. E. C. and Johnson, J. L. (1982) Taxonomy of the genus *Veillonella* Prevot. *International Journal of Systematic Bacteriology*, **32**, 28–36

Phillips, K. D. (1976) A simple and sensitive technique for determining the fermentation reactions of nonsporing anaerobes. *Journal of Applied Bacteriology*, **41**, 325–328

Phillips, K. D. (1986) A sporulation medium for *Clostridium perfringens*. *Letters in Applied Microbiology*, **3**, 77–79

Phillips, K. D., Brazier, J. S., Levett, P. N. and Willis, A. T. (1985) Clostridia. In *Isolation and Identification of Micro-organisms of Medical and Veterinary Importance* (eds C. H. Collins and J. M. Grange), Society for Applied Microbiology Technical Series No. 21, Academic Press, London, pp. 215–236

Willis, A. T. (1977) *Anaerobic Bacteriology*, 3rd edn, Butterworths, London

Chapter 29

Mycobacterium

These organisms are acid-fast: if they are stained with a strong phenolic solution of a dye, e.g. carbol fuchsin, they retain the stain when washed with dilute acid. Other organisms are decolorized. There are no grounds for the commonly held belief that some mycobacteria are acid- and alcohol-fast but that others are only acid-fast. These properties vary with the technique and the organism's physiological state.

The most important members of this genus are obligate parasites. They include the three 'species' of tubercle bacilli, *Mycobacterium tuberculosis, M. bovis* and *M.africanum; M. paratuberculosis* (Johne's bacillus) and the leprosy (Hansen's) bacillus *M. leprae*. Several species that appear to be free living are opportunist pathogens of humans and animals. Several free-living species are also opportunist pathogens of man and animals.

The tubercle bacillus, and in some countries certain other species of mycobacteria (see national lists) are Risk/Hazard Group 3 pathogens and infectious by the airborne route. All manipulations that might produce aerosols should be done in microbiological safety cabinets in Biosafety/Containment Level 3 laboratories.

Tubercle bacilli

Direct microscopic examination

Sputum
This is the most common material suspected of containing tubercle bacilli. It is often viscous and difficult to manipulate. To make direct films use disposable 10 μg plastic loops. These avoid the use of bunsens in safety cabinets. Discard them into disinfectant. Ordinary bacteriological loops are unsatisfactory and unsafe. Spread a small portion of the most purulent part of the material carefully on a slide, avoiding hard rubbing which releases infected airborne particles. Allow the slides to dry in the safety cabinet then remove them for fixing and staining. This is quite safe; aerosols are released during the spreading, not from the dried film, although these should not be left unstained for any length of time. Fixing may not kill the organisms and dried material is easily detached.

For direct microscopy of centrifuged deposits after homogenizing sputum for culture spread a 10-μl loopful over an area of about 1 cm^2. Dry slowly and handle with care. The material tends to float off the slide while it is being stained.

CSF
Make two parallel marks 2–3 mm apart and 10-mm long in the middle of a microscope slide. Place a loopful of the centrifuged deposit between the marks. Allow to dry and superimpose another loopful. Do this several times before drying, fixing and staining. Examine the whole area.

Urine
Direct microscopy of the centrifuged deposit is unreliable because urine frequently contains acid-fast bacilli which may resemble tubercle bacilli but which have entered the specimen from the skin or the environment.

Aspirated fluids, pus
Centrifuge if possible and make films from the deposit. Otherwise prepare thin films. Thick films float off during straining. Material containing much blood is difficult to examine and may give false positives.

Gastric lavage
Direct microscopy is unreliable because environmental mycobacteria are frequently present in food and hence in the stomach contents.

Laryngeal swabs
Direct microscopy is unrewarding.

Blood, faeces, milk
Direct microscopy is unreliable because acid-fast artefacts and saprophytic mycobacteria may be present.

Tissue
Cut the tissue into small pieces. Work in a safety cabinet. Remove caseous material with a scalpel and scrape the area between caseation and soft tissue. Spread this on a slide. It is more likely to contain the bacilli than other material.

Staining

Stain direct films of sputum or other material by the ZN method. The bacilli may be difficult to find and it is often necessary to spend several minutes examining each film. Fluorescence microscopy, using the auramine phenol (AP) method, is popular in some clinical laboratories. A lower-power objective may be used and more material examined in less time. False positives are not uncommon with this method and the presence of acid-fast bacilli should be confirmed by overstaining the film with ZN stain. These staining methods are described in Chapter 6.

Reporting

Report 'acid-fast bacilli seen/not seen'. If they are present report the number per high power field. Table 29.1 gives a reporting system in common use in Europe for sputum films.

Table 29.1 Scale for reporting acid-fast bacilli in sputum smears examined microscopically

No. of bacilli observed	Report
0 per 300 fields	Negative of AFB
1–2 per 300 fields	(±) Repeat test
1–10 per 100 fields	+
1–10 per 10 fields	++
1-10 per field	+++
10 or more per field	++++

Note that this is a logarithmic scale; this facilitates plotting.

False positives

Single acid-fast bacilli should be regarded with caution. They may be environmental contaminants, e.g. from water. They may be transferred from one slide to another during staining or if the same piece of blotting paper is used for more than one slide.

Isolation from pathological material

Mycobacteria are often scanty in pathological material and not uniformly distributed. Many other organisms may be present and the plating methods used in other bacteriological examinations are useless. Acid-fast bacilli usually grow at a very much slower rate than other bacteria and are overgrown on plate cultures. Isolation methods depend on homogenization of the suspected specimens with a reagent that is less lethal for mycobacteria than for other organisms and that reduces the viscosity of the preparation so that it may be centrifuged. No ideal reagent is known; there is a choice between 'hard' and 'soft' reagents.

Hard reagents are recommended for specimens that are heavily contaminated with other organisms. Exposure times are critical or the mycobacteria are also killed. Soft reagents are to be preferred when there are fewer other organisms, e.g. in freshly collected sputum. The reagent can be left in contact with the specimen for several hours without significantly reducing the numbers of mycobacteria. Soft reagents may permit the recovery of mycobacteria other than tubercle bacilli, which are often killed by hard reagents.

The methods described below are intended to reduce the number of manipulations, particularly in neutralization and centrifugation, so that the hazards to the operator are minimized.

Culture of the centrifuged deposit after treatment gives optimum results, but is avoided by some workers because of the dangers of centrifuging tuberculous liquids. Sealed centrifuge buckets (safety cups) overcome this problem (see Chapter 1). These reduce considerably the hazards of aerosol formation should a bottle leak or break in the centrifuge. They should be opened in a safety cabinet.

If material is not centrifuged, more tubes of culture media should be inoculated and a liquid medium should be included. This should contain an antibiotic mixture (p.350).

Sputum culture

Encourage the sputum from the specimen container into the preparation bottle with a pipette made from a piece of sterile glass tubing (200 × 5 mm) with one end left rough to cut through sputum strands and the other smoothed in a flame to accept a rubber teat.

Hard method

Add about 1 ml of sputum to 2 ml of 4% NaOH solution in a 25-ml screw-capped bottle. Stopper securely, place in a self-sealing plastic bag to prevent accidental dispersal of aerosols and shake mechanically, but not vigorously, for not less than 15 min but not more than 30 min. Thin specimens require the shorter time. Incubation does not help.

Remove the bottle from the plastic bag and add 3 ml of 14% (approximately 1 M) dipotassium hydrogen orthophosphate (KH_2PO_4) solution containing enough phenol red to give a yellow colour. This neutralizing fluid should be dispensed ready for use and sterilized in small screw-capped bottles. It should not be pipetted from a stock bottle.

The colour change of the phenol red from yellow to orange pink indicates correct neutralization. Re-stopper, mix gently and centrifuge with the bottles in sealed centrifuge buckets ('safety cups') for 15 min at 3000 rev/min. Pour off the supernatant fluid carefully into disinfectant. Wipe the neck of the bottle with a piece of filter-paper, which is then discarded into disinfectant. Culture the deposit.

Soft method

To 2–4 ml of sputum in a screw-capped bottle, add an equal volume of 23% trisodium orthophosphate solution. Mix gently and stand at room temperature or in a refrigerator for 18 h. Add the contents of a 10-ml bottle of sterile distilled water to reduce the viscosity, centrifuge, decant and culture the deposit using the methods and precautions described above.

Laryngeal and other swabs

Place the swab in a tube containing 1 N sodium hydroxide solution for 5 min. Remove, drain and place in another tube containing 14% potassium dihydrogen orthophosphate (KH_2PO_4) solution for 5 min. Drain and inoculate culture media.

Urine culture

Use fresh, early morning mid-stream specimens. Do not bulk several specimens as this often leads to contamination of cultures. Centrifuge 25–50 ml at 3000 rev/min and assess the number of organisms in the deposit by examining a Gram-stained film. Treat the deposit with 2 ml of 4% sulphuric acid for 15–40 min depending on the load of other organisms. Neutralize with 15 ml of distilled water, centrifuge and culture the deposit.

Alternatively plate some of the urine on blood agar, incubate overnight and if this is sterile pass 50–100 ml of the urine through a membrane filter. Cut the filter into strips and place each strip on the surface of the culture media in screw-capped bottles.

Culture of CSF, pleural fluids, pus, etc.

Plate original specimen or centrifuged deposit on blood agar and incubate overnight. Keep the remainder of specimen in a deep-freeze. If the blood agar culture is sterile, inoculate media for mycobacteria without further treatment. Use as much material and inoculate as many tubes as possible. If the blood agar culture shows the presence of other organisms, proceed as for sputum and/or culture directly into media containing antibiotics (see below). Half fill the original container with Kirchner medium and incubate. Some tubercle bacilli may adhere to the glass or plastic.

Automated culture
A commercial instrument (BACTEC: Johnston Laboratories) has been developed for detecting the early growth of mycobacteria by a radiometric method.

Sputum or other homogenates, decontaminated if necessary, are added to Middlebrook broth medium containing antibiotics to discourage the growth of other organisms, and ^{14}C-labelled palmitic acid. The medium is prepared commercially (Becton Dickinson) in rubber sealed bottles and is inoculated with a syringe and hypodermic needle. If growth occurs, $^{14}CO_2$ is evolved. The air space above the medium in each bottle is sampled automatically at fixed intervals and the amount of radioactive gas is estimated and recorded. Growth of mycobacteria may be detected in 2–12 days, but positive results require further tests to distinguish between tubercle bacilli and other mycobacteria. In the BACTEC, *p*-nitro-α-acetylamino-β-propiophenone (NAP) is used and this takes another 2–5 days. Mycobacteria that grow in media containing this substance are subcultured and identified by traditional methods. (See Damato *et al.*, 1983; Roberts *et al.*, 1983.)

Culture of tissue
Grind very small pieces of tissue, e.g. endometrial curettings, in Griffith's tubes or emulsify them in sterile water with glass beads on a Vortex mixer. Homogenize larger specimens with a stomacher or blender. Check the sterility and proceed as for CSF, etc. Keep some of the material in a deep freeze in case the cultures are contaminated and the tests need repeating.

Isolation from the environment

This may be done to trace the sources of opportunist and saprophytic mycobacteria which contaminate clinical material and laboratory reagents.

Water
Pass up to 2 litres of water from cold and hot taps through membrane filters (as for water bacteriology p.233). Drain the membranes and place them in 3% sulphuric or oxalic acid for 3 min and then in sterile water for 5 min. Cut them into strips and place each strip on the surface of the culture medium in screw-capped bottles. Use Lowenstein–Jensen medium and Middlebrook 7H11 medium containing antibiotics (see below).

Swab the insides of cold and hot water taps and treat them as laryngeal swabs.

Milk
Centrifuge at least 100 ml from each animal. Examination of bulk milk is useless. Treat cream and deposit separately by the NaOH method and culture deposit on several tubes of medium.

Dust and soil
Place 2–5 g samples in 1 litre of sterile distilled water containing 0.5% Tween 80. Mix gently to avoid too much froth. Pass through a coarse filter to remove large particles and allow to settle in a refrigerator for 24 h. Decant the supernatant and

pass it through membrane filters as described above for waters. Centrifuge the sediment and treat it with sodium hydroxide as for sputum. Neutralize, centrifuge and inoculate several tubes of medium.

Culture media

Egg media are most commonly used and there seems little to choose between any of them. In the UK, Lowenstein–Jensen medium is most popular, but in the USA bacteriologists seem to favour ATS or Piezer media. Both glycerol medium (which encourages the growth of the human tubercle bacillus) and pyruvate medium (which encourages the bovine bacillus) should be used.

Pyruvate medium should not be used alone as some opportunist mycobacteria grow very poorly on it.

Some workers prefer agar-based media e.g. Middlebrook 7H10 or 7H11. Kirchner liquid medium is also useful for fluid specimens that cannot be centrifuged or when it is desirable to culture a large amount of material.

The antibiotic media of Mitchison *et al.* (1987) are useful for contaminated specimens, even after treatment with NaOH, etc. They offer a safety net for non-repeatable specimens such as biopsies. These media should be used as well as, not in place of, egg media for tissues, fluids and urines. They are made as follows.

Add the following to complete Kirchner or Middlebrook 7H11 media: Polymyxin, 200 units/ml: carbenicillin, 100 mg/l; trimethoprim, 10 mg/l; amphotericin, 10 mg/l. Dispense fluid medium in 10 ml and make slopes of solid medium.

Inoculating culture media

Use an inoculum of at least 0.2 ml (not a loopful) for each tube of medium. Inoculate two tubes each of egg medium, one containing glycerol and the other pyruvic acid and other media as indicated above.

Incubation

Incubate all cultures at 35–37°C and cultures from superficial lesions also at 30–33°C for at least 8 weeks. Prolonging the incubation for a further 2–4 weeks may result in a small increment of positives, especially from tissues. Examine weekly.

Identification of tubercle bacilli

Most mycobacteria cultured from pathological material will be tubercle bacilli. No growth will be evident on egg media until 10–14 days. Colonies of the human tubercle bacillus are cream coloured, dry and look like breadcrumbs or cauliflowers. Those of the bovine tubercle bacillus are smaller, whiter and flat. Growth of this organism may be very poor; it grows much better on pyruvate medium than on glycerol medium.

On Middlebrook agar media, colonies of both types are flat and grey. In Kirchner fluid media, colonies are round, granular and may adhere to one another in strings or masses. They grow at the bottom of the tube and settle rapidly after the fluid has been shaken.

Make ZN films to check acid-fastness. Some yeasts and coryneform organisms grow on egg media and their colonies may resemble those of tubercle bacilli. Make the films in a drop of saturated mercuric chloride as a safety measure against dispersing live bacilli in aerosols, and spread the films gently. Note whether the organisms are difficult or easy to emulsify. Check the morphology microscopically. Tubercle bacilli may be arranged in serpentine cords, are usually uniformly stained and usually 3–4 μm, in length.

Make suspensions of the organisms as follows. Prepare small, glass screw-capped bottles containing two wire nails shorter than the diameter of the base of the bottles, a few glass beads 2–3 mm in diameter and 2 ml of phosphate buffer, pH 7.2–7.4 (the nails will rust if water is used; wash both nails and beads in dilute hydrochloric acid and then in water before use). Sterilize by autoclaving. Place several colonies of the organisms in a bottle and mix on a magnetic stirrer. Allow large particles to settle and use the supernatant.

Inoculate two tubes of egg medium and one tube of the same medium containing 500 μg/ml of p-nitrobenzoic acid (PNBA medium). To make the p-nitrobenzoic acid stock solution, dissolve 0.5 g of the compound in 50 ml of water to which a small volume of 1 N sodium hydroxide solution has been added. Carefully neutralize with hydrochloric acid and make up to 100 ml with water. This solution keeps for several months in a refrigerator. Incubate one egg slope at 25±0.5°C, the other in an incubator with an internal light at 35–37°C. Incubate the p-nitrobenzoic acid egg slope at 35–37°C. If an internally illuminated incubator is not available, grow the organisms for 14 days and then expose to the light of a 25-W lamp at a distance of 2–3 ft for several hours and re-incubate for 1 week.

Tubercle bacilli have the following characteristics:

(1) Not easily emulsified in water.
(2) Regular morphology, 3–4 μm in length and usually showing serpentine cords.
(3) Relatively slow growing, taking 10 days or more to show visible growth.
(4) Produce no yellow, orange or red pigment.
(5) Fail to grow in the presence of 500 μg/ml of p-nitrobenzoic acid (the bovine organism may show a trace of growth).
(6) Fail to grow at 25°C (the bovine organism may show a trace of growth).

Acid-fast organisms that grow in 1 week or less, or yield yellow or red colonies, or emulsify easily, or are morphologically small, coccoid, or long, thin and beaded or irregularly stained, are unlikely to be tubercle bacilli.

Variants of tubercle bacilli and BCG

Although the identification of human, bovine and 'African' variants of *M. tuberculosis* is of no clinical value it may be necessary to recognize them for epidemiological work (Collins *et al.*, 1982). It is also useful to identify BCG, particularly if it may have been responsible for disseminated disease as a result of vaccination of neonates or the treatment of malignancy.

Use suspensions prepared as described above and inoculate media for the following tests and incubate at 37°C.

TCH susceptibility
Use Lowenstein–Jensen medium containing 5 μg/ml of thiophen-2-carboxylic acid hydrazide. Read at 18 days.

Pyruvate preference
Compare the growth on Lowenstein–Jensen medium containing glycerol with that containing pyruvate after 18 days' incubation.

Nitratase test
Use Middlebrook 7H10 broth and test after 18 days by method (3) described on p.105.

Oxygen preference
Use Kirchner medium made semisolid with 0.2% agar. Inoculate with about 0.02 ml of suspension. Mix gently to avoid air bubbles and incubate undisturbed for 18 days. Aerobic growth occurs at or near the surface; microaerophilic growth as a band 1–3 cm below the surface, sometimes extending upwards.

Pyrazinamide susceptibility
Use the method described on p.363.

Cycloserine susceptibility
Use Lowenstein–Jensen medium containing 20µg/ml cycloserine. Read after 18 days.

See Table 29.2. Identification of a strain as 'human', 'bovine', 'Asian' or 'African' does not indicate origin of strain or ethnic group of the patient. The variants are widely distributed.

Table 29.2 Variants of tubercle bacilli

	TCH	Nitratase	Oxygen preference	Pyrazinamide
Classic bovine	S	−	M	R
African I[a]	S	−	M	S
African II[a]	S	+	M	S
Asian human	S	+	A	S
Classic human	R	+	A	S
BCG[b]	S	−	A	R

TCH, thiophen-2-carboxylic acid hydrazide; S, sensitive; R, resistant; M, microaerophilic; A, aerobic
[a]both variants are known as *M. africanum*
[b]Differs from others in being resistant to cycloserine

The niacin test

It was originally claimed that this test distinguished between human tubercle bacilli, which synthesize niacin, and bovine tubercle bacilli, which do not. Unfortunately some human strains give negative niacin results and some other mycobacteria do synthesize niacin. It is an unreliable test and uses a hazardous chemical (cyanogen bromide). It has little to commend it.

Species of tubercle bacilli

M. tuberculosis

This species grows well (eugonic) on egg media containing glycerol or pyruvate. Colonies resemble breadcrumbs and are cream coloured. Films show clumping and cord formation – the bacilli are orientated in ropes or cords, especially on moist medium. There is no growth at 25 or 37°C or on PNBA medium. It is usually resistant to TCH (the Asian variants are sensitive), is nitratase positive, aerobic and usually susceptible to pyrazinamide. It causes tuberculosis in humans and may infect domestic and wild animals (usually directly or indirectly from humans). Simians are particularly likely to become infected.

M. bovis

Is also known as the bovine variant of *M. tuberculosis*. It grows poorly (dysgonic) on egg medium containing glycerol but growth is enhanced on pyruvate medium. Colonies are flat and grey or white; growth may be effuse. Cords are present in films. There is no or very poor growth at 25°C and on PNBA medium compared with a control slope at 37°C. It is sensitive to TCH, nitratase negative, microaerophilic and resistant to pyrazinamide. This is the classic bovine tubercle bacillus, common in milch cows before eradication schemes were introduced. It is still occasionally associated with human disease both pulmonary and extrapulmonary, but such infections are now rarely associated with the consumption of infected milk. Most disease represents reactivation of infections acquired much earlier.

M. africanum

This name is given to the African group of strains with properties intermediate between those of the human and bovine tubercle bacillus. (See Collins *et al.*, 1982.) Growth is usually dysgonic, enhanced by pyruvate. It is susceptible to TCH, microaerophilic and susceptible to pyrazinamide. The nitratase test may be negative (most West African strains), or positive (most East African strains). It causes tuberculosis, clinically indistinguishable from that caused by the other tubercle bacilli.

BCG

This is the bacillus of Calmette and Guerin, an organism of attenuated virulence used in immunization against tuberculosis. It is eugonic; growth is not enhanced by pyruvate. It is susceptible to TCH, nitratase negative, aerobic, resistant to pyrazinamide and (unlike other strains in this group) is resistant to cycloserine.

Animal inoculation tests

Cultural methods have now largely superseded animal inoculation. They are generally more reliable and much less expensive. Tubercle bacilli that are resistant to isoniazid usually show limited or no virulence to guinea pigs, and so do some Asian or South Indian strains. Reasons for ending the routine guinea pig test were given by Marks (1972).

Opportunist mycobacteria

Some mycobacteria, normally present in the environment, may, under some circumstances, cause or be associated with human disease. Such diseases are correctly called mycobacterioses, not tuberculosis. Before specific names were given to these organisms and their taxonomic position was clarified, they were known as 'atypical' or 'anonymous' mycobacteria. Various temporary systems of classification were proposed and were used until species were delineated.

A number of other mycobacteria, not so far known to be associated with human disease, are frequently cultured in clinical laboratories. These are usually contaminants of the specimen or are introduced during collection and/or laboratory processing. Mostly they come from water but they may also be present in dust, soil and food. Occasionally they are seen in direct films but fail to grow on cultures incubated at 35–37°C.

It is also difficult to assess the significance of opportunists. Single isolations may be regarded with suspicion. Whenever possible, attempts should be made to recover the organisms from subsequent specimens.

Most of these mycobacteria may be identified with the aid of the following tests and with Table 29.3.

Inoculation
Unless otherwise stated, inoculate the media with one loopful or a 10-μl drop of a suspension prepared as described on p.362.

Pigment production
Inoculate two egg medium slopes. Incubate both at 35–37°C, one exposed to light and the other in a light-proof box. Examine at 14 days. Photochromogens show a yellow pigment only when exposed to light. Scotochromogens are pigmented in light and dark, but the culture exposed to light is usually deeper in colour. False photochromogenicity may arise if the inoculum is too heavy and pigment precursor is carried over. Continue to incubate the light slope and look for orange-coloured crystals of carotene pigment which may form in the growth.

Temperature tests
Inoculate egg medium slopes with a single streak down the centre and incubate at the following (exact) temperatures: 20, 25, 42 and 44°C. Examine for growth at 3 and 7 days and thereafter weekly for 3 weeks.

Thiacetazone susceptibility
Inoculate egg medium slopes containing 20 μg/ml of thiacetazone (TZ). Make a stock 0.1% solution of this drug in formdimethylamide (*caution*). This solution keeps well in a refrigerator. Incubate slopes for 18–21 days and observe the growth in comparison with that on an egg slope without the drug.

Nitratase test
Use method (3) on p.105.

Sulphatase test
Use the method described on p.106. If the organism is a rapid grower, add the ammonia after 3 days.

Catalase test
Prepare the egg medium in butts in screw-capped tubes (20 × 150 mm). Inoculate and incubate at 35–37°C for 14 days with the cap loosened and then add 1 ml of a mixture of equal parts of 30% hydrogen peroxide (*caution*) and 10% Tween 80. Allow to stand for a few minutes and measure the height in millimetres of the column of bubbles. More than 40 mm of froth is a strong positive result (+ + +) (Wayne's test).

Table 29.3 Usual properties of some opportunist mycobacteria and others that may be encountered in clinical material

Species	Pigment	TZ	Nitratase	Tween hydrolysis	Growth at (°C)					Sulphatase[a]		Catalase[b]	Tellurite[c] reduced	Growth on N-medium	Rapid growth
					20	25	33	42	44	3 days	21 days				
M. kansasii	P	S	+	+	-	+	+	v	-		+	+++	-	-	-
M. marinum	P	R	-	+(late)	+	+	+	-	-		++	++	-	v	+
M. xenopi	-/S	R	-	-	-	-	+	+	+		+++	-	-	-	-
M. avium-intracellulare	-/S	R	-	-	v	+	+	+	v		v	-	+	-	-
M. scrofulaceum	S	v	+	+(late)	v	+	+	-	-		v	-	-	-	-
M. malmoense	-	R	-	-	-	+	+	-	-		-	++	-	-	-
M. simiae	P	R	+	+	-	+	+	-	-		+	+	-	-	-
M. szulgai	S/P[d]	R	+	-	+	+	+	v	-		+++	v	-	+	+
M. fortuitum	-	R	+	+	+	+	+	-	-	+++	+++	v	+	+	+
M. chelonei	-	R	-	+	+	+	+	-	-	+++	+++	++	+	-	-
M. gordonae	S	R	-	+	+	+	+	-	-		+	+		-	-
M. flavescens	S	R	+	+	+	+	+	-	-		+	++	+	+	+
M. gastri	-	R	-	+	+	+	+	-	-		+	+++	-	-	-
M. terrae	-	R	+	+	v	+	+	-	-		+	+	-	-	+
M. triviale	-	R	+	+	-	+	+	-	-		+	+++	-	-	-
M. nonchromogenicum	-	R	-	+	v	+	+	-	-		+	+++	-	-	-
M. smegmatis	-	R	+	+	+	+	+	+	+	-	+	+++	+	+	+
M. phlei	S	R	+	+	+	+	+	+	+	-	+	+++	+	+	+
M. ulcerans	-/S	R	-	-	-	-	+	-	-		-	-	-	-	-

TZ, thiacetazone; P, photochromogen; S, scotochromogen; +, usually positive; -, usually negative or none; v, variable

[a] Sulphatase: +++, deep pink; ++, pink; +, pale pink
[b] Catalase: Amount of foam, +++, more than 20 mm; ++, 10–20 mm; +, 5–10 mm; -, less than 5 mm
[c] Test not done on pigmented strains
[d] M. szulgai: At 25°C = photochromogen; at 37°C = scotochromogen

Tween hydrolysis
Add 0.5 ml of Tween 80 and 2.0 ml of a 0.1% aqueous solution of neutral red to 100 ml of M/15 phosphate buffer. Dispense this in 2-ml lots and autoclave at 115°C for 10 min. This solution keeps for about 2 weeks in the dark and should be a pale straw or amber colour. Add a large loopful of culture from solid medium and incubate at 35–37°C. Examine at 5 and 10 days for a colour change from amber to deep pink.

Tellurite reduction
Grow the organisms in Middlebrook 7Ha medium for 14 days or until there is a heavy growth. Add 4 drops of a sterile 0.02% aqueous solution of potassium tellurite (*caution*) and incubate for a further 7 days. If tellurite is reduced there will be a black deposit of metallic tellurium. Ignore grey precipitates. This test is of no value for highly pigmented organisms.

Growth in N medium
Inoculate the medium with a straight wire and incubate. Rapid growers give a turbidity with or without a pellicle in 3 days. 'Frosting', i.e. adherence of a film of growth to the walls of the tube above the surface of the medium, is often seen. Ignore a very faint turbidity.

Resistance to antibiotics
Methods of doing these tests are described below. Some mycobacteria exhibit consistent patterns. All opportunists are resistant to PAS and pyrazinamide, and all but a few strains of *M. xenopi* are moderately susceptible to INAH.

Aerial hyphae
Some Nocardia spp. are partially acid-fast and resemble rapidly growing mycobacteria. Do a slide culture (p.117) and look for aerial hyphae. In very young cultures of mycobacteria, mycelium may be seen, but it fragments early into bacilli. Aerial hyphae are not formed except by *M. xenopi*.

Emulsifiability and morphology
When making films for microscopic examination, note if the organisms emulsify easily. Note the morphology, whether very short, coccobacilli, long, poorly stained filaments or beaded forms.

Species of opportunist and other mycobacteria

M. kansasii
This is a photochromogen (but see below). Optimum pigment production is observed if the inoculum is not too heavy, if there is a plentiful supply of air (loosened cap) and if exposure to light is continuous. Young cultures grown in the dark and then exposed to light for 1 h will develop pigment, but old cultures that have reached the stationary phase may not show any pigment after exposure. Continuous incubation in the light results in the formation of orange-coloured crystals of carotene. The morphology is distinctive; the bacilli are long (5–6 μm) and are beaded. *M. kansasii* is nitratase positive, sensitive to 20 μg/ml of thiacetazone, hydrolyses Tween 80 rapidly and grows at 25°C but not at 20°C; some

stains grow poorly at 42°C but not at 44°C. The sulphatase test is weakly positive and the catalase strongly positive (more than 40 mm in the Wayne test). It is resistant to streptomycin, isoniazid and PAS but susceptible to ethionamide, ethambutol and rifampicin.

Some strains are susceptible to amikacin and erythromycin by the disc technique.

Occasional non-chromogenic or scotochromogenic strains have been reported but these may be recognized by their biochemical reactions and morphology.

This organism is an opportunist pathogen, associated with pulmonary infection. It is rarely significant when isolated from other sites. It has been found in water supplies.

M. xenopi

Pigment, if any, is more obvious on cultures incubated at 42–44°C, which become pale yellow. Morphologically it is easily recognized; it stains poorly and the bacilli are long (5-6 μm) and filamentous. It is nitratase negative, resistant to thiacetazone and does not hydrolyse Tween 80. It is a thermophile, growing well at 44°C, slowly at 35°C, when growth is effuse, and not at 25°C. (This is the only opportunist mycobacterium which fails to grow at 25°C.) The sulphatase test is strongly positive, the catalase test negative. It does not reduce tellurite. It is more susceptible to INAH than other opportunist mycobacteria, usually giving a resistance ratio of 4. It is susceptible to ethionamide, but its susceptibility to other antituberculous drugs varies between strains. Some strains are susceptible to amikacin and erythromycin by the disc method.

It is an opportunist pathogen in human lung disease and is rarely significant in other sites. It is a frequent contaminant of pathological material, especially in urines, and has been found in hospital hot water supplies. 'Outbreaks' of laboratory contamination are not uncommon. This organism is very common in south-east England and north-west France.

M. avium–intracellulare

The two species in this group may be separated by agglutination serology but as this is not done by most reference laboratories they are usually grouped together as the MAI bacilli, after the initial letters of its components. A feeble yellow pigment, not influenced by light, is produced by some species. The bacilli are small, almost coccoid. Some strains are susceptible to thiacetazone. All are nitratase negative, catalase negative or weakly positive, reduce tellurite and do not hydrolyse Tween 80. The sulphatase reaction varies from strongly positive to negative. There is growth on egg medium at 25°C, growth at 20°C and 42°C is variable. Resistance to antituberculous drugs is usual, but some strains are susceptible to ethionamide.

MAI bacilli are opportunist pathogens of humans, associated with cervical adenitis (scrofula), especially in young children, but pulmonary infections also occur. They are a frequent cause of disseminated disease in patients with acquired immune deficiency diseases (AIDS). They are also opportunist pathogens of pigs and birds (M. avium was once described as the avian tubercle bacillus). They have also been found in soil and water.

M. scrofulaceum

This, the 'scrofula scotochromogen', is phenetically similar to the MAI bacilli but is scotochromogenic and fails to reduce tellurite to tellurium. It may also be differentiated by agglutination serology.

M. marinum (formerly M. balnei)

This is missed in clinical laboratories that do not culture material from superficial lesions at 30–33°C. Primary cultures do not grow at 35–37°C. It is a photochromogen. Beading or banding similar to that of *M. kansasii* may be evident. It is nitratase negative, resistant to thiacetazone and hydrolyses Tween 80. After laboratory subculture its temperature range is modified. It will grow at 25°C and at 37°C but not at 44°C. The catalase and sulphatase tests are weakly positive and growth may occur in N medium. Resistance to streptomycin, isoniazid and PAS is usual; resistance to other drugs is variable. Some strains are susceptible to cotrimoxazole, erythromycin and amikacin by the disc method. This is an opportunist pathogen responsible for superficial infections, known as swimming pool granuloma, fish tank granuloma or fish-fanciers' finger. It is found in sea-bathing pools and in tanks where tropical fish are kept, and is a pathogen of some fish. (See the review by Collins *et al.*, 1985a.)

M. gordonae

This is also known as the tap water scotochromogen, although other scotochromogens are found in water supplies. Organism in this group grow slowly, produce a deep orange pigment in light and usually a yellow pigment in the dark. No crystals of carotene are formed. If such crystals are seen, the organisms may be a scotochromogenic strain of *M. kansasii*. Morphology is not distinctive. They are nitratase negative, resistant to thiosemicarbazone and hydrolyse Tween 80. Growth occurs at 20°C but not at 44°C. Some strains are psychrophilic. The sulphatase reaction is weak and the catalase test is usually strongly positive. Growth does not occur in N medium. They are usually resistant to isoniazid and PAS but susceptible to streptomycin and other antituberculous drugs.

The organisms in this group are very rarely associated with human disease and may be reported as non-significant or environmental scotochromogens. They are not infrequent contaminants of pathological material but usually appear as single colonies on egg medium. They may be found in tap water, dust and soil.

M. szulgai

This scotochromogen differs from *M. gordonae* in being nitratase positive, strongly sulphatase positive and giving a weak catalase reaction. It is a rare human opportunist pathogen.

Rapidly-growing scotochomogens

There are many species in this group, e.g. *M. flavescens*, *M. gilvum*, *M. duvalii* and *M. vaccae*. With very rare exceptions they are not known to cause human disease but occur in the environment and sometimes contaminate pathological material.

M. fortuitum and M. chelonei

These two non-pigmented rapid growers are conveniently considered together. There have been taxonomic problems: *M. fortuitum* has also been called *M. ranae*, *M. giae* and *M. peregrinum; M. chelonei* was formerly named *M. abcessus, M.*

runyonii and *M. borstelense*. Growth occurs in most media in 3 days. The bacilli tend to be rather fat and solidly stained. The nitratase test is positive (*M. fortuitum*) or negative (*M. chelonei*). Tween 80 is not hydrolysed. There is growth at 20°C and in some strains at 42°C. Psychrophilic strains, which fail to grow at 37°C, are not uncommon. The sulphatase test is positive at 3 days and the catalase reaction is strongly positive. Tellurite is reduced rapidly. There is growth in N medium in 3 days. There is a general resistance to all antituberculous drugs, except that *M. fortuitum* is usually susceptible to ethionamide and the quinolones but *M. chelonei* is resistant.

These are opportunist pathogens, occurring in superficial infections (e.g. injection abcesses) and occasionally as secondary agents in pulmonary disease. They are common in the environment and frequently appear as laboratory contaminants.

M. smegmatis and M. phlei
These two rapidly growing saprophytes are more common in textbooks than in clinical laboratories. Growth occurs in 3 days. The bacilli have no distinguishing features. They are nitratase positive and hydrolyse Tween 80 in 10 days. They grow at 20°C and at 44°C, *M. phlei* grows at 52°C. The 3-day sulphatase test is negative, but longer incubation gives positive reactions. The catalase test is strongly positive. Tellurite is reduced. There is growth in N medium.

They are not associated with humans or animal disease.

They have contributed to the myth of acid-fast versus acid- and alcohol-fast bacilli (p.345).

Slowly-growing nonchromogens
There are at least four species, *M. terrae*, *M. nonchromogenicum*, *M. triviale* and *M. gastri*. All grow at 25°C but poorly or not at all at 42°C. All hydrolyse Tween 80 and none reduce tellurite, which permits differentiation from MAI bacilli. *M. terrae* and *M. triviale* give a positive nitratase test but *M. nonchromogenicum* and *M. gastri* are negative. *M. gastri* is occasionally photochromogenic and is usually susceptible to thiacetazone and may therefore be confused with *M. kansasii* to which, though non-pathogenic, it is genetically closely related. Susceptibility to anti-tuberculous drugs is variable.

M. ulcerans
This is likely to be missed by clinical laboratories because it does not grow at 37°C. It grows only between 31 and 34°C. Growth is very slow, taking 10–12 weeks to give small colonies resembling those of *M. bovis*. It is biochemically inert and undistinguished. It causes Buruli ulcer, a serious skin infection.

M. simiae
Originally isolated from monkeys, this is also known as *M. habana*. It is a photochromogen, is nitratase negative and does not hydrolyse Tween 80. Pulmonary and disseminated infections have been reported.

M. malmoense
This resembles the MAI organisms, grows very slowly and is sometimes confused with *M. bovis* by inexperienced workers. This suggests that it is commoner than might be expected. It is difficult to identify except by lipid chromatography. It causes pulmonary disease in adults and cervical adenopathy in children.

M. haemophilum

Another rare species, possibly missed because it grows poorly or not at all on media that do not contain iron or haemin. It grows at 30°C on LJ medium containing 2% ferric ammonium citrate and on Middlebrook 7H11 medium plus 60 µg/ml of haemin. All the usual tests are negative. It has been isolated from the skin and subcutaneous tissues of immunologically compromised patients.

Drug susceptibility tests

Most cases of tuberculosis yield tubercle bacilli that are sensitive to the commonly used antituberculous drugs. Susceptibility tests are, however, frequently required by physicians at the commencement of and during treatment, particularly if the patient's condition does not improve. Susceptibility tests for opportunist mycobacteria are of questionable value as patients often respond to drug regimens despite *in vitro* resistance.

It is customary for physicians to treat patients with combinations of three drugs. Treatment with one drug only usually results in the emergence of resistant organisms.

The drugs in general use are INAH (isoniazid), rifampicin, ethambutol and pyrazinamide. Streptomycin, being an injectable drug, is now usually reserved for special purposes.

Susceptibility tests with mycobacteria are complex and are usually done in specialist and reference laboratories. There are four principal methods:

(1) the absolute concentration method which is popular in Europe,
(2) the proportion method, also used in Europe and popular in America,
(3) the resistance ratio method used in the UK,
(4) the radiometric method.

Different techniques, are, however, used for pyrazinamide susceptibility tests.

The absolute concentration method

Carefully measured amounts of standardized inocula are placed on control media and on media containing varying amounts of drugs and the lowest concentration that will inhibit all, or nearly all, of the growth is reported. This method gives different results in different laboratories. It is not possible to use a medium that must be heated after the addition of the drug (e.g. egg medium) because some drugs are partially heat labile and heating times are rarely constant even in the same laboratory. Middlebrook 7H10 or 7H11 media are used but this method has not found much favour in the UK.

The proportion method

Several dilutions of the inoculum are made and media containing no drug and standard concentration of drugs are inoculated. The number of colonies growing on the control from suitable dilutions of the inoculum is counted and also the number growing on drug-containing medium receiving the same inoculum. Comparison of the two shows the proportion of organisms that are resistant. This is usually expressed as a percentage.

This method is popular in the USA and in Europe but it is technically very difficult and as it is usually done in petri dishes we regard it as highly hazardous. There are also more risks attacked to standardizing the inocula than with the resistance ratio method.

The resistance ratio method

The minimal inhibitory concentrations of the drug that inhibit test strains are divided by those which inhibit control strains to give the resistance ratios. Ratios of 1 and 2 are considered susceptible those of 4 or more resistant (see below). These results usually correlate with the clinical findings. This method is used extensively in the UK and is described below.

Each of the above methods may be used for 'direct' tests on sputum homogenates that contain enough tubercle bacilli to give positive direct films as well as for 'indirect' tests on cultures. The latter are more reliable and reproducible.

For background information about these tests, see Canetti et al. (1969) and Vestal (1975). For a discussion on the design of resistance ratio tests, see Marks (1961) and Collins et al. (1985b).

Technique of resistance ratio method

Dilutions are incorporated in Lowenstein–Jensen medium, which is then inspissated. As some of these drugs are affected by heat, it is essential that the inspissation procedure is standardized. The inspissator must have a large circulating fan so that all tubes are raised to the same temperature in the same time. The load (i.e. number of tubes) and the time of exposure must be constant. To ensure that the drug-containing medium is not overheated, the machine should be raised to its correct temperature *before* it is loaded and racks can be devised that allow loading in a few seconds. A period of 45 min at 80°C is usually sufficient to coagulate this medium and no further heat is necessary if it is prepared with a reasonable aseptic technique.

Control strains
The resistance ratio method compares the MIC of the unknown strain with that of control strains on the same batch of medium.

Some workers use the H37Rv strain of *M. tuberculosis* as the control strain, but the susceptibility of this to some drugs does not parallel that of wild tubercle bacilli and may give misleading ratios. It is better to use the modal resistance method of Leat and Marks (1970). The unknown strains are compared with the modal resistance (i.e. that which occurs most often) of a number of known susceptible strains.

Drug concentrations
Each laboratory must determine its own range, as these will vary slightly according to local conditions. Initially use those suggested in Table 29.4 and inoculate at least 12 sets with known susceptible organisms to arrive at a baseline for future work. A drug-free control slope must be included.

Stock solutions of the drugs are conveniently prepared as 1% solutions in water (except for rifampicin which should be dissolved in formdimethylamide (*caution*)). All stock solutions keep well at −4°C.

Table 29.4 TB susceptibility tests: suggested concentrations of drugs in Lowenstein–Jensen medium

	Final concentration (µg/ml)						
INAH	0.007	0.015	0.03	0.06	0.125	0.25	0.5
Ethambutol	0.07	0.15	0.31	0.62	1.25	2.5	5.0
Rifampicin	0.53	1.06	3.12	6.25	12.5	25	50
Streptomycin	0.53	1.06	3.12	6.25	12.5	25	50

These are for the preliminary titration. Choose six that give confluent or near confluent growth in the two lowest and no growth in the next four.

Bacterial suspension

Smooth suspensions must be used. Large clumps or rafts of bacilli give irregular results and make readings difficult.

Sterilize 7 ml (bijou) screw-capped bottles containing a wire nail (shorter than the diameter of the bottle), a few glass beads and 2 ml of phosphate buffer (13.3 g of anhydrous Na_2HPO_4 and 3.5 g of KH_2PO_4 in 2 litres of water; pH 7.4). Into each bottle place a scrape of growth equal to about three or four large colonies and place the bottle on a magnetic stirrer for 3–4 min. Allow to stand for a further 5 min for any lumps to settle and use the supernatant to inoculate culture media.

Methods for inoculation

Place the bottle containing the suspension on a block of Plasticine at a suitable angle. Inoculate each tube with a 3-mm loopful of the suspension, withdrawn edgewise. Plastic disposable 10 µl loops are best.

A better and quicker method uses an automatic micropipette (the Jencon Micro-Repette is ideal) fitted with a plastic pipette tip. Deposit 10 µl of the suspension near to the top of each slope of medium. As it runs down it spreads out to give a suitable inoculum. The plastic tips are sterilized individually in small, capped test-tubes and after use are deposited in a jar containing glutaraldehyde, which is subsequently autoclaved. The tips may be re-used.

Incubation

Incubate at 37°C for 18–21 days.

Reading results

Examine tubes with a hand lens and record as follows: confluent growth, CG; innumerable discrete colonies, IC; between 20 and 100 colonies, +; less than 20 colonies, 0.

The drug-free control slope must give growth equal to CG or IC. Less growth invalidates the test. The *modal* resistance, i.e. the MIC occurring most frequently in the susceptible control strains, should give readings of CG or IC on the first two tubes (Table 29.5). The range must therefore be adjusted to give this result during the preliminary exercises. Once the range is set it is seldom necessary to vary it by more than one dilution.

Interpretation

The resistance ratio is found by dividing the MIC of the test strain by the modal MIC of the control strains. When the readings are all CG or IC, this is easy, but when there are tubes showing +, care and experience may be required in interpretation. In general, a resistance ratio of two or less can be reported as *susceptible*, 4 as *resistant* and 8 as *highly resistant*. A ratio of 3 is *borderline* except for

Table 29.5 'Modal resistance'

Tube no.	Drug concentration					
	1	2	3	4	5	6
Strain A	CG	CG	0	0	0	0
B	CG	IC	0	0	0	0
C	CG	CG	+	0	0	0
D	IC	IC	0	0	0	0
E	CG	CG	+	0	0	0
Mode	CG	CG	0	0	0	0

ethambutol, when it probably indicates resistance. Mixed *susceptible* and *resistant* strains occur. Examples of these findings are shown in Table 29.6.

The radiometric method

The BACTEC instrument (p.349) may be used to obtain rapid results, e.g. within 7 days (Laszlo *et al.*, 1983). Appropriate amounts of each drug are added to the vials, followed by the inoculum.

Pyrazinamide sensitivity tests

Pyrazinamide acts upon tubercle bacilli in the lysosomes, which have a pH of about 5.2. For reliable susceptibility tests the medium should therefore be at this pH. Tubercle bacilli do not grow well on egg medium at pH 5.2 and agar medium (e.g. Middlebrook 7H11) are frequently used. We have had good and consistent results with Yates' (1984) modification of Marks (1964) stepped pH method. This is a Kirchner semisolid medium layered on to butts of Lowenstein–Jensen medium.

Pyrazinamide stock solution
Dissolve 0.22 g of dry powder in 100 ml of distilled water and sterilize by filtration or steaming.

Table 29.6 Interpretation of TB drug susceptibility

Tube no.	Drug concentration						Resistance ratio	Interpretation
	1	2	3	4	5	6		
Mode	CG	IC	0	0	0	0		
Strain A	CG	0	0	0	0	0	0.5	Susceptible
B	CG	CG	0	0	0	0	1	Susceptible
C	CG	IC	+	0	0	0	1	Susceptible
D	CG	CG	CG	0	0	0	2	Susceptible
E	CG	CG	CG	+	0	0	3	Borderline[a]
F	CG	CG	CG	IC	0	0	4	Resistant
G	CG	CG	CG	CG	IC	0	8	Highly resistant
H	CG	CG	CG	CG	CG	CG	16	Highly resistant
I	CG	CG	CG	+	+	+	–	Mixed susceptible and resistant

[a]Probably resistant with ethambutol

Solid medium

Use Lowenstein–Jensen medium containing only one half the usual concentration of malachite green (at an acid pH less dye is bound to the egg and the higher concentration of 'free' dye is inhibitory to tubercle bacilli). Adjust 600 ml of medium to pH 5.2 with N HC1. To 300 ml add 9 ml of water (control); to the other 300 ml add 9 ml of the stock pyrazinamide solution (test). Tube each batch in 1-ml amounts and inspissate upright, to make butts, at 87°C for 1 h.

Semisolid medium

Add 1 g of agar and 3 g of sodium pyruvate to 1 litre of Kirchner medium. Adjust pH to 5.2 with 5N HCl. To 500 ml add 15 ml of water (control); to the other 500 ml 500 ml add 15 ml of the stock pyrazinamide solution. Steam both bottles to dissolve the agar. Cool to 40°C and add to each bottle 40 ml of OADC supplement.

Final medium

Layer 2 ml of the control semisolid medium on butts of the control Lowenstein–Jensen medium. Do the same with the test media. Store in a refrigerator and use within 3 weeks. (It may keep longer; we have never tried it.)

Bacterial suspensions

Two inocula are required. Use one as described above for other susceptibility tests and also a 1:10 dilution of it in sterile water.

Inoculation

With a suitable pipette (e.g. the Jencons MicroRepette and plastic tip) add 20 μl (approximately) of (a) the undiluted, and (b) the diluted suspension to pairs of test and control media.

Interpretation

Two concentrations of inoculum are used because

(1) some strains require a heavy inoculum to give growth at an acid pH, even in the control medium, and
(2) a heavy inoculum of some other strains may overcome the inhibitory action of pyrazinamide.

Colonies of tubercle bacilli should be distributed throughout the semisolid medium in one or both controls. If there is no growth in the test bottle from either the heavy or light inoculum report the strain as *susceptible*. If there is growth in the test bottle that received the heavy inoculum but not in that which received the diluted suspension, report the strain as *susceptible*. If there is growth in both test bottles, comparable with that in their respective controls report the strain as *resistant*.

NB Human strains of tubercle bacilli from untreated patients are invariably susceptible. 'Classic bovine' strains and all other species of mycobacteria found in clinical material are naturally resistant.

References

Canetti, G. and seven others (1969) Advances in techniques of testing mycobacterial drug sensitivity and the use of sensitivity tests in tuberculosis control programmes. *Bulletin of the World Health Organisation,* **41,** 21–43

Collins, C. H., Yates, M. D. and Grange, J. M. (1982) Subdivision of *Mycobacterium tuberculosis* into five variants for epidemiological purposes: methods and nomenclature. *Journal of Hygiene (Cambridge),* **89,** 235–242

Collins, C. H., Grange, J. M. and Yates, M. D. (1985a) *Mycobacterium marinum* infections in man. *Journal of Hygiene (Cambridge),* **94,** 135–149

Collins, C. H., Grange, J. M. and Yates, M. D. (1985b) *Organization and Practice in Tuberculosis Bacteriology.* Butterworths, London, pp. 90–97

Damato, J. J., Collins, M. T., Rothlauf, M. V. and MacLatchy, J. K. (1983) Detection of mycobacteria by radiometric and standard plate procedures. *Journal of Clinical Microbiology,* **17,** 1066–1073

Laszlo, A. and four others. (1983) Conventional and radiometric drug susceptibility testing of *Mycobacterium tuberculosis* complex. *Journal of Clinical Microbiology,* **18,** 1335–1339

Leat, J. L. and Marks, J. (1970) Improvements in drug sensitivity tests on tubercle bacilli. *Tubercle,* **51,** 68–73

Marks, J. (1961) The design of sensitivity tests on mycobacteria. *Tubercle,* **42,** 314–316

Marks, J. (1964) A 'stepped pH' technique for the estimation of pyrazinamide sensitivity. *Tubercle,* **45,** 47–50

Marks, J. (1972) Ending the routine guinea pig test. *Tubercle,* **53,** 31–34

Mitchison, D. A., Allen, B. J. and Manickavasagar, D. (1987) Selective Kirchner medium for the culture of specimens other than sputum for mycobacteria. *Journal of Clinical Pathology,* **36,** 1357–1361

Roberts, G. D. *et al.* (1983) Evaluation of the BACTEC radiometric method for the recovery for *Mycobacterium tuberculosis* from acid-fast smear positive specimens. *Journal of Clinical Microbiology,* **18,** 689–696

Vestal, A. L. (1975) *Procedures for the Isolation and Identification of Mycobacteria,* DHEW (CDC) 75-8230, Government Printing Office, Washington DC

Yates, M. D. (1984) The differentiation and epidemiology of the tubercle bacilli and a study on the identification of other mycobacteria. MPhil Thesis, University of London

Chapter 30

Actinomyces, nocardia, actinomadura and streptomyces

Actinomyces, nocardias and streptomyces are Gram-positive mycelial bacteria. The mycelium may branch or fragment into bacillary forms. Actinomyces are anaerobic, microaerophilic or capnophilic. Nocardias and streptomyces are aerobic. Species of medical and veterinary importance occur in pus and discharges as colonies or granules.

Actinomyces

Isolation

Wash pus gently with saline in a sterile bottle or petri dish. Look for 'sulphur granules', about the size of a pinhead or smaller. Aspirate granules with a pasteur pipette and crush one between two slides and stain by the Gram method. Look for mycelium and large club forms radiating from the centre of the granule.

Crush granules with a sterile glass rod in a small sterile tube containing a drop of broth. Inoculate blood agar plates, brain heart infusion agar and enriched thioglycollate medium. These may be made selective for Gram-positive non spore-bearers by adding 30 µg/ml nalidixic acid and 10 µg/ml metronidazole. Do cultures in triplicate and incubate:

(1) anaerobically plus 5% carbon dioxide,
(2) in a 5% carbon dioxide atmosphere, and
(3) aerobically at 37°C for 2–7 days.

Subculture any growth in the broth media to solid media and incubate under the same conditions. Look for colonies resembling 'spiders' or 'molar teeth'.

Identification

Do catalase tests on colonies from anaerobic or microaerophilic cultures that show colonies of Gram-positive coryneform or filamentous organisms. Actinomyces are catalase negative; corynebacteria and bifidobacteria are usually catalase positive. It is often difficult to distinguish actinomyces from bifidobacteria, arachnia and propionibacteria by simple tests. Gas-liquid chromatography (GLC) may be necessary. The spot indole test applied to colonies is useful; *Propionibacterium acnes* is usually positive.

Table 30.1 Species of *Actinomyces*

Species	Nitrates reduced	Acid from			Starch hydrolysis
		Glucose	*Mannitol*	*Xylose*	
A. bovis	−	+	−	v	+
A. israelii	v	+	+	+	−
A. naeslundii	+	+	−	−	−
A. odontolyticus	+	+	−	v	−
A. meyer	−	+	−	+	−

Subculture for nitratase test and test for acid production from glucose, mannitol and xylose and for starch utilization (see Table 30.1). The API ZYM system may be useful in distinguishing between actinomyces and related genera.

Species of *Actinomyces*

Actinomyces bovis
Colonies at 48 h are pinpoint and smooth, later becoming white and shining with an entire edge. Spider and molar tooth colonies are rare. In thioglycollate broth, growth is usually diffuse but occasionally there are colonies that resemble bread-crumbs. Microscopically the organisms are coryneform, rarely branching.

Nitrates are not reduced, acid is produced from glucose but not from mannitol or sometimes xylose. Starch is hydrolysed.

This is a pathogen of animals, usually of bovines, causing lumpy jaw.

A. israelii
Colonies at 48 h are microscopic, with a spider appearance, later becoming white and lobulated with the appearance of molar teeth. In thioglycollate broth, there are distinct colonies with a diffuse surface growth and a clear medium. The colonies do not break when the medium is shaken. Microscopically the organisms are coryne-form, with branching and filamentous forms.

Nitrates may be reduced, acid is produced from glucose, mannitol and xylose but starch is not hydrolysed.

This species causes human actinomycosis and may be responsible for intrauterine infections in women fitted with IUCDs.

A. naeslundii
Colonies at 48 h are similar to those of *A. israelii;* spider forms are common, but not molar tooth forms. In thioglycollate broth, growth is diffuse and the medium is turbid. Microscopically the organisms are irregular and branched, with mycelial and diphtheroid forms.

Nitrates are reduced, acid is produced from glucose but not from mannitol or xylose and starch is not hydrolysed. This is a facultative aerobe.

It is not known to be a human pathogen but has been found in human material.

A. odontolyticus
Colonies at 48 h are 1–2 mm and grey but they may develop a deep reddish colour after further incubation. CO_2 is required for growth. The organisms are coryne-form, rarely branched. Nitrate is reduced; acid is produced from glucose, some-times from xylose but not from mannitol. Starch is not usually hydrolysed.

The normal habitat is the human mouth but the organisms have been isolated from the tear ducts.

A. meyer

Colonies at 48 h are pinpoint, greyish white and rough. CO_2 is required for both aerobic and anaerobic growth. The organisms are short and coryneform but branching is rare. Nitrate is not usually reduced; acid is produced from glucose and xylose but not from mannitol; starch is not hydrolysed.

The normal habitat is the human mouth but the organism has been isolated from abscesses.

Nocardia, Actinomadura and Streptomyces

Isolation

Wash pus to recover granules as described on p.366. Treat sputum by the 'mild' method used for culturing mycobateria (p.348) or use one of the sputum digesting digesting agents. Culture on blood agar and on duplicate Lowenstein–Jensen (LJ) slopes. Incubate at 37°C for several days. Incubate one of the LJ slopes at 45°C at which nocardias may grow while other organisms are discouraged.

Identification

Subculture aerobic growth on blood agar and incubate at 37°C for 3–10 days. Colonies vary in size and may be flat or wrinkled, sometimes 'star shaped'. They may be pigmented (pink, red, green, brown or yellow) and often covered with a whitish, downy or chalky aerial mycelium. Colonies of nocardias are usually on the surface; those of streptomyces may be embedded in the medium.

Examine Gram- and ZN-stained films (minimum decolorization with the latter). Do slide culture (see below) and inoculate lysozyme broth: dissolve 20 mg lysozyme in 2 ml of 50% ethanol in water and add 50 μl to 2 ml of nutrient broth at pH 6.8). Test for the hydrolysis of casein, xanthine, hypoxanthine and tyrosine (see Table 30.2). Test also for acid production from arabinose and xylose. The API ZYM system may be used to distinguish between nocardias, streptomyces and related organisms.

Slide cultures

Melt a tube of malt agar, dilute with an equal volume of distilled water and cool to 45°C. Suck about 1 ml of this mixture into a pasteur pipette, followed by a drop of a

Table 30.2 *Nocardia, Rhodococcus, Actinomadura* **and** *Streptomyces*

	Acid fast	Aerial mycelium	Lysozyme	Hydrolysis of		
				Casein	Xanthine	Tyrosine
N. asteroides	+[a]	+	R	−	−	−
N. brasiliensis	+[a]	+	R	+	−	+
N. otitidiscaviarum	+[a]	+	R	−	+	−
Actinomadura	−	+	S	+	v	+
Streptomyces	−	+	S	v	v	v
Rhodococcus	−	−	S	−	−	−

R, resistant; S, sensitive; v, variable
[a]Rarely complete. Acid-fast elements may be few in number or absent.

thin suspension of the organisms. Run this over the surface of a slide in a moist chamber. No cover-glass is needed. Examine daily with a low-power microscope until a mycelium is seen. Spores may be observed as powdery spots on the surface of the medium.

Partially acid-fast, Gram-positive mycelium which does not branch and which fragments early in slide cultures may be tentatively identified as nocardia. Non-acid fast Gram-positive mycelium which branches with aerial mycelium and conidiophores abstricted in chains suggest streptomyces.

Nocardia

Two (possibly three) species of medical importance. Acid-fast elements may be rare or occur in cultures at different times and are best observed on Middlebrook media. Nocardias are resistant to lysozyme and may be differentiated by hydrolysis of casein, xanthine and tyrosine (Table 30.2).

N. asteroides
The most common; causes severe pulmonary infections, brain abscesses and occasionally cutaneous infections.

N. brasiliensis
Mostly confined to North America and the southern hemisphere and usually causes cutaneous infections but may be associated with systemic disease.

A. otitidiscaviarum (N. caviae)
Rare cause of human infections.

Actinomadura and Streptomyces

A. madurae causes madura foot, *A. pelleteri* and *S. somaliensis* cause mycetomas. They are difficult to distinguish bacteriologically although *A. madurae* is said to hydrolyse aesculin while *S. somaliensis* does not. Unlike the saprophytic streptomyces they do not produce acid from lactose and xylose. Cell wall analysis is necessary to sort out these and associated species.

Rhodococcus

Several species, none of which are pathogenic for humans although one, *R. equi*, usually known as *Corynebacterium equi*, is a pathogen of horses (p.321). Rhodococci are important only in that they may be confused with nocardias, actinomaduras and *S. somaliensis*.

For more information on these organisms see Wilson (1983); Goodfellow (1986); Goodfellow and Lechevalier (1986); Gordon (1988) and Collins *et al.* (1988).

References

Collins, C. H., Uttley, A. H. C. and Yates, M. D. (1988) Presumptive identification of nocardias in a clinical laboratory. *Journal of Applied Bacteriology*, **65**, 55–59

Goodfellow, M. (1986) Genus *Rhodococcus* Zopf 1891. In *Bergey's Manual of Determinative Bacteriology*, Vol 2 (eds P. H. A. Sneath, N. S. Mair, M. E. Sharpe and J. G. Holt), William and Wilkins, Baltimore, pp. 1472–1475, 1460–1514

Goodfellow, M. and Lechevalier, M. P. (1986) Genus *Nocardia* Trevisan 1889. In *Bergey's Manual of Determinative Bacteriology*, Vol. 2 (eds P. H. A. Sneath, N. S. Mair, M. E. Sharpe and J. G. Holt), William and Wilkins, Baltimore, pp. 1460–1514

Gordon, M. A. (1988) Aerobic pathogenic Actinomycetaceae. In *Manual of Clinical Microbiology*, 4th edn (eds. E. H. Lennette, A. Balows, W. J. Hausler and H. J. Shadomy), American Society for Microbiology, Washington, pp. 249–262

Wilson, G. S. (1983) *Actinomyces, Nocardia* and *Actinobacillus*. In *Topley and Wilson's Principles of Bacteriology and Immunity*, Vol. 2 (ed. M. T. Parker), Edward Arnold, London, pp. 40–43

Chapter 31

Spirochaetes

Three genera that are of medical importance are considered briefly here: *Borrelia, Treponema, Leptospira.*

Borrelia

Borrelia duttonii and *B. recurrentis* cause relapsing fever, transmitted by lice and ticks.

Lyme disease (inflammatory arthropathy), first described in Lyme, Connecticut, USA and now known in Europe, is caused by *B. burgdorferi* (Burgdorfer, 1984, 1985). It is a tick-borne zoonosis and small woodland rodents and wild deer are reservoirs. Infection is also transmitted to domesticated and farm animals. Unlike the relapsing fever spirochaetes, this species is antigenically stable.

Identification

Take blood during a febrile period. For the relapsing fever species examine wet preparations by dark field using a high power dry objective. Stain thick and thin films with Giemsa stain.

Method
Fix smear in 10% methyl alcohol for 30 s. Stain with 1 part Giemsa stock stain and 49% Sorensen's buffer at pH 7 in a Coplin jar for 45 min. Wash off Sorensen's buffer. Dry in air and examine.

Inoculate two laboratory rats with 1–2 ml of fresh or refrigerated defibrinated blood and examine the animal's blood daily between the second and seventh days.

Morphology
They form shallow, coarse, irregular, highly motile coils 0.25–0.5 × 8–16 μm.

Lyme disease
Diagnosis is made by fluorescence antibody methods (Technicon kit), ELISA or silver staining of biopsies.

Treponema

The species are *T. pallidum, T. pertenue, T. carateum* and *T. vincentii (Borrelia vincentii).*

Identification

In suspected syphilis
Examine exudate (free from blood and antiseptics) from lesion by dark field or fluorescence microscopy. For detailed methods see Sequira (1987).

T. pallidum
Is the causative organism of syphilis, transmitted by direct contact.

Morphology
It forms tightly wound slender coils $0.1–0.2 \times 6–20$ μm with pointed ends each having three axial fibrils. They are sluggishly motile with drifting flexuous movements.

In suspected Vincent's angina
Stain smears with dilute carbol fuchsin. Large numbers of spirochaetes are seen together with fusobacteria.

T. vincentii
Is the causative organism of Vincent's angina (ulcerative lesions of the mouth or genitals) and pulmonary infections. It can be spread by direct contact.

Culture
Inoculate peptone yeast extract medium with added serum or ascitic fluid under anaerobic conditions at 37°C. Small, while colonies 12–15 mm in diameter having the appearance of a slight haze are visible after 2 weeks.

Morphology
It is a loosely wound single contoured spirochaete $0.2–0.6$ μm $\times$ 7–18 μm.

Other treponemes

T. pertenue
This organism causes yaws and is transmitted by direct contact. It is morphologically indistinguishable from *T. pallidum*. Laboratory tests are unhelpful.

T. carateum
Causes pinta. The mode of spread is, in common with yaws, by direct contact. It is morphologically indistinguishable from *T. pallidum*. Diagnosis is by silver impregnation of tissue.

Leptospira

Identification is important only for epidemiological purposes. *Leptospira interrogans* (19 subgroups) is primarily an animal parasite but in man causes an acute febrile illness with or without jaundice, conjunctivitis and meningitis (Weil's disease). *L. biflexa* is non-pathogenic.

Isolation and identification

Examine blood during the first week of illness and urine thereafter. Add 9 parts of blood to 1 part of 1% sodium oxalate in phosphate buffer at pH 8.1 and centrifuge 15 min at 1500 rev/min. Examine clear plasma under dark-field microscopy using low power magnification. If this is negative, centrifuge the remainder of the plasma at 10 000 rev/min for 20 min and examine the sediment. Direct films rarely show leptospira either in dark field or Giemsa-stained preparations. Centrifuge urine and examine deposit by dark field within 15 min.

Morphology
Leptospiras are short, fine, closely wound spirals, 0.25×6–20 µm, resembling a string of beads. The ends are bent at right angles to the main body to form hooks.

Culture
The methods given here are adapted from those of Waitkins (1985).

Add 2 drops of fresh blood, CSF or urine (at pH 8) to 5 ml of EMJH/5FU medium (p.69) and make five serial dilutions in the same medium to dilute out antibody. Incubate at 30°C, examine daily for 1 week, then weekly for several weeks. Use dark-field, phase contrast or fluorescence microscopy.

Identification
If leptospiras are seen subculture to EMJH medium and incubate at 30°C and 13°C. Subculture also to EMJH medium containing 225 mg azoguanine per litre.

L. interrogans grows at 30°C, not at 13°C and not in azoguanine medium.

L. biflexa grows at 13 and 30°C and also in azoguanine medium.

Serological diagnosis

Macroslide agglutinations may be done with patients' sera and (commercial) genus-specific antigen but further agglutination tests are best done in reference laboratories.

There is a rapid test kit (Leptese: Bradsure Biologicals).

For more information about leptospires, culture, serology and epidemiology see Waitkins (1985).

References

Burgdorfer, W. (1984) Discovery of the Lyme spirochaete and its relationship to tick vectors. _Yale Journal of Biology and Medicine_, **57**, 165–168

Burgdorfer, W. (1985) _Borrelia_. In _Manual of Clinical Microbiology_, 4th edn (eds E. H. Lennette, A. Balows, W. J. Hauser and H. J. Shadomy), Association of American Microbiologists, Washington, pp. 154–175

Sequira, P. J. L. (1987) Syphilis. In _Sexually Transmitted Diseases_ (ed. A. E. Jephcott), Public Health Laboratory Service, London, pp. 6–22

Waitkins, S. A. (1985) Leptospiras and leptospirosis. In _Isolation and Identification of Micro-organisms of Medical and Veterinary Importance_ (eds C. H. Collins and J. M. Grange), Society for Applied Bacteriology Technical Series No. 21, Academic Press, London, pp. 251–296

Chapter 32

Yeasts

Yeasts are fungi whose main growth form is unicellular and which usually replicate by budding. Many can also grow in the hyphal form and the distinction between a yeast and a mould is one of convention only. Some yeast-like fungi such as *Acremonium, Geotrichum* and the 'black yeasts' *Exophiala, Aureobasidium* and *Phialophora* are traditionally excluded from texts on yeasts (see Chapter 33).

The yeasts are not a natural group. Many are Ascomycotina, a few are Basidiomycotina and others are asexual forms of these groups (Deuteromycotina or Fungi Imperfecti). Some genera are even 'convenience groups': for example both *Candida* and *Torulopsis* contain asexual forms of Ascomycotina and Basidiomycotina. In addition, as these two genera are distinguished only by their ability to produce pseudohyphae some workers group them together (as *Candida*).

Identification

The methods given in Chapter 8 will yield information about morphology, sugar fermentation and assimilation of carbon and nitrogen sources. The morphological key (Table 32.1) will allow most yeasts isolated in clinical and food laboratories to be identified to genus level. Specific identification may then be made by reference to Table 32.2 (see also Campbell *et al.*, 1985).

Candida
Candidas are commensals in the human gut. The principal species of medical importance is *Candida albicans*, which causes infections of skin and mucous membranes, particularly in immunosuppressed patients.

C. parapsilosis is a normal skin commensal and frequently infects nails. It is also a common cause of candida endocarditis.

Other species rarely causing human disease include *C. tropicalis, C. kefyr (C. pseudotropicalis), C. krusei, C. guillermondii, C. lusitaniae* and *C. zeylanoides*.

All produce creamy, white, smooth colonies on malt or peptone agar, except *C. krusei* which has a ground glass appearance (see Table 32.2).

Cryptococcus
Like other encapsulated yeasts *(Rhodotorula, Trichosporon)*, these are the asexual forms of Basidiomycetes although the sexual forms are not normally encountered.

All species are large, round, non-fermenting, urease-positive yeasts, found on fruit and in bird droppings.

Table 32.1 Key to identification of genera of yeasts

Budding cells only (corn meal and coverslip preparation); no ascospores

(1) Cells small (2-3 μm), bottle-shaped, with broad base to daughter bud   *Pityrosporum*
(2) Cells oval with prominent lateral 'spur'; colony pinkish, depositing mirror
    image of colony on lid of inverted petri dish   *Sporobolomyces*
(3) Cells not as in 1 or 2
    (a) Urease negative   *Torulopsis*
    (b) Urease positive; colonies pink/red; cells large, encapsulated   *Rhodotorula*
    (c) Urease positive; colonies white or cream; cells large; encapsulated   *Cryptococcus*

Budding cells, some with ascospores

(1) Ascospores liberated from parent cells, kidney-shaped or elongate   *Kluyveromyces*
(2) Ascospores liberated, round with or without disc-like flange; nitrate
    assimilated   *Hansenula*
(3) Ascospores as in 2; nitrate not assimilated   *Pichia*
(4) Ascospores retained in parent cells, round, smooth   *Saccharomyces*
(5) Ascospores retained, round, warty and ridged   *Debaromyces*

Budding cells borne on pseudomycelium; no ascospores   *Candida* (most
                                                                              spp.)

Budding cells borne on pseudomycelium which grows out to true mycelium after
3–4 days

(1) Chlamydospores present   *C. albicans*
(2) No chlamydospores   *C. tropicalis*

Extensive true mycelium, fragmenting to form arthrospores

(1) Budding on short side branches   *Trichosporum*
(2) No budding   *Geotrichum*

Table 32.2 Biochemical differentiation of some common yeasts

	Fermentation of				Assimilation of								
	glu	ma	su	la	glu	ma	su	la	mn	ra	ce	er	NO₃
Candida albicans	+	+	−	−	+	+	+	−	+	−	−	−	−
C. tropicalis	+	+	+	−	+	+	+	−	+	−	+	−	−
C. kefyr	+	−	+	+	+	−	+	+	+	+	+	−	−
C. parapsilosis	+	−	−	−	+	+	+	−	+	−	−	−	−
C. guillermondii	+	−	+	−	+	+	+	−	+	+	+	−	−
C. krusei	+	−	−	−	+	−	−	−	−	−	−	−	−
Torulopsis glabrata	+	−	−	−	+	−	−	−	−	−	−	−	−
T. candida	±	−	±	−	+	+	+	+	+	+	+	±	−
Cryptococcus neoformans	−	−	−	−	+	+	+	−	+	+ʷ	+ʷ	±	−
C. albidus	−	−	−	−	+	+	+	±	+	+ʷ	+	−	+
C. laurentii	−	−	−	−	+	+	+	+	−	−	−	−	−
Trichosporum cutaneum	−	−	−	−	+	+	+	+	±	±	±	±	−
T. capitatum	−	−	−	−	+	−	−	−	−	−	−	−	−
Saccharomyces cerevisiae	+	+	+	−	+	+	+	−	±	±	−	−	−

glu, glucose   ra, raffinose   +, gas produced or growth occurs
ma, maltose   ce, cellobiose   −, no gas produced or no growth occurs
su, sucrose   er, erythritol   w, week
la, lactose   NO₃, nitrate
mn, mannitol

One species, *C. neoformans*, is pathogenic for humans, causing meningitis (which may be fulminant or slowly progressive and is common in AIDS patients) and subcutaneous or deep granulomata.

In advanced infections agglutination tests may be negative because the capsular material is poorly metabolized by the body, and the accumulation of it causes a high dose immunological tolerance. This antigen can be detected in serum or in CSF by the agglutination of latex sensitized with γ-globulin from an immune rabbit serum as described on p.121. For details, see MacKenzie *et al.* (1980).

Torulopsis

This is a large group of asexual yeasts. The commonest species in clinical mycology is *T. glabrata*, a small-celled species commensal on the skin. It occasionally causes peritonitis and urinary tract infections in debilitated patients. *T. candida* is a large-celled type, occasionally found in similar sites.

Rhodotorula

These are frequent skin contaminants. They form red or orange colonies on malt or glucose peptone agar. Morphologically they resemble cryptococci and may be capsulated. They are of no clinical significance.

Spororobolomyces

These yeasts live on leaf surfaces and reproduce by ballistospores which are ejected into the air. The salmon pink colonies of *S. roseus* produce mirror images on the lids of petri dishes if left undisturbed. They are of no clinical significance.

Trichosporon

T. beigelli is the one most frequently isolated in clinical work. It is a skin contaminant but occasionally causes deep infections in immunocompromised patients. *T. capsitatum* causes 'white piedra' in which masses of yeasts grow on the outer surfaces of axillary and other hair which is kept permanently moist.

The colonies of these species are whitish, wrinkled, tough, and often slightly 'hairy' on the surface, due to the presence of aerial hyphae.

Sexual yeasts

Saccharomyces

This genus contains the brewing ('pitching') and baking yeasts. *S. cerevisiae* is the 'top yeast' used in making beer, and *S. carlsbergensis* the 'bottom yeast' used in making lager. There are many special strains and variants used for particular purposes.

S. pastorianus, S. rouxii and S. mellis

A bottom yeast with long sausage-shaped cells, which produces unpleasant flavours in beer. *S. rouxii* and *S. mellis* are 'osmophilic yeasts' which will grow in concentrated sugar solution (*cf.* aspergilli of the glaucus group) and spoil honey and jam.

Pichia and Hansenula

Occur as contaminants in alcoholic liquors. They can use alcohol as a source of carbon, and are a nuisance in the fermentation industry. They grow as a dry pellicle

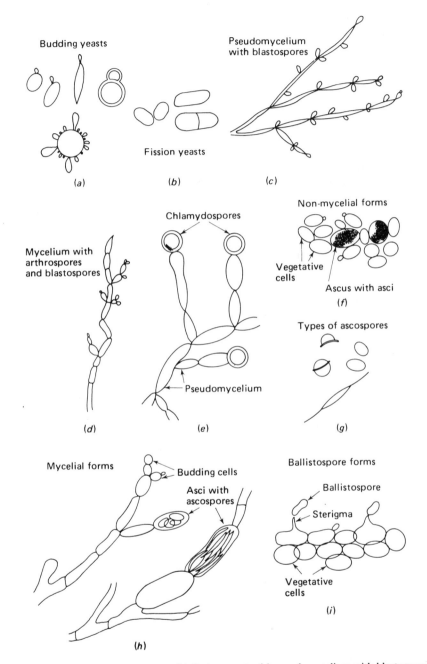

Figure 32.1 (a) Budding yeasts; (b) fission yeasts; (c) pseudomycelium with blastospores; (d) mycelium with arthrospores and blastospores; (e) pseudomycelium with chlamydospores; (f) yeasts with asci; (g) types of ascospores; (h) mycelial forms with asci; (i) ballistospore forms

on the surface of the substrate. Some species occur as the 'flor' in sherry and in certain French wines, to which they give a distinctive flavour. They also cause pickle spoilage, particularly in low-salt (10–15%) brines.

Debaromyces

These are also 'pellicle forming' yeasts. They are found in animal products such as cheese, glue and rennet, and are also responsible for spoilage in high-salt (20–25%) pickle brines.

Apiculate yeasts

These lemon-shaped yeasts are common contaminants in food. Perfect forms are placed in the genus *Hanseniaspora* and imperfect forms in the genus *Kloeckera*.

References

Campbell, C. K., Davis, C. and Mackenzie, D. W. R. (1985) Detection and isolation of pathogenic fungi. In *Isolation and Identification of Micro-organisms of Medical and Veterinary Importance* (eds C. H. Collins and J. M. Grange), Society for Applied Bacteriology Technical Series No. 21, Academic Press, London, pp. 329–343

Campbell, M. C. and Stewart, J. L. (1980) *The Medical Mycology Handbook,* J. Wiley & Sons, New York and Chichester

Kreger-Van RIJ (ed.) (1984) *The Yeasts, A Taxonomic Study,* 3rd edn, Elsevier, Amsterdam

Mackenzie, D. W. R., Philpot, C. M. and Proctor, A. G. J. (1980) *Basic Serodiagnostic Methods for Diseases Caused by Fungi.* Public Health Laboratory Service Monograph No. 12, HMSO, London

Chapter 33

Common moulds

Many common moulds occur as contaminants on ordinary culture media. Most of them are saprophytes, causing spoilage of food and other commodities; others are plant pathogens. Some are described in Chapter 34 because they are also recognized causes of human disease, especially in immunosuppressed patients. Each isolation of a mould from human material must therefore be assessed individually and some reliance must be placed on the observation of hyphae by direct microscopy. The preliminary identification guide (Table 33.1) therefore includes some of the 'pathogens' described in Chapter 34.

Alternaria and Ulocladium
Large genera of plant parasites and saprophytes. Without knowledge of the host plant speciation is almost impossible. Rare cases of subcutaneous granulomata in humans have been reported.

Colonies form a high, densely fluffy mat, white at first, becoming dark grey, deepest in colour in the centre. On some media the whole colony is black, with less aerial growth. Spores are large, brown pigmented, club shaped, multicellular with some septa longitudinal or oblique. In *Alternaria* (Figure 33.1a) the spores are in chains: in *Ulocladium* they are single or only occasionally joined together. A tape impression mount will demonstrate this difference.

Arthrinium
Common contaminants of no medical importance. Most species are from plant material, on which they look quite unlike the *in vitro* growth.

The growth on laboratory media is dense, vigorous and at first pure white in colour. The centres of older cultures become grey because of the abundance of spores which are unicellular, rounded and black.

Chaetomium
Saprophytes on plant material. Some are troublesome pests of paper and cellulosic materials. There are no medical implications.

Colonies are flat with low, often sparse aerial growth, with black or greenish discrete bodies standing on the agar surface. Low power examination shows that these are covered with long black spines (Figure 33.1b). Under the high power these are seen to be straight, spiral or irregularly branched according to species. Inside the black mass is a definite cell wall enclosing many dark brown, ovoid spores.

Table 33.1 Colour guide to mould cultures

Black or dark brown mycelium spores or both

Alternaria	Aspergillus (niger)
Arthrinium	Cladosporium
Aureobasidium	Exophiala
Botrytis	Fonseceae
Chaetomium	Madurella
	Phialophora
	Sprorothrix

Some shade of green predominant

Trichoderma	Aspergillus
Chaetomium	Penicillium

Some shade of red predominant

Monascus	Trichophyton
Paecilomyces (lilacinus)	Acremonium
Aureobasidium	Fusarium

Colonies white or cream

Chrysosporium	Dermatophytes
Geomyces	Histoplasma
	Coccidioides
	Blastomyces
	Mucor

Colonies sandy-brown to khaki

Paecilomyces (varioti)	Aspergillus terreus
Scopuraliopsis	Epidermophyton floccosum

The fungi in the left-hand column are described in Chapter 33; those in the right-hand column in Chapter 34

Aureobasidium

Occurs on damp cellulosic materials and grows on painted surfaces.

Colonies are at first yeast-like, often mucoid, with pink, white and black areas. In time the black areas increase in size and the colony becomes drier and rough surfaced. There may be aerial mycelium.

Microscopy shows a bizarre mixture of oval to long budding yeasts and pigmented, irregularly-shaped hyphal cells. Under low power the edge of the colony shows yeasts clustered in round masses along the submerged hyphae. This is best seen in a slide culture or needle mount (Figure 33.1c).

A. pullulans is the commonest of the saprophytic species in this group.

Botrytis

A group of plant pathogens, of which the common *B. cinerea* has a very wide host range and is also a successful saprophyte.

Growth is vigorous, producing a high, grey turf which develops discrete black bodies (stroma) at the agar surface, especially at the edge of the culture vessel. The large, unicellular spores are produced in grape-like clusters on long, robust conidiophores (Figure 33.1d).

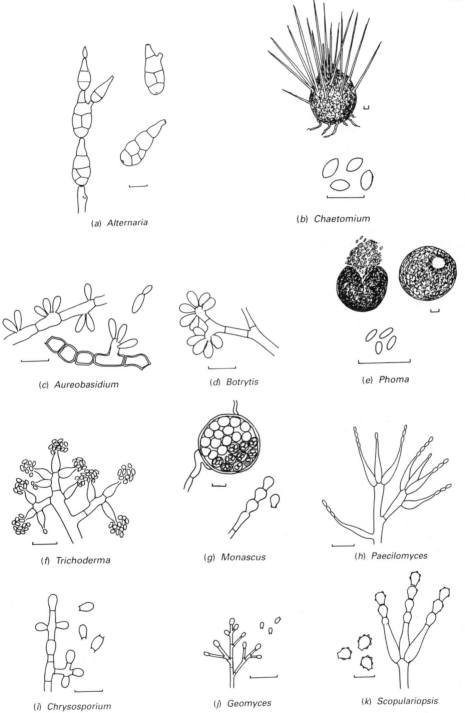

(a) *Alternaria*

(b) *Chaetomium*

(c) *Aureobasidium*

(d) *Botrytis*

(e) *Phoma*

(f) *Trichoderma*

(g) *Monascus*

(h) *Paecilomyces*

(i) *Chrysosporium*

(j) *Geomyces*

(k) *Scopulariopsis*

Figure 33.1 (a) *Alternaria;* (b) *Chaetomium;* (c) *Aureobasidium;* (d) *Botrytis;* (e) *Phoma;* (f) *Trichoderma;* (g) *Monascus;* (h) *Paecilomyces;* (i) *Chrysosporium;* (j) *Geomyces;* (k) *Scopulariopsis* (Bar = 10 μm)

Phoma

This is a representative of a large number of pycnidial genera, all plant parasites, any of which may be encountered as contaminants. It is included here because it is probably one of the commonest and many others were once classified with it.

Some species produce low, flat colonies with areas of black mixed with pink (spore masses). Others may give dense, fluffy growths with black or grey shading. Microscopically they all show pycnidia – spherical structures containing large numbers of small conidia, often released in a wet, worm-like mass through a round ostiole. These are best seen by examining the growing colony *in situ* under low power (Figure 33.1e).

Trichoderma

Saprophytes which produce several interesting metabolites.

Growth is white, rapidly spreading and cobweb-like, with irregular patches of dark green spores. These are abundant, small, ovoid, arising in ball-like groups on short right-angled branches (Figure 33.1f). This fungus may be mistaken for a *Penicillium*.

Monascus

A saprophyte. Colonies are flat, rather granular with a deep red pigment. The characteristic ascospores are produced inside spherical ascocarps. These are in that each develops from a single (or a very few) hyphae and they have a thin wall (Figure 33.1g). Chains of conidia superficially resembling *Scopulariopsis* may also be present.

Paecilomyces

Usually seen as a spoilage fungus but one species, *P. variotti*, has been found in rare, deep infections in immunosuppressed patients.

It resembles *Penicillium*, producing either pale purple or yellowish olive colonies depending on species. This feature, and the long, tapering tips to the spore-producing cells of the 'penicillus' distinguish it from *Penicillium* (Figure 33.1h).

Chrysosporium

Soil dwellers, many of which specialize in utilizing keratinaceous substrates.

This genus is close to the dermatophytes and like them is unaffected by cyclcohexamide in the medium. Colonies of most species are white or cream coloured and flat with a felt-like surface. Spores are similar to those of dermatophyte microconidia but mostly longer than 5 μm (Figure 33.1i).

Geomyces

Saprophytes. One species, *G. pannorus*, a soil organism, is often mistaken for a dermatophyte.

Colonies are restricted, often heaped up, with a very thin white crust of aerial spores. These resemble dermatophyte microconidia but are smaller than most (< 1.5 μm long) and produced on minute, acutely-branched Christmas tree-like structures (Figure 33.1j).

Scopulariopsis

One species, *S. brevicaulis*, is a saprophyte, living in soil but does seem to be adapted to utilize keratin and is one of the commoner invaders of human nail tissue. It is also notable for releasing gaseous arsines from the arsenical pigment Paris green.

Colonies are often folded and heaped up, but they may be flat. When sporing they are cinnamon-brown in colour and have a powdery surface. The spores are large, rounded, with a flat detachment scar and are produced in long chains from a penicillium-like branching structure. In most strains the spore surfaces are distinctly roughened (Figure 33.1k).

Reference

Onions, A. M. S., Allsop, D. and Eggins, H. O. W. (1981) *Smith's Introduction to Industrial Mycology*, 7th edn, Edward Arnold, London

Chapter 34

Pathogenic moulds

The following fungi are considered in this chapter: Dermatophytes, causing infections of epidermal tissues; opportunist deep-tissue pathogens (*Aspergillus* and *Zygomycotina*); agents of subcutaneous infections; agents of coccidiomycosis, histoplasmosis and blastomycosis.

Dermatophytes

The nature of the disease assists in identification as many of these fungi have characteristic infection patterns (Table 34.1). This is particularly helpful in identifying strains with abnormal morphology.

Table 34.1 Characteristic infection patterns of dermatophytes

Patient	Body site	Geographical origin	Organism
Adult	Feet, hands, groin	Any	T. rubrum
			T. interdigitale
			E. floccosum
Adult or child	Scalp, face, arms, chest, legs	Any	M. canis
			M. gypseum
			T. verrucosum
			T. mentagrophytes
Adult or child	Scalp, face, arms, chest, legs	N Africa, E and SE Asia	T. violaceum
Child	Scalp	Any	T. tonsurans
Child	Scalp	W and Central Africa, Caribbean	M. audouinii
Child	Scalp	N and E Africa	T. soudananse
Child	Scalp	N and E Africa, Middle East	T. schoenleinii

Appearance in clinical material

In nail and skin cleared with caustic soda fungal hyphae may be seen, some of them breaking into arthrospores, but specific identification cannot be made from their arrangement.

Some dermatophyte-infected hairs fluoresce under a Wood's lamp (examine in a darkened room).

Table 34.2 shows the appearance of infected hairs and Figure 34.1 shows various types of infected hairs, skin and nails.

There are three genera of common dermatophytes, classified according to the type of macroconidia they produce:

Microsporum – spindle-shaped and roughened
Trichophyton – cylindrical and smooth
Epidermophyton – club-shaped and smooth

Microsporum and *Epidermophyton* are easily recognized by their macroconidia but *Trichophyton* rarely produces them.

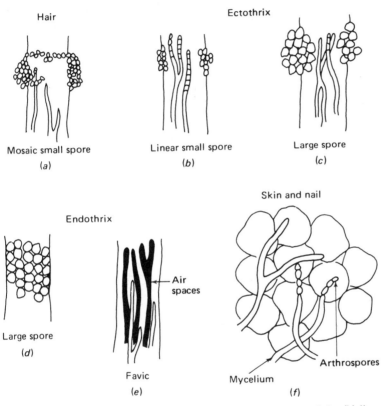

Figure 34.1 Infected hair, skin and nail: (a) mosaic small-spore ectothrix; (b) linear small-spore ectothrix; (c) large-spore ectothrix; (d) large-spore endothrix; (e) favic-type endothrix; (f) skin and nail appearance

Table 34.2 Appearance of infected hairs

Spore arrangement	Fluorescence	Organism
Small spore ectothrix	+	*M. canis*
	+	*M. audouinii*
	−	*T. mentagrophytes*
Large spore ectothrix	−	*M. gypseum*
	−	*T. verrucosum*
Endothrix	−	*T. soudanense*
	−	*T. violaceum*
'Favic' hairs	weak	*T. schoenleinii*

Identification

Identification to species level depends on colonial and microscopical appearances, both of which are influenced by the growth medium.

Use a good brand of Sabouraud dextrose agar which has been shown to give adequate pigmentation and spore production.

Microsporum audouinii

This is a scalp fungus with no animal or soil reservoir. It spreads from child to child in schools, and infection usually resolves at puberty. The organism is slow growing, forming a tenacious colony with little aerial mycelium, sometimes having a light pink colour on the back. Macroconidia are rare (Figure 34.2a).

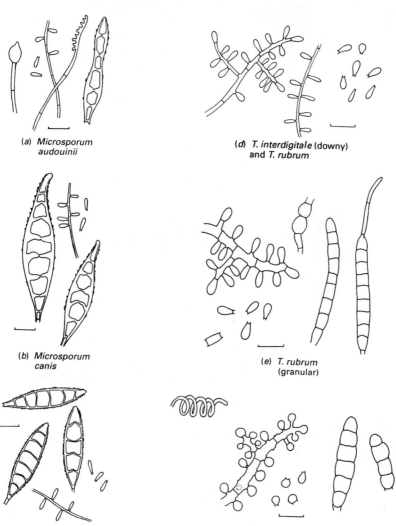

(a) *Microsporum audouinii*

(d) *T. interdigitale* (downy) and *T. rubrum*

(b) *Microsporum canis*

(e) *T. rubrum* (granular)

(c) *Microsporum gypseum*

(f) *T. mentagrophytes* and *T. interdigitale* (granular)

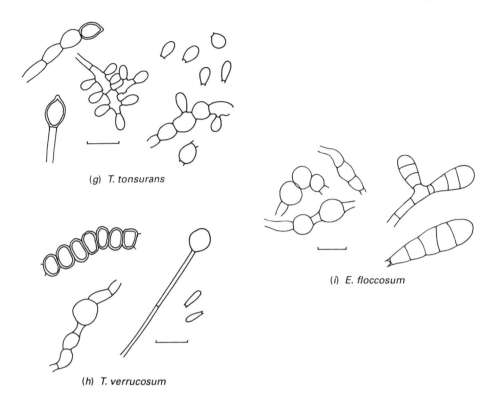

(g) T. tonsurans

(i) E. floccosum

(h) T. verrucosum

Figure 34.2 *Microsporum audouinii:* (b) *Microsporum canis*; (c) *Microsporum gypseum*; (d) *T. interdigitale* (downy) and *T. rubrum*; (e) *T. rubrum* (granular); (f) *T. mentagrophytes* and *T. interdigitale* (granular); (g) *T. tonsurans*; (h) *T. verrucosum*; (i) E. floccosum

M. canis

As its name suggests, this is a common parasite of dogs and cats, where the infection is often difficult to detect. Children are commonly infected from pets. This organism, which is fast growing, usually produces a vivid yellow pigment and fairly abundant macroconidia. Occasionally, strains are non-pigmented and slow growing. To differentiate them from *M. audouinii*, grow on rice (5 g of rice and 20 ml of water in a 100-ml flask, autoclaved at 115°C for 20 min and cooled). *M. canis* produces aerial mycelium; *M. audouinii* does not (Figure 34.2b).

M. gypseum

This occurs as a saprophyte in soil and has a low virulence for humans. The occasional case which occurs usually shows a solitary lesion, related to contact with the soil. The fungus grows fast, and is soon covered with a buff-coloured granular coating of spores. These tend to be arranged in radial strands on the surface, rather like a spider's web. Macroconidia are abundant (Figure 34.2c).

Trichophyton rubrum

Commonest dermatophyte seen in dermatology clinics in developed countries, but the less common *T. interdigitale* is probably more widespread in the general population.

There are many varieties of *T. rubrum*; that most often seen gives a downy, white colony with a red-brown pigment and a narrow white edge on reverse. Microscopy shows only club-shaped microconidia (Figure 34.2d).

Other varieties are 'melanoid' in which a dark brown pigment diffuses into the medium; 'yellow', which is a non-sporing pale yellow form; and 'granular' which has the usual dark red reverse but powdery aerial growth with patches of pink. This variety has larger microconidia and usually produces macroconidia (Figure 34.2e).

Non-pigmented strains occur which resemble the downy variety of *T. interdigitale* (Figure 34.2d). To distinguish these, subculture on urea agar. *T. rubrum* does not usually produce the colour change by 7 days but most other dermatophytes do.

Trichophyton mentagrophytes complex

The commonest form in this group is the anthropophilic *T. interdigitale*, which causes athlete's foot.

Colonies resemble those of the zoophilic *T. mentagrophytes* (see below), with a uniformly powdery, cream-coloured surface made up of an abundance of nearly spherical microconidia. Spiral hyphae and macroconidia may be present (Figure 34.2f). The downy variety forms a pure white cottony colony and has elongated microconidia resembling those of *T. rubrum*.

The zoophilic (usually from rodents) *T. mentagrophytes* has coarsely granular colonies with areas of agar surface showing between radiating zones of white or cream sporing growth. Reverse pigmentation often shows dark brown 'veins'.

T. tonsurans

A cause of scalp ringworm. Its colony is slow growing and velvety, often with a folded centre. Macroconidia are large, long, oval and arranged along wide, often empty hyphae. (Figure 34.2g).

T. verrucosum

The organism of cattle ringworm. It causes scalp, beard or nail infections in farm workers, and other people such as slaughtermen and veterinary surgeons who come into contact with cattle. In culture it grows very poorly; suspected cultures should be examined carefully with a lens after 14 days' incubation, as colonies are often submerged and minute. This, together with the large, thin walled, balloon-like chlamydospores at the end of a straight hypha, and chains of thick-walled cells (Figure 34.2h) is sufficient to identify the organism.

T. schoenleinii

The cause of favus. The colony is slow growing, hard and leathery, with a surface like white suede. The only microscopic features are thick, knobbly hyphae, like arthritic fingers – the so-called 'favic chandeliers' – and chlamydospores.

T. violaceum

Another cause of scalp ringworm, microscopically similar to *T. verrucosum*, although the hyphae are usually thinner. The characteristic feature is a deep red-purple pigment, which appears as a spot in the centre of a young colony, and gradually spreads to the edge.

Epidermophyton
There is only one species, *E. floccosum*. This is one cause of tinea cruris ('dhobi itch') and 'jungle rot'. It also occasionally infects feet. Its growth on malt agar is khaki in colour with a powdery surface due to masses of macronidia whose shape is very characteristic (Figure 34.2i).
    For further information about dermatophytes see Mackenzie and Philpot (1981).

Aspergillus and *Penicillium*

These genera are closely related and there are several species that are intermediate in morphology. Both produce long, dry chains of conidia by repeated budding through the end of a bottle-shaped cell, the phialide.
    In *Penicillium* the phialides are clustered at the tips of tree-like conidiophores and develop unevenly.
    In *Aspergillus* they are clustered on a club-shaped or spherical vesicle and develop synchronously (Figure 34.3).
    Intermediate forms are unbranched but have only a slight swelling at the stalk apex. By convention those in which the vesicle is less than one and a half times the diameter of the stalk are grouped in the section Aspergilloidea of *Penicillium*.
    In both groups the mycelium is colourless (white) and any colony colour is due to the coloured spores (conidia). *Penicillium* spores are all some shade of green: *Aspergillus* spores may be green, brown, yellow or black.
    Some *Penicillium* spp. are plant parasites but the majority of both genera occur on decaying vegetation because of their wide range of secondary metabolites several are of importance in industrial fermentation and synthetic processes. They are also important in that they may produce mycotoxins in foods.

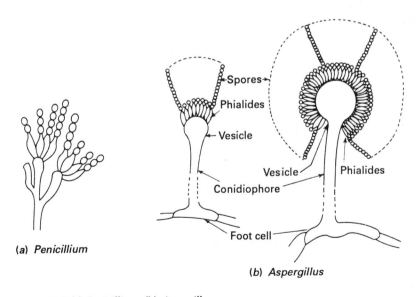

Figure 34.3 (a) *Penicillium;* (b) *Aspergillus*

A. glaucus group

Members of this group flourish in conditions of physiological dryness, e.g. on the surface of jam, on textiles and on tobacco. They are very distinctive in appearance, with bluish green or grey-green conidia and large yellow perithicia.

A. restrictus group

Like *A. glaucus*, these will grow on dry materials, such as slightly dry textiles. As their name suggests, they are slow-growing organisms.

A. fumigatus

So-called from its smoky green colour. It is a common mould in compost, and can grow at temperatures above 40°C and hence it is pathogenic for birds. It spores more readily at 37°C than at 26°C. It is now notorious as a human pathogen, underlying pulmonary eosinophilia. In this condition, the fungus may be found in the mucous plugs that are expectorated. It also invades old tuberculous cavities and gives rise to aspergilloma of the lung. When isolated from such cases, the organism often does not produce spores; incubation at 42°C may encourage it to do so. Patients who harbour this organism develop precipitating antibody, which is rather weak in pulmonary eosinophilia but strong in cases of aspergilloma. For methods, see p.118. If *A. fumigatus* is examined with a lens, the heads are seen to be columnar – the whole effect is like a test-tube brush.

A. niger

On the other hand, this organism has round heads which are large enough to appear discrete to the naked eye. This, together with their black colour, makes them easy to recognize.

The Flavus–Oryzae group

Members of this group are notorious for the production of aflatoxin in groundnuts and other foods such as millet when badly stored (see Moss *et al.* 1990). They are also of value commercially, as a source of diastatic enzymes such as takadiastase. The specific name, which means yellow, is an error; the colony colour in this series is green. The heads are round, but smaller than those of *A. niger*.

Table 34.3 shows the morphological properties of some *Aspergillus* species/ groups.

Table 34.3   Morphology of *Aspergillus* species

	A. fumigatus	A. flavus	A. niger	A. terreus	A. versicolor	A. nidulans	A. glaucus
Finely roughened stalks	–	+	–	–	–	–	–
Stalks pale brown	–	–	–	–	–	+	–
Green colony	+	+	–	–	+	+	+
Black colony	–	–	+	–	–	–	–
Sand-brown colony	–	–	–	+	–	–	–
Metulae present	–	+/–	+	+	+	+	–

Zygomycotina

Fungi with non-septate, wide hyphae in which sexual fusion results in a thick-walled resting zygospore. These are seldom seen, however, and distinction of genera and species is largely based on the sexual structures. The group contains two main orders. Many of them grow rapidly and produce aerial mycelium resembling sheep wool.

Mucorales

Reproduced by many-spored sporangia. The majority belong to three genera: *Mucor, Rhizopus* and *Absidia*.

Rhizopus   Columella enlarges greatly after rupture of the sporangium and collapses to form a characteristic mushroom shape. Spores are angular and delicately striated. Root-like rhizoids may occur at the base of sporangium stalk (Figure 34.4b).

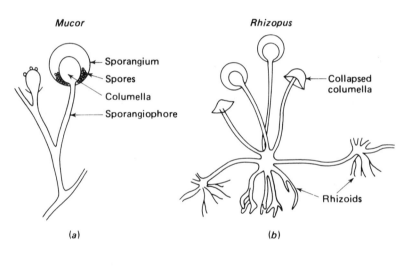

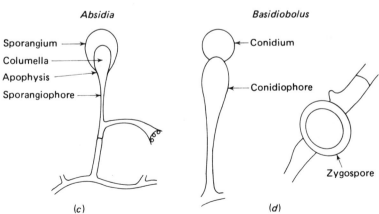

Figure 34.4 Zygomycetes: (a) *Mucor;* (b) *Rhizopus;* (c) *Absidia;* (d) *Basidiobolus*

Absidia   Stalk trumpet-shaped and sporangium more or less continuous (Figure 34.4c).

The absence of features characteristic of the latter two genera suggests *Mucor* spp.

The common pathogenic species, all causing mucormycosis, are *M. pusillus*, *M. corymbifera* and *R. oryzae*. They are thermophilic: check their identity by culturing at 37°C.

Entomophthorales

Insect pathogens, rarely seen in clinical material. The asexual spore form is a single, relatively large 'conidium' which is forcefully discharged.

Colonies are wrinkled, cream coloured and waxy. A haze of discharged spores forms a mirror image of the colony on the lid of the petri dish.

Basidiobolus   Cause of tropical subcutaneous zygomycosis.

Conidiobolus coronatus (Entomophthora coronata) Causes tropical rhinoentomophthoromycosis of the nasal mucosa.

Miscellaneous pathogenic moulds

Madurella

M. mycetomatis and *M. grisea* cause black grain mycetoma. The former produces flat or wrinkled, yellowish colonies, often releasing melanoid pigments into the medium. The latter grows as domed, densely fluffy, mouse-grey to black colonies. Neither form conidia.

Exophiala

Colonies are dark grey to black, intensely fluffy or wet and yeast-like depending on degree of conidial budding, strain and growth conditions. All produce conidia from small, tapering nipples which grow in length as more conidia are cut off (Figure 34.5a).

E. jeanselmei is a cause of mycetoma, *E. werneckii* of tinta nigra, *E. dermatitidis* of chromomycosis and *E. spinifera* subcutaneous cysts.

Phialophora

Black moulds in which the spore apex has a definite collarette or funnel-shaped cup in which the conidia are formed (Figure 34.5b).

P. verrucosa is one cause of chromomycosis; *P. parasitica* is occasionally found in subcutaneous cysts.

Fonsecaea

Black moulds in which the conidia are formed in short chains by apical budding of the cell below. Differs from *Cladosporium* only in that the chains consist of no more than four spores (Figure 34.5c).

F. pedrosi is the common cause of chromomycosis; *F. compacta* is less common.

Cladosporium

Many species, mostly of no clinical significance. Long-chains of spores are

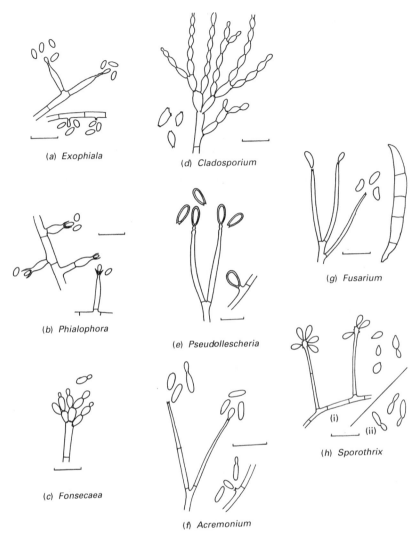

Figure 34.5 (a) *Exophiala;* (b) *Phialophora;* (c) *Fonsecaea;* (d) *Cladosporium;* (d) *Pseudoallescheria;* (f) *Acremonium;* (g) *Fusarium;* (h) *Sporothrix;* (i) Mycelial form, (ii) tissue form

produced by acropetal budding and may branch where a spore develops two buds (Figure 34.5d).

C. *carionii* causes chromomycosis; C. *bantianum* cerebral infections.

Pseudoallescheria boydii
Colonies fluffy, white to light grey, producing oval, slightly yellowish spores in succession at the tips of long, straight hyphae. Each spore has a basal scar. Occasional strains may produce black ascomata beneath the agar (Figure 34.5e).

Variously named *Allescheria, Petriellidium* or *Monosporium.* Causes pale grain mycetoma and miscellaneous other infections (otitis externa in the UK).

Acremonium (Cephalosporium)

A very large group of white or pink moulds with wet heads of conidia produced one by one from the tips of straight hyphae or lateral nipples. In some, the conidia bud secondary spores and the colony becomes yeast-like (Figure 34.5f).

A. kilense is a cause of pale grain mycetoma; other species invade nail tissue.

Fusarium

Colonies often loose and fluffy with a pink or purple pigment. Characterized by curved, multiseptate macroconidia, but these may be absent. Many strains produce microconidia similar to those of Acremonium but usually somewhat curved (Figure 34.5g).

F. solani and F. oxysporum cause a variety of infections including mycetomas and keratitis.

Sporothrix

At low temperatures (25–30°C) colonies are grey to black, dry, membranous and wrinkled. Hyaline to pigmented spores are produced initially as rosettes on minute, teeth-like points on expanded apical knobs (Figure 34.5h).

Differentiation from non-pathogenic species requires conversion to yeast form by culture at 37°C.

Causes sporotrichosis in which it exists as a budding, spindle-shaped yeast.

Blastomyces, Histoplasma and Coccidioides

These cause systemic or generalized infections resulting from the inhalation of airborne conidia. Heavily sporing mould cultures are more hazardous than those in the yeast-like phase. All manipulations that might release conidia should be done in microbiological safety cabinets in Containment Level 3 laboratories.

A useful cultural feature of these organisms is that they are not inhibited by cyclohexamide.

Blastomyces dermatitidis

Causes North American blastomycosis. Infection is almost always by inhalation and skin lesions are usually secondary to a pulmonary infection, even a mild one. The mycelial form, at 26°C, is a moist colony at first, developing a fluffy serial mycelium with microconidia. At 37°C or in tissue, large oval multinucleate yeasts are seen (Figure 34.6a).

Histoplasma capsulatum

Causes histoplasmosis, which is very common in the southern half of the USA although it is also found in other countries. Most people recover rapidly: only an occasional patient develops progressive disease. The route of infection is inhalation, and the pathology similar to tuberculosis (Figure 34.6b). See caution above.

At 26°C, the organism grows with a white aerial mycelium, which becomes light brown. The spores are striking and characteristic. At 37°C, small, oval budding yeasts are found. In tissue, these yeasts are intracellular, except in necrotic material.

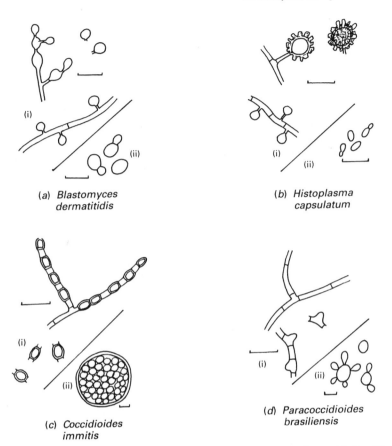

(a) *Blastomyces dermatitidis*

(b) *Histoplasma capsulatum*

(c) *Coccidioides immitis*

(d) *Paracoccidioides brasiliensis*

Figure 34.6 (a) *Blastomyces dermatitidis;* (b) *Histoplasma capsulatum;* (c) *Coccidioides immitis;* (d) *Paracoccidioides brasiliensis;* (i) Mycelial form, (ii) tissue form

Coccidioides immitis
Causes coccidiomycosis. This disease is limited to a few special areas in the south west of the USA and in Mexico and Venezuela. It is inhaled in desert dust. Again, most people suffer only a mild, transient, 'flu-like' illness. Occasionally, cases progress to generalized destructive lesions (Figure 34.6c). See caution above.

At 26°C, the colony is at first moist, then develops an aerial mycelium, becoming brown with age and breaking up into arthrospores. There is no second form in artificial culture. In the body, however, large, thick-walled cells are seen. When mature, these are full of spores.

Paracoccidioides braziliensis
Causes South American blastomycosis. The primary lesion is almost always in the oral mucosa, possibly because of the habit of cleaning the teeth with wood splinters and chewing bark in the places where the disease is found. The spread, again, is through the lymphatics. The organism is very slow growing, white and featureless at 26°C. At 37°C on rich media, characteristic yeast cells are formed (Figure 34.6d).

References

Campbell, M. C. and Stewart, J. C. (1980) *The Medical Mycology Handbook,* J. Wiley & Sons, New York and Chichester

Clayton, Y. and Midgeley, G. (1985) *Medical Mycology*, Gower Medical Publishing, London and New York

Mackenzie, D. W. R. and Philpot, C. M. (1981) *Isolation and Identification of Ringworm Fungi*, Public Health Laboratory Service Monograph No. 15, HMSO, London

Moss, M. O., Jarvis, B. and Skinner, F. A. (1989) *Filamentous Fungi in Foods and Feeds*, Journal of Applied Bacteriology Symposium Supplement No 18, **67**, 1S–144S

Index